Lecture Notes in Computer Science   14832

Founding Editors

Gerhard Goos
Juris Hartmanis

The series Lecture Notes in Computer Science (LNCS), including its subseries Lecture Notes in Artificial Intelligence (LNAI) and Lecture Notes in Bioinformatics (LNBI), has established itself as a medium for the publication of new developments in computer science and information technology research, teaching, and education.

LNCS enjoys close cooperation with the computer science R & D community, the series counts many renowned academics among its volume editors and paper authors, and collaborates with prestigious societies. Its mission is to serve this international community by providing an invaluable service, mainly focused on the publication of conference and workshop proceedings and postproceedings. LNCS commenced publication in 1973.

Leonardo Franco · Clélia de Mulatier ·
Maciej Paszynski · Valeria V. Krzhizhanovskaya ·
Jack J. Dongarra · Peter M. A. Sloot
Editors

Computational Science – ICCS 2024

24th International Conference
Malaga, Spain, July 2–4, 2024
Proceedings, Part I

Springer

Editors
Leonardo Franco 🆔
University of Malaga
Malaga, Spain

Clélia de Mulatier 🆔
University of Amsterdam
Amsterdam, The Netherlands

Maciej Paszynski 🆔
AGH University of Science and Technology
Krakow, Poland

Valeria V. Krzhizhanovskaya 🆔
University of Amsterdam
Amsterdam, The Netherlands

Jack J. Dongarra 🆔
University of Tennessee
Knoxville, TN, USA

Peter M. A. Sloot 🆔
University of Amsterdam
Amsterdam, The Netherlands

ISSN 0302-9743   ISSN 1611-3349 (electronic)
Lecture Notes in Computer Science
ISBN 978-3-031-63748-3   ISBN 978-3-031-63749-0 (eBook)
https://doi.org/10.1007/978-3-031-63749-0

Preface

Welcome to the proceedings of the 24th International Conference on Computational Science (https://www.iccs-meeting.org/iccs2024/), held on July 2–4, 2024 at the University of Málaga, Spain.

In keeping with the new normal of our times, ICCS featured both in-person and online sessions. Although the challenges of such a hybrid format are manifold, we have always tried our best to keep the ICCS community as dynamic, creative, and productive as possible. We are proud to present the proceedings you are reading as a result.

ICCS 2024 was jointly organized by the University of Málaga, the University of Amsterdam, and the University of Tennessee.

Facing the Mediterranean in Spain's Costa del Sol, Málaga is the country's sixth-largest city, and a major hub for finance, tourism, and technology in the region.

The University of Málaga (Universidad de Málaga, UMA) is a modern, public university, offering 63 degrees and 120 postgraduate degrees. Close to 40,000 students study at UMA, taught by 2500 lecturers, distributed over 81 departments and 19 centers. The UMA has 278 research groups, which are involved in 80 national projects and 30 European and international projects. ICCS took place at the Teatinos Campus, home to the School of Computer Science and Engineering (ETSI Informática), which is a pioneer in its field and offers the widest range of IT-related subjects in the region of Andalusia.

The International Conference on Computational Science is an annual conference that brings together researchers and scientists from mathematics and computer science as basic computing disciplines, as well as researchers from various application areas who are pioneering computational methods in sciences such as physics, chemistry, life sciences, engineering, arts, and the humanities, to discuss problems and solutions in the area, identify new issues, and shape future directions for research.

The ICCS proceedings series have become a primary intellectual resource for computational science researchers, defining and advancing the state of the art in this field.

We are proud to note that this 24th edition, with 17 tracks (16 thematic tracks and one main track) and close to 300 participants, has kept to the tradition and high standards of previous editions.

The theme for 2024, "Computational Science: Guiding the Way Towards a Sustainable Society", highlights the role of Computational Science in assisting multidisciplinary research on sustainable solutions. This conference was a unique event focusing on recent developments in scalable scientific algorithms; advanced software tools; computational grids; advanced numerical methods; and novel application areas. These innovative novel models, algorithms, and tools drive new science through efficient application in physical systems, computational and systems biology, environmental systems, finance, and others.

ICCS is well known for its excellent lineup of keynote speakers. The keynotes for 2024 were:

- David Abramson, University of Queensland, Australia
- Manuel Castro Díaz, University of Málaga, Spain
- Jiří Mikyška, Czech Technical University in Prague, Czechia
- Takemasa Miyoshi, RIKEN, Japan
- Coral Calero Muñoz, University of Castilla-La Mancha, Spain
- Petra Ritter, Berlin Institute of Health & Charité University Hospital Berlin, Germany

This year we had 430 submissions (152 to the main track and 278 to the thematic tracks). In the main track, 51 full papers were accepted (33.5%); in the thematic tracks, 104 full papers (37.4%). The higher acceptance rate in the thematic tracks is explained by their particular nature, whereby track organizers personally invite many experts in the field to participate. Each submission received at least 2 single-blind reviews (2.6 reviews per paper on average).

ICCS relies strongly on our thematic track organizers' vital contributions to attract high-quality papers in many subject areas. We would like to thank all committee members from the main and thematic tracks for their contribution to ensuring a high standard for the accepted papers. We would also like to thank Springer, Elsevier, and Intellegibilis for their support. Finally, we appreciate all the local organizing committee members for their hard work in preparing this conference.

We hope the attendees enjoyed the conference, whether virtually or in person.

July 2024

Leonardo Franco
Clélia de Mulatier
Maciej Paszynski
Valeria V. Krzhizhanovskaya
Jack J. Dongarra
Peter M. A. Sloot

Organization

Conference Chairs

General Chair

Valeria Krzhizhanovskaya   University of Amsterdam, The Netherlands

Main Track Chair

Clélia de Mulatier   University of Amsterdam, The Netherlands

Thematic Tracks Chair

Maciej Paszynski   AGH University of Krakow, Poland

Thematic Tracks Vice Chair

Michael Harold Lees   University of Amsterdam, The Netherlands

Scientific Chairs

Peter M. A. Sloot   University of Amsterdam, The Netherlands
Jack Dongarra   University of Tennessee, USA

Local Organizing Committee

Leonardo Franco (Chair)   University of Malaga, Spain
Francisco Ortega-Zamorano   University of Malaga, Spain
Francisco J. Moreno-Barea   University of Malaga, Spain
José L. Subirats-Contreras   University of Malaga, Spain

Thematic Tracks and Organizers

Advances in High-Performance Computational Earth Sciences: Numerical Methods, Frameworks & Applications (IHPCES)

Takashi Shimokawabe	University of Tokyo, Japan
Kohei Fujita	University of Tokyo, Japan
Dominik Bartuschat	FAU Erlangen-Nürnberg, Germany

Artificial Intelligence and High-Performance Computing for Advanced Simulations (AIHPC4AS)

Maciej Paszynski	AGH University of Krakow, Poland

Biomedical and Bioinformatics Challenges for Computer Science (BBC)

Mario Cannataro	University Magna Graecia of Catanzaro, Italy
Giuseppe Agapito	University Magna Graecia of Catanzaro, Italy
Mauro Castelli	Universidade Nova de Lisboa, Portugal
Riccardo Dondi	University of Bergamo, Italy
Rodrigo Weber dos Santos	Federal University of Juiz de Fora, Brazil
Italo Zoppis	University of Milano-Bicocca, Italy

Computational Diplomacy and Policy (CoDiP)

Roland Bouffanais	University of Geneva, Switzerland
Michael Lees	University of Amsterdam, The Netherlands
Brian Castellani	Durham University, UK

Computational Health (CompHealth)

Sergey Kovalchuk	Huawei, Russia
Georgiy Bobashev	RTI International, USA
Anastasia Angelopoulou	University of Westminster, UK
Jude Hemanth	Karunya University, India

Computational Optimization, Modelling, and Simulation (COMS)

Xin-She Yang   Middlesex University London, UK
Slawomir Koziel   Reykjavik University, Iceland
Leifur Leifsson   Purdue University, USA

Generative AI and Large Language Models (LLMs) in Advancing Computational Medicine (CMGAI)

Ahmed Abdeen Hamed   State University of New York at Binghamton, USA
Qiao Jin   National Institutes of Health, USA
Xindong Wu   Hefei University of Technology, China
Byung Lee   University of Vermont, USA
Zhiyong Lu   National Institutes of Health, USA
Karin Verspoor   RMIT University, Australia
Christopher Savoie   Zapata AI, USA

Machine Learning and Data Assimilation for Dynamical Systems (MLDADS)

Rossella Arcucci   Imperial College London, UK
Cesar Quilodran-Casas   Imperial College London, UK

Multiscale Modelling and Simulation (MMS)

Derek Groen   Brunel University London, UK
Diana Suleimenova   Brunel University London, UK

Network Models and Analysis: From Foundations to Artificial Intelligence (NMAI)

Marianna Milano   Università Magna Graecia of Catanzaro, Italy
Giuseppe Agapito   University Magna Graecia of Catanzaro, Italy
Pietro Cinaglia   University Magna Graecia of Catanzaro, Italy
Chiara Zucco   University Magna Graecia of Catanzaro, Italy

Numerical Algorithms and Computer Arithmetic for Computational Science (NACA)

Pawel Gepner	Warsaw Technical University, Poland
Ewa Deelman	University of Southern California, Marina del Rey, USA
Hatem Ltaief	KAUST, Saudi Arabia

Quantum Computing (QCW)

Katarzyna Rycerz	AGH University of Krakow, Poland
Marian Bubak	Sano and AGH University of Krakow, Poland

Simulations of Flow and Transport: Modeling, Algorithms, and Computation (SOFTMAC)

Shuyu Sun	King Abdullah University of Science and Technology, Saudi Arabia
Jingfa Li	Beijing Institute of Petrochemical Technology, China
James Liu	Colorado State University, USA

Smart Systems: Bringing Together Computer Vision, Sensor Networks and Artificial Intelligence (SmartSys)

Pedro Cardoso	University of Algarve, Portugal
João Rodrigues	University of Algarve, Portugal
Jânio Monteiro	University of Algarve, Portugal
Roberto Lam	University of Algarve, Portugal

Solving Problems with Uncertainties (SPU)

Vassil Alexandrov	Hartree Centre – STFC, UK
Aneta Karaivanova	IICT – Bulgarian Academy of Science, Bulgaria

Teaching Computational Science (WTCS)

Evguenia Alexandrova   Hartree Centre – STFC, UK
Tseden Taddese   UK Research and Innovation, UK

Reviewers

Ahmed Abdelgawad   Central Michigan University, USA
Samaneh Abolpour Mofrad   Imperial College London, UK
Tesfamariam Mulugeta Abuhay   Queen's University, Canada
Giuseppe Agapito   University of Catanzaro, Italy
Elisabete Alberdi   University of the Basque Country, Spain
Luis Alexandre   UBI and NOVA LINCS, Portugal
Vassil Alexandrov   Hartree Centre – STFC, UK
Evguenia Alexandrova   Hartree Centre – STFC, UK
Julen Alvarez-Aramberri   Basque Center for Applied Mathematics, Spain
Domingos Alves   Ribeirão Preto Medical School, University of São
                                      Paulo, Brazil
Sergey Alyaev   NORCE, Norway
Anastasia Anagnostou   Brunel University London, UK
Anastasia Angelopoulou   University of Westminster, UK
Rossella Arcucci   Imperial College London, UK
Emanouil Atanasov   IICT – Bulgarian Academy of Sciences, Bulgaria
Krzysztof Banaś   AGH University of Krakow, Poland
Luca Barillaro   Magna Graecia University of Catanzaro, Italy
Dominik Bartuschat   FAU Erlangen-Nürnberg, Germany
Pouria Behnodfaur   Curtin University, Australia
Jörn Behrens   University of Hamburg, Germany
Adrian Bekasiewicz   Gdansk University of Technology, Poland
Gebrail Bekdas   Istanbul University, Turkey
Mehmet Ali Belen   Iskenderun Technical University, Turkey
Stefano Beretta   San Raffaele Telethon Institute for Gene Therapy,
                                      Italy
Anabela Moreira Bernardino   Polytechnic Institute of Leiria, Portugal
Eugénia Bernardino   Polytechnic Institute of Leiria, Portugal
Daniel Berrar   Tokyo Institute of Technology, Japan
Piotr Biskupski   IBM, Poland
Georgiy Bobashev   RTI International, USA
Carlos Bordons   University of Seville, Spain
Bartosz Bosak   PSNC, Poland
Lorella Bottino   University Magna Graecia of Catanzaro, Italy

Roland Bouffanais	University of Geneva, Switzerland
Marian Bubak	Sano and AGH University of Krakow, Poland
Aleksander Byrski	AGH University of Krakow, Poland
Cristiano Cabrita	Universidade do Algarve, Portugal
Xing Cai	Simula Research Laboratory, Norway
Carlos Calafate	Universitat Politècnica de València, Spain
Victor Calo	Curtin University, Australia
Mario Cannataro	University Magna Graecia of Catanzaro, Italy
Karol Capała	AGH University of Krakow, Poland
Pedro J. S. Cardoso	Universidade do Algarve, Portugal
Eddy Caron	ENS-Lyon/Inria/LIP, France
Stefano Casarin	Houston Methodist Hospital, USA
Brian Castellani	Durham University, UK
Mauro Castelli	Universidade Nova de Lisboa, Portugal
Nicholas Chancellor	Durham University, UK
Thierry Chaussalet	University of Westminster, UK
Sibo Cheng	Imperial College London, UK
Lock-Yue Chew	Nanyang Technological University, Singapore
Pastrello Chiara	Krembil Research Institute, Canada
Su-Fong Chien	MIMOS Berhad, Malaysia
Marta Chinnici	enea, Italy
Bastien Chopard	University of Geneva, Switzerland
Maciej Ciesielski	University of Massachusetts, USA
Pietro Cinaglia	University of Catanzaro, Italy
Noelia Correia	Universidade do Algarve, Portugal
Adriano Cortes	University of Rio de Janeiro, Brazil
Ana Cortes	Universitat Autònoma de Barcelona, Spain
Enrique Costa-Montenegro	Universidad de Vigo, Spain
David Coster	Max Planck Institute for Plasma Physics, Germany
Carlos Cotta	University of Málaga, Spain
Peter Coveney	University College London, UK
Alex Crimi	AGH University of Krakow, Poland
Daan Crommelin	CWI Amsterdam, The Netherlands
Attila Csikasz-Nagy	King's College London, UK/Pázmány Péter Catholic University, Hungary
Javier Cuenca	University of Murcia, Spain
António Cunha	UTAD, Portugal
Pawel Czarnul	Gdansk University of Technology, Poland
Pasqua D'Ambra	IAC-CNR, Italy
Alberto D'Onofrio	University of Trieste, Italy
Lisandro Dalcin	KAUST, Saudi Arabia

Bhaskar Dasgupta	University of Illinois at Chicago, USA
Clélia de Mulatier	University of Amsterdam, The Netherlands
Ewa Deelman	University of Southern California, Marina del Rey, USA
Quanling Deng	Australian National University, Australia
Eric Dignum	University of Amsterdam, The Netherlands
Riccardo Dondi	University of Bergamo, Italy
Rafal Drezewski	AGH University of Krakow, Poland
Simon Driscoll	University of Reading, UK
Hans du Buf	University of the Algarve, Portugal
Vitor Duarte	Universidade NOVA de Lisboa, Portugal
Jacek Długopolski	AGH University of Krakow, Poland
Wouter Edeling	Vrije Universiteit Amsterdam, The Netherlands
Nahid Emad	University of Paris Saclay, France
Christian Engelmann	ORNL, USA
August Ernstsson	Linköping University, Sweden
Aniello Esposito	Hewlett Packard Enterprise, Switzerland
Roberto R. Expósito	Universidade da Coruna, Spain
Hongwei Fan	Imperial College London, UK
Tamer Fandy	University of Charleston, USA
Giuseppe Fedele	University of Calabria, Italy
Christos Filelis-Papadopoulos	Democritus University of Thrace, Greece
Alberto Freitas	University of Porto, Portugal
Ruy Freitas Reis	Universidade Federal de Juiz de Fora, Brazil
Kohei Fujita	University of Tokyo, Japan
Takeshi Fukaya	Hokkaido University, Japan
Wlodzimierz Funika	AGH University of Krakow, Poland
Takashi Furumura	University of Tokyo, Japan
Teresa Galvão	University of Porto, Portugal
Luis Garcia-Castillo	Carlos III University of Madrid, Spain
Bartłomiej Gardas	Institute of Theoretical and Applied Informatics, Polish Academy of Sciences, Poland
Victoria Garibay	University of Amsterdam, The Netherlands
Frédéric Gava	Paris-East Créteil University, France
Piotr Gawron	Nicolaus Copernicus Astronomical Centre, Polish Academy of Sciences, Poland
Bernhard Geiger	Know-Center GmbH, Austria
Pawel Gepner	Warsaw Technical University, Poland
Alex Gerbessiotis	NJIT, USA
Maziar Ghorbani	Brunel University London, UK
Konstantinos Giannoutakis	University of Macedonia, Greece
Alfonso Gijón	University of Granada, Spain

Jorge González-Domínguez	Universidade da Coruña, Spain
Alexandrino Gonçalves	CIIC – ESTG – Polytechnic University of Leiria, Portugal
Yuriy Gorbachev	Soft-Impact LLC, Russia
Pawel Gorecki	University of Warsaw, Poland
Michael Gowanlock	Northern Arizona University, USA
George Gravvanis	Democritus University of Thrace, Greece
Derek Groen	Brunel University London, UK
Loïc Guégan	UiT the Arctic University of Norway, Norway
Tobias Guggemos	University of Vienna, Austria
Serge Guillas	University College London, UK
Manish Gupta	Harish-Chandra Research Institute, India
Piotr Gurgul	SnapChat, Switzerland
Oscar Gustafsson	Linköping University, Sweden
Ahmed Abdeen Hamed	State University of New York at Binghamton, USA
Laura Harbach	Brunel University London, UK
Agus Hartoyo	TU Kaiserslautern, Germany
Ali Hashemian	Basque Center for Applied Mathematics, Spain
Mohamed Hassan	Virginia Tech, USA
Alexander Heinecke	Intel Parallel Computing Lab, USA
Jude Hemanth	Karunya University, India
Aochi Hideo	BRGM, France
Alfons Hoekstra	University of Amsterdam, The Netherlands
George Holt	UK Research and Innovation, UK
Maximilian Höb	Leibniz-Rechenzentrum der Bayerischen Akademie der Wissenschaften, Germany
Huda Ibeid	Intel Corporation, USA
Alireza Jahani	Brunel University London, UK
Jiří Jaroš	Brno University of Technology, Czechia
Qiao Jin	National Institutes of Health, USA
Zhong Jin	Computer Network Information Center, Chinese Academy of Sciences, China
David Johnson	Uppsala University, Sweden
Eleda Johnson	Imperial College London, UK
Piotr Kalita	Jagiellonian University, Poland
Drona Kandhai	University of Amsterdam, The Netherlands
Aneta Karaivanova	IICT-Bulgarian Academy of Science, Bulgaria
Sven Karbach	University of Amsterdam, The Netherlands
Takahiro Katagiri	Nagoya University, Japan
Haruo Kobayashi	Gunma University, Japan
Marcel Koch	KIT, Germany

Harald Koestler	University of Erlangen-Nuremberg, Germany
Georgy Kopanitsa	Tomsk Polytechnic University, Russia
Sotiris Kotsiantis	University of Patras, Greece
Remous-Aris Koutsiamanis	IMT Atlantique/DAPI, STACK (LS2N/Inria), France
Sergey Kovalchuk	Huawei, Russia
Slawomir Koziel	Reykjavik University, Iceland
Ronald Kriemann	MPI MIS Leipzig, Germany
Valeria Krzhizhanovskaya	University of Amsterdam, The Netherlands
Sebastian Kuckuk	Friedrich-Alexander-Universität Erlangen-Nürnberg, Germany
Michael Kuhn	Otto von Guericke University Magdeburg, Germany
Ryszard Kukulski	Institute of Theoretical and Applied Informatics, Polish Academy of Sciences, Poland
Krzysztof Kurowski	PSNC, Poland
Marcin Kuta	AGH University of Krakow, Poland
Marcin Łoś	AGH University of Krakow, Poland
Roberto Lam	Universidade do Algarve, Portugal
Tomasz Lamża	ACK Cyfronet, Poland
Ilaria Lazzaro	Università degli studi Magna Graecia di Catanzaro, Italy
Paola Lecca	Free University of Bozen-Bolzano, Italy
Byung Lee	University of Vermont, USA
Mike Lees	University of Amsterdam, The Netherlands
Leifur Leifsson	Purdue University, USA
Kenneth Leiter	U.S. Army Research Laboratory, USA
Paulina Lewandowska	IT4Innovations National Supercomputing Center, Czechia
Jingfa Li	Beijing Institute of Petrochemical Technology, China
Siyi Li	Imperial College London, UK
Che Liu	Imperial College London, UK
James Liu	Colorado State University, USA
Zhao Liu	National Supercomputing Center in Wuxi, China
Marcelo Lobosco	UFJF, Brazil
Jay F. Lofstead	Sandia National Laboratories, USA
Chu Kiong Loo	University of Malaya, Malaysia
Stephane Louise	CEA, LIST, France
Frédéric Loulergue	University of Orléans, INSA CVL, LIFO EA 4022, France
Hatem Ltaief	KAUST, Saudi Arabia
Zhiyong Lu	National Institutes of Health, USA

Stefan Luding	University of Twente, The Netherlands
Lukasz Madej	AGH University of Krakow, Poland
Luca Magri	Imperial College London, UK
Anirban Mandal	Renaissance Computing Institute, USA
Soheil Mansouri	Technical University of Denmark, Denmark
Tomas Margalef	Universitat Autònoma de Barcelona, Spain
Arbitrio Mariamena	Consiglio Nazionale delle Ricerche, Italy
Osni Marques	Lawrence Berkeley National Laboratory, USA
Maria Chiara Martinis	Università Magna Graecia di Catanzaro, Italy
Jaime A. Martins	University of Algarve, Portugal
Paula Martins	CinTurs – Research Centre for Tourism Sustainability and Well-being; FCT-University of Algarve, Portugal
Michele Martone	Max-Planck-Institut für Plasmaphysik, Germany
Pawel Matuszyk	Baker-Hughes, USA
Francesca Mazzia	University di Bari, Italy
Jon McCullough	University College London, UK
Pedro Medeiros	Universidade Nova de Lisboa, Portugal
Wen Mei	National University of Defense Technology, China
Wagner Meira	Universidade Federal de Minas Gerais, Brazil
Roderick Melnik	Wilfrid Laurier University, Canada
Pedro Mendes Guerreiro	Universidade do Algarve, Portugal
Isaak Mengesha	University of Amsterdam, The Netherlands
Wout Merbis	University of Amsterdam, The Netherlands
Ivan Merelli	ITB-CNR, Italy
Marianna Milano	Università Magna Graecia di Catanzaro, Italy
Magdalena Misiak	Howard University College of Medicine, USA
Jaroslaw Miszczak	Institute of Theoretical and Applied Informatics, Polish Academy of Sciences, Poland
Dhruv Mittal	University of Amsterdam, The Netherlands
Fernando Monteiro	Polytechnic Institute of Bragança, Portugal
Jânio Monteiro	University of Algarve, Portugal
Andrew Moore	University of California Santa Cruz, USA
Francisco J. Moreno-Barea	Universidad de Málaga, Spain
Leonid Moroz	Warsaw University of Technology, Poland
Peter Mueller	IBM Zurich Research Laboratory, Switzerland
Judit Munoz-Matute	Basque Center for Applied Mathematics, Spain
Hiromichi Nagao	University of Tokyo, Japan
Kengo Nakajima	University of Tokyo, Japan
Philipp Neumann	Helmut-Schmidt-Universität, Germany
Sinan Melih Nigdeli	Istanbul University – Cerrahpasa, Turkey

Fernando Nobrega Santos	University of Amsterdam, The Netherlands
Joseph O'Connor	University of Edinburgh, UK
Frederike Oetker	University of Amsterdam, The Netherlands
Arianna Olivelli	Imperial College London, UK
Ángel Omella	Basque Center for Applied Mathematics, Spain
Kenji Ono	Kyushu University, Japan
Hiroyuki Ootomo	Tokyo Institute of Technology, Japan
Eneko Osaba	TECNALIA Research & Innovation, Spain
George Papadimitriou	University of Southern California, USA
Nikela Papadopoulou	University of Glasgow, UK
Marcin Paprzycki	IBS PAN and WSM, Poland
David Pardo	Basque Center for Applied Mathematics, Spain
Anna Paszynska	Jagiellonian University, Poland
Maciej Paszynski	AGH University of Krakow, Poland
Łukasz Pawela	Institute of Theoretical and Applied Informatics, Polish Academy of Sciences, Poland
Giulia Pederzani	Universiteit van Amsterdam, The Netherlands
Alberto Perez de Alba Ortiz	University of Amsterdam, The Netherlands
Dana Petcu	West University of Timisoara, Romania
Beáta Petrovski	University of Oslo, Norway
Frank Phillipson	TNO, The Netherlands
Eugenio Piasini	International School for Advanced Studies (SISSA), Italy
Juan C. Pichel	Universidade de Santiago de Compostela, Spain
Anna Pietrenko-Dabrowska	Gdansk University of Technology, Poland
Armando Pinho	University of Aveiro, Portugal
Pietro Pinoli	Politecnico di Milano, Italy
Yuri Pirola	Università degli Studi di Milano-Bicocca, Italy
Ollie Pitts	Imperial College London, UK
Robert Platt	Imperial College London, UK
Dirk Pleiter	KTH/Forschungszentrum Jülich, Germany
Paweł Poczekajło	Koszalin University of Technology, Poland
Cristina Portalés Ricart	Universidad de Valencia, Spain
Simon Portegies Zwart	Leiden University, The Netherlands
Anna Procopio	Università Magna Graecia di Catanzaro, Italy
Ela Pustulka-Hunt	FHNW Olten, Switzerland
Marcin Płodzień	ICFO, Spain
Ubaid Qadri	Hartree Centre – STFC, UK
Rick Quax	University of Amsterdam, The Netherlands
Cesar Quilodran Casas	Imperial College London, UK
Andrianirina Rakotoharisoa	Imperial College London, UK
Celia Ramos	University of the Algarve, Portugal

Robin Richardson	Netherlands eScience Center, The Netherlands
Sophie Robert	University of Orléans, France
João Rodrigues	Universidade do Algarve, Portugal
Daniel Rodriguez	University of Alcalá, Spain
Marcin Rogowski	Saudi Aramco, Saudi Arabia
Sergio Rojas	Pontifical Catholic University of Valparaiso, Chile
Diego Romano	ICAR-CNR, Italy
Albert Romkes	South Dakota School of Mines and Technology, USA
Juan Ruiz	University of Buenos Aires, Argentina
Tomasz Rybotycki	IBS PAN, CAMK PAN, AGH, Poland
Katarzyna Rycerz	AGH University of Krakow, Poland
Grażyna Ślusarczyk	Jagiellonian University, Poland
Emre Sahin	Science and Technology Facilities Council, UK
Ozlem Salehi	Özyeğin University, Turkey
Ayşin Sancı	Altinay, Turkey
Christopher Savoie	Zapata Computing, USA
Ileana Scarpino	University "Magna Graecia" of Catanzaro, Italy
Robert Schaefer	AGH University of Krakow, Poland
Ulf D. Schiller	University of Delaware, USA
Bertil Schmidt	University of Mainz, Germany
Karen Scholz	Fraunhofer MEVIS, Germany
Martin Schreiber	Université Grenoble Alpes, France
Paulina Sepúlveda-Salas	Pontifical Catholic University of Valparaiso, Chile
Marzia Settino	Università Magna Graecia di Catanzaro, Italy
Mostafa Shahriari	Basque Center for Applied Mathematics, Spain
Takashi Shimokawabe	University of Tokyo, Japan
Alexander Shukhman	Orenburg State University, Russia
Marcin Sieniek	Google, USA
Joaquim Silva	Nova School of Science and Technology – NOVA LINCS, Portugal
Mateusz Sitko	AGH University of Krakow, Poland
Haozhen Situ	South China Agricultural University, China
Leszek Siwik	AGH University of Krakow, Poland
Peter Sloot	University of Amsterdam, The Netherlands
Oskar Slowik	Center for Theoretical Physics PAS, Poland
Sucha Smanchat	King Mongkut's University of Technology North Bangkok, Thailand
Alexander Smirnovsky	SPbPU, Russia
Maciej Smołka	AGH University of Krakow, Poland
Isabel Sofia	Instituto Politécnico de Beja, Portugal
Robert Staszewski	University College Dublin, Ireland

Magdalena Stobińska	University of Warsaw, Poland
Tomasz Stopa	IBM, Poland
Achim Streit	KIT, Germany
Barbara Strug	Jagiellonian University, Poland
Diana Suleimenova	Brunel University London, UK
Shuyu Sun	King Abdullah University of Science and Technology, Saudi Arabia
Martin Swain	Aberystwyth University, UK
Renata G. Słota	AGH University of Krakow, Poland
Tseden Taddese	UK Research and Innovation, UK
Ryszard Tadeusiewicz	AGH University of Krakow, Poland
Claude Tadonki	Mines ParisTech/CRI – Centre de Recherche en Informatique, France
Daisuke Takahashi	University of Tsukuba, Japan
Osamu Tatebe	University of Tsukuba, Japan
Michela Taufer	University of Tennessee, USA
Andrei Tchernykh	CICESE, Mexico
Kasim Terzic	University of St Andrews, UK
Jannis Teunissen	KU Leuven, Belgium
Sue Thorne	Hartree Centre – STFC, UK
Ed Threlfall	United Kingdom Atomic Energy Authority, UK
Vinod Tipparaju	AMD, USA
Pawel Topa	AGH University of Krakow, Poland
Paolo Trunfio	University of Calabria, Italy
Ola Tørudbakken	Meta, Norway
Carlos Uriarte	University of the Basque Country, BCAM – Basque Center for Applied Mathematics, Spain
Eirik Valseth	University of Life Sciences & Simula, Norway
Rein van den Boomgaard	University of Amsterdam, The Netherlands
Vítor V. Vasconcelos	University of Amsterdam, The Netherlands
Aleksandra Vatian	ITMO University, Russia
Francesc Verdugo	Vrije Universiteit Amsterdam, The Netherlands
Karin Verspoor	RMIT University, Australia
Salvatore Vitabile	University of Palermo, Italy
Milana Vuckovic	European Centre for Medium-Range Weather Forecasts, UK
Kun Wang	Imperial College London, UK
Peng Wang	NVIDIA, China
Rodrigo Weber dos Santos	Federal University of Juiz de Fora, Brazil
Markus Wenzel	Fraunhofer Institute for Digital Medicine MEVIS, Germany

Contents – Part I

ICCS 2024 Main Track Full Papers

Effects of Wind on Forward and Turning Flight of Flying Cars Using Computational Fluid Dynamics

Taiga Magata[1]([✉]), Ayato Takii[2], Masashi Yamakawa[1], Yusei Kobayashi[1], Shinichi Asao[3], and Seiichi Takeuchi[3]

[1] Kyoto Institute of Technology, Matsugasaki, Sakyo-ku, Kyoto 606-8585, Japan
m3623028@kit.edu.ac.jp
[2] RIKEN Center for Computational Science, 7-1-26 Minatojima-minami-machi, Chuo-ku, Kobe 650-0047, Hyogo, Japan
[3] College of Industrial Technology, 1-27-1, Amagasaki 661-0047, Hyogo, Japan

Abstract. We have been using various environments and spaces to meet high transportation demands. However, traffic congestion, deteriorating transportation infrastructure, and environmental pollution have become current social problems. To solve these problems, flying vehicles that use near ground space (NGS) are attracting attention. In order to develop such vehicles efficiently, highly accurate computer simulation technology is required. In this study, computer simulations are performed by coupling fluid and rigid body motions using two calculation methods. One is the moving computational domain method, in which the object and computational domain are moved as a single unit to represent the motion of the object and the flow around the body, and the other is the multi-axis sliding mesh method, in which physical quantities are transferred at the boundaries to reproduce the motion of objects with different motions, such as rotating parts. Because the flying car in the development stage is small and has a shape that obtains thrust from multiple propellers, the insertion of disturbances was considered because of the possible effects of wind on the aircraft during actual flight. In this study, we attempted to clarify the effect of wind on the flying car by performing flight simulations in six patterns, one with no wind and the other with a disturbance inserted that causes a headwind during forward flight and a crosswind during turning flight.

Keywords: CFD · Flying Car · turn flight

1 Introduction

To date, we have used a variety of spaces and environments – above ground, above water, underwater, in the air, and underground – to meet the growing transportation demands of technological development. However, as populations increase around the world, urban traffic congestion, infrastructure deterioration, and environmental pollution have become problems. To address these problems, development of the next generation

L. Franco et al. (Eds.): ICCS 2024, LNCS 14832, pp. 3–18, 2024.
https://doi.org/10.1007/978-3-031-63749-0_1

of air mobility, the flying car, is underway around the world. This flying car is said to use near ground space (NGS), which has not been used in the past, and to significantly reduce noise and exhaust emissions inside the vehicle by installing electric rotors [1]. In addition, various shapes of flying car currently under development are envisioned, most of which are fixed-wing aircraft and rotary-wing aircraft (VTOL) [2]. In particular, many rotary-wing aircraft (VTOLs) in the development stage have many environmental considerations, such as fully electric-driven and hybrid types [3]. Thus, the flying car is a new vehicle that solves current problems.

Flying cars, which are expected to become as common to us as cars, ships, trains, and airplanes in the future, require high safety standards to fly over towns and urban areas. In other words, the design and development of a new flying car airframe requires enormous cost and time. Therefore, various studies on flying cars have been conducted in order to develop actual aircraft more efficiently. Some studies include not only fluid dynamics but also control characteristics, such as wind tunnel tests at high incidence angles using actual propellers [4] and the relationship between VTOL rotors and electric motors in propeller pitch maneuvering [5]. Furthermore, not only research related to actual aircraft testing, but also research related to CFD assuming actual aircraft, such as research on aerodynamic interaction in a two-rotor system with a front rotor and rear rotor assuming a multi-rotor type aircraft [6], and research involving aerodynamic interaction caused by the positional configuration of the propeller [7]. However, most of the studies using CFD have focused on one part of the aircraft or one function, and there have been few reports on numerical flight simulation of eVTOL flight. In response to this current situation, a turning flight simulation [8], which solves the problems of complex attitude changes, acceleration, and long-distance travel that had been considered difficult to solve, and an investigation of the aerodynamic effects on the aircraft's attitude in flight by bringing the propeller to an abrupt stop [9] were conducted. Because these studies were conducted under no-wind conditions, and because many flying cars in the development stage obtain thrust from the rotation of each propeller and are relatively small in length compared to conventional aircraft, there is a new concern that wind will have a large effect during actual flight. Regarding the effect of wind, which has been raised as a new problem, studies have been conducted on stabilization [10] under the influence of wind disturbance for helicopters, which are conventional aircraft. However, the effect of wind on the airframe of an eVTOL, whose rotor diameter is smaller than that of a helicopter, has not been studied because it is still in the early stages of development. Therefore, in this study, forward and turning flight simulations are performed by inserting wind as a disturbance expected during actual flight, in contrast to the previous study of turning flight simulations [8]. In this study, six types of winds up to strong winds on the Beaufort scale (a measure of wind speed) were given as disturbances. This will attempt to clarify the effects of wind on the flight of flying car, which have not been studied before. We believe that this simulation will enable us to understand the effects of wind on airframes in advance of the development of actual airframes, leading to significant reductions in design time and development costs.

This study uses an octorotor-type airframe model with four double-reversing propeller units as the flying car. The flying car flight simulation uses the forward Euler method and Crank-Nicolson method of discretization. At each time step, the motion of the aircraft is calculated using the governing equations of the rigid body, followed by a weakly coupled simulation in which the flow field around the flying car is solved using the governing equations of the fluid. In addition, the moving computational domain (MCD) method [12] based on the unstructured moving grid finite volume method [11], which not only allows for complex motions but also eliminates the need for space constraints. To compute the relative motion between objects with this MCD method, physical quantities are transferred at the boundaries, and the sliding mesh method [13] is used in combination to reproduce the motion of objects with different motions, such as rotating parts. This method not only reproduces the rotation of a flying car's propeller, but also enables the transmission of physical quantities generated by the propeller's rotational motion. The computations were conducted under the unstructured parallel computational environment [14].

The objective of this study is to establish computer simulation as a new method to solve the problem of developing a new flying car by using CFD to reproduce realistic phenomena.

2 Numerical Approach

2.1 Fundamental Equation of Fluid

The three-dimensional Euler equation, which is the basic equation for a non-viscous compressible fluid, is used as the Fundamental equation of fluid from the viewpoint of computational efficiency. The three-dimensional Euler equation:

$$\frac{\partial q}{\partial t} + \frac{\partial E}{\partial x} + \frac{\partial F}{\partial y} + \frac{\partial G}{\partial z} = 0, \tag{1}$$

$$q = \begin{pmatrix} \rho \\ \rho u \\ \rho v \\ \rho w \\ e \end{pmatrix}, E = \begin{pmatrix} \rho u \\ \rho u^2 + p \\ \rho uv \\ \rho uw \\ u(e+p) \end{pmatrix}, F = \begin{pmatrix} \rho v \\ \rho uv \\ \rho v^2 + p \\ \rho vw \\ v(e+p) \end{pmatrix}, G = \begin{pmatrix} \rho v \\ \rho uv \\ \rho v^2 + p \\ \rho vw \\ v(e+p) \end{pmatrix} \tag{2}$$

where q is the conserved quantity vector and E, F, G are the inviscid flux vectors in the x, y, z directions. ρ is the density of the fluid, u, v, w are the velocity components in the x, y, z directions, and e is the total energy per unit volume. However, the above equation is dimensionless. The p in Eq. (2) is the pressure of the gas, which can be obtained by the equation of state of an ideal gas shown in Eq. (3) when the gas is assumed to be an ideal gas. Note that γ is the specific heat ratio.

$$p = (\gamma - 1)\{e - \tfrac{1}{2}\rho(u^2 + v^2 + w^2)\} \tag{3}$$

In this study, the specific heat ratio was set to 1.4 because the air temperature was set to 10[°C].

2.2 Fundamental Equation of Rigid Body

The three-dimensional Newton-Euler equation is used as the governing fundamental equation of rigid body to represent translational and rotational motion in three-dimensional space. The three-dimensional Newton-Euler equation:

$$m\frac{d\ddot{r}}{dt^2} = f \tag{4}$$

$$I\frac{d\dot{\omega}}{dt} + \omega \times I\omega = T \tag{5}$$

where m is the mass of the rigid body and I is the inertia tensor. In addition, r, f, ω and T represent the position vector, external force vector, angular velocity vector, and torque vector in three-dimensional space, respectively. However, because the flying car used in this research has no moving parts other than the propeller, the aircraft is assumed to be a rigid body with a constant center of gravity. The equations in the previous section and the above equation are computed coupled to simulate the six degrees of freedom.

2.3 Moving Computational Domain Approach

The calculations are performed using a moving computational domain method [12] based on the unstructured moving lattice finite volume method [11]. A schematic diagram of the moving computational domain method is shown in Fig. 1. This method makes it possible to move the computational domain along with the object, thus eliminating restrictions due to the size of the computational domain. In other words, this method allows objects to move in an infinite region. In addition, it satisfies each of the following two problems that are often encountered when dealing with moving boundary problems.

- Strictly satisfying the conservation law of the fluid on all moving computational grids.
- To reproduce the moving boundary while maintaining the computational grid geometry for highly accurate fluid calculations.

As shown above, the moving computational domain method enables calculations that strictly satisfy the geometric conservation law even when the lattice moves.

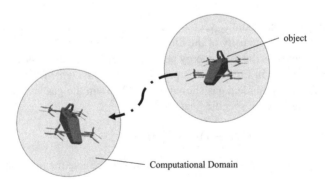

Fig. 1. Moving Computational Domain Method

2.4   Multi-axis Sliding Mesh Approach

The MCD method enables the representation of free motion of an object by moving the computational domain along with the object. However, the flying car used in this study is an octorotor-type airframe model with four double-reversing propeller units, so it is necessary to represent the relative motion between multiple objects on a boundary-fitted grid. Therefore, the sliding mesh method [13] has been used to arbitrarily divide a region and slide the regions on the boundary surface. In particular, the multi-axis sliding mesh method corresponds to the region segmentation with multiple rotation axes of the sliding mesh method. The simplicity of the algorithm of this method reduces the computational load and satisfies the user without overlapping computational domains. Figure 2 shows a schematic of the multi-axis sliding mesh method, which divides a region with multiple rotation axes. In this study, the multi-axis sliding mesh method shown in Fig. 2 was used to reproduce the rotation of the rotor. Specifically, as shown in Fig. 3, eight rotor regions are provided for the fuselage region to make up the overall calculation region. In addition, this research can not only reproduce the rotational motion of a flying car by the multi-axis sliding mesh method, but also realize the physical quantity transfer obtained by the rotational motion.

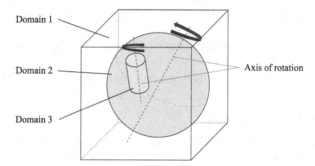

Fig. 2. Conceptual Diagram of Multi-axis Sliding Mesh Method

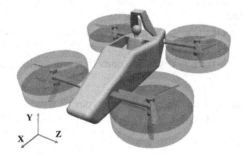

Fig. 3. Flying car model and Sliding mesh interface

3 Simulation Summary

3.1 Flight Simulation Conditions

In this study, a total of seven forward and turning flight simulations were performed under no-wind conditions and with six different winds inserted as expected disturbances during actual flight. Two types of flight simulation conditions are shown below. The first type is a simulation overview of an Acceleration turning flight when the aircraft continues to accelerate its speed after transitioning from forward flight to a turn. The second type is constant velocity turning flight, in which the aircraft is controlled to track the target speed after transitioning from forward flight to turning flight. In both simulations, the wind is set to be headwind during forward flight. Since the turning flight is directed toward the negative direction of the z-axis, the crosswind is in the direction that agitates the turning motion during the turning flight.

• Acceleration turning flight.

1) The initial conditions are either no wind or a constant wind blowing in the negative direction along the x-axis.
2) The target speed of 60 km/h is given in the x-axis direction.
3) As soon as the difference between the forward speed of the aircraft and the target speed is less than 2 km/h, the target angle in the roll direction is set to 30° and the aircraft begins to turn.
4) The acceleration during forward flight is continued after the transition to turning flight.
5) The aircraft performs a turning flight and ends the turning flight when the yaw angle exceeds 90°.

Constant velocity turning flight.

1) The initial conditions are either no wind or a constant wind blowing in the negative direction along the x-axis.
2) The target speed of 60 km/h is given in the x-axis direction.
3) As soon as the difference between the forward speed of the aircraft and the target speed is less than 2 km/h, the target angle in the roll direction is set to 30° and the aircraft begins to turn.
4) After the turning flight transition, the aircraft continues to maintain the target forward speed of 60 km/h.
5) The aircraft performs a turning flight and ends the turning flight when the yaw angle exceeds 90°.

However, the initial conditions in (1) are six types of wind given as disturbances in 10 km/h increments from 10 km/h to 60 km/h, and no wind conditions, for a total of seven patterns.

3.2 Computation Model

Figure 4 shows the computational domain of the airframe and the area around the airframe. The spherical computational domain is approximately 30 times larger than the total length of the aircraft, which is the characteristic length, to account for the downwash effect caused by the propellers. As shown in Fig. 4, the grids around the aircraft are finely detailed so that the flow field around the aircraft can be obtained with high accuracy. Figure 5(a) shows the computational model of the oct-rotor machine with four pairs of eight double-reversing rotors used in this simulation. Figure 5(b) shows the computational grid corresponding to the rotor area propeller locations shown in Fig. 3. However, this model was created based on the Japanese SkyDrive's SD-03[15]. The model was created with an overall length of 4 m and a weight of 400 kg. The overall length of the aircraft was used as a characteristic length for the calculations in the simulations. As shown in Fig. 4, the computational model used in this study is an unstructured lattice and was created using MEGG3D [16, 17]. In addition, the total lattice count is about 3 million, and each propeller lattice count is about 120 thousand. The minimum lattice width on the calculated lattice is approximately 6 [mm] of the rotor surface.

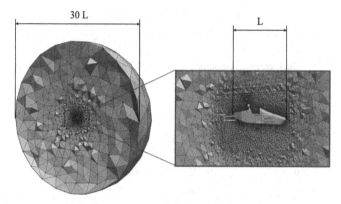

Fig. 4. Computational domain and grid size around the Flying car model

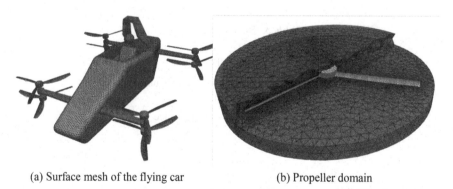

(a) Surface mesh of the flying car   (b) Propeller domain

Fig. 5. Surface grid of Flying car model

3.3 Calculation Conditions

Table 1 shows characteristic values for this simulation. Characteristic length, density, and velocity are shown as characteristic values. The characteristic length is the total length of the aircraft as shown in Fig. 4, the characteristic density is the density of the air, and the characteristic speed is the speed of sound. Table 2 shows the wind speeds u, v and w given as initial conditions in the x, y and z axes. The wind speeds v and w on the y and z axes were 0 km/h. The wind speed u on the x axis was given as no wind (0 km/h) or six different wind speeds from 10 km/h to 60 km/h in the negative direction of the x-axis at 10 km/h intervals, for a total of seven simulation patterns. Table 3 shows the boundary conditions for this simulation. Three types of boundary conditions are set: a slip wall boundary, a Riemann invariant boundary condition, and a sliding mesh interface using the sliding mesh method.

Table 1. Characteristic Values

Density of the air	1.247 [kg/m^3]
Characteristic velocity	340.29 [m/s]
Characteristic length	4.0 [m]

Table 2. Initial condition

u	0, -10 ~ -60 [km/h]
v	0 [km/h]
w	0 [km/h]

Table 3. Boundary condition

Aircraft surface	Slip wall condition
Outer boundary	Riemann invariant boundary condition
Other boundary	Sliding mesh interface

3.4 Attitude Control

The model used in this simulation represents the rotation of the aircraft in pitch and roll direction by the difference in thrust between the front-back and left-right directions. The yaw rotation of the aircraft is expressed by using the counter-torque generated by the rotation of the propeller [18]. Therefore, control is performed by the difference in propeller speed. Figure 6 shows how the posture is controlled by the different number of rotations to reproduce each operation. However, the throttle, aileron, rudder, and elevator shown in Fig. 6 represent operations to raise and lower the aircraft and to promote rolling, yawing, and pitching of the aircraft. The model used in this study, which was created based on the Japanese SkyDrive SD-03 [15], is based on a rotational speed of 1930 rpm when hovering. Figure 7 shows the direction of rotation of each propeller (clockwise: CW, counter-clockwise: CCW).

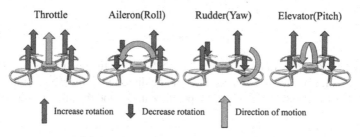

Fig. 6. Attitude control by rotation speed

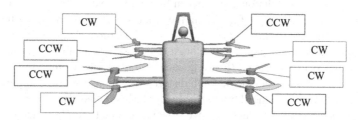

Fig. 7. Rotation direction of each propeller

4   Simulation Results

4.1   Forward Flight Simulation

The results of the aircraft's forward flight toward the target forward speed of 60 km/h are shown in this section. Figure 8 shows a graph of the relationship between time and aircraft forward speed. Simulation results of (a) an acceleration turning flight and (b) a constant velocity turning flight to follow the target speed of 60 km/h are shown in (a) and (b), respectively. The simulation results are for the no-wind condition and for a constant wind speed of 10 km/h to 60 km/h in 10 km/h increments along the negative x-axis, respectively. Figure 8 shows that the time to reach the target speed of 60 km/h in the forward direction is slowed down as the headwind is increased in both results. Figure 8(b) shows that the target speed is maintained near 60 km/h compared to (a), indicating that the constant velocity turning flight can be reproduced.

Figures 9(a) and (b) show the results for no wind and a constant wind speed of 60 km/h for the negative x-axis, respectively, and the visualization results during forward flight with the velocity isosurface of 25 m/s displayed for each result. Figure 9 is presented in dimensionless quantities for consistency with other papers. The dimensioned quantities can be calculated using the characteristic values in Table 1. From the comparison in Fig. 9, it can be confirmed that the flow in front of the fuselage in (b) is unstable due to disturbance, even though the fuselage pitch angle is controlled to be 15 ° during forward flight and the projected area from the front of the fuselage is equal.

Furthermore, in order to quantitatively evaluate the effect of wind during forward flight, a comparison of the drag force acting on the airframe during forward flight is performed for each simulation result. The drag force F:

$$F = T \sin \theta - F_x \tag{6}$$

where T is the thrust generated by the eight propellers, θ is the pitch angle of the aircraft in forward flight, and F_x is the force applied in the direction of the x axis of the aircraft. The thrust T is obtained by the propeller specific value A and the propeller speed n as in Eq. (7).

$$T = An^2 \tag{7}$$

Figure 10 shows the relationship between time and x-axis force F_x during forward flight. Figure 10 shows that the force F_x in the x-axis direction decreases as the disturbance is increased. Table 4 shows the average values of the force F_x in the x-axis direction from 4 to 8[s], the average values of the thrust force $T\sin\theta$ from 4 to 8[s], and the drag force F calculated from formula (6) using each average value. Comparing the results for the no-wind condition and the 60 km/h disturbance, Table 4 shows that the force F_x in the x-axis direction is reduced by 26.8% due to the headwind. This decrease in force F_x in the x-axis direction is thought to have slowed the time to reach the target speed as shown in Fig. 9. It was also confirmed that the drag force exerted on the aircraft during forward flight in no-wind conditions acts 349[N]. From the drag force during forward flight in no wind, the force F_{wind} exerted by the wind on the aircraft is calculated using Eq. (8).

$$F_{wind} = F_i - F_0 \tag{8}$$

F_i is the drag force F calculated by substituting the force F_x for the constant wind disturbance of 10 km/h to 60 km/h in the negative direction of the x-axis into Eq. (6), and F_0 is the drag force F calculated from the force F_x in the no wind condition into Eq. (6). Table 5 shows the force F_{wind} exerted on the aircraft by the wind calculated from Eq. (8). From Table 5, the increase in force to force F_{60}, calculated from the force F_{10} when subjected to a disturbance of 10 km/h, was obtained as 374%. The above confirms that the wind exerts a pressure of 56.6[Pa] on a projected area of 2.42[m^2] from the forward direction during high winds in the Beaufort scale.

4.2 Acceleration Turning Flight

As soon as the difference between the forward speed of the aircraft and the target speed is less than 2 km/h, the target angle in the roll direction is set to 30° and the aircraft begins

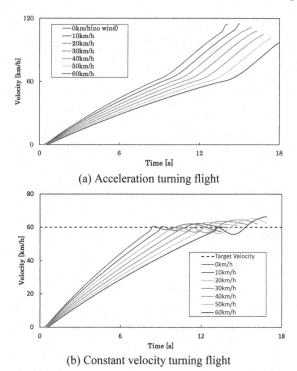

(a) Acceleration turning flight

(b) Constant velocity turning flight

Fig. 8. Time and aircraft forward speed

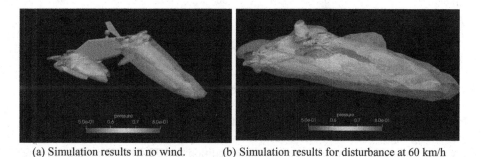

(a) Simulation results in no wind.   (b) Simulation results for disturbance at 60 km/h

Fig. 9. Visualization of velocity isosurfaces during forward flight

to turn. Accelerated turning flight simulation results are shown in this section when acceleration is continued after turning flight as it was during forward flight. Figure 11 shows a graph of the flight trajectory of a turning aircraft viewed from above with the x-axis on the horizontal axis and the z-axis on the vertical axis. Figure 11 shows the simulation results when no wind or a constant wind speed of 10 km/h to 60 km/h is applied as a disturbance in the negative x-axis direction, respectively. In addition, the trajectory of the turning flight in Fig. 11 shows a forward flight in the positive direction from 0 on the x-axis and then a turning flight in the negative direction on the z-axis.

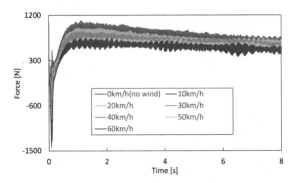

Fig. 10. Force in the forward direction of the aircraft.

Table 4. Various Forces Generated by Each Simulation

v[km/h]	0	10	20	30	40	50	60
$T\sin\theta$ [N]	1115	1101	1086	1070	1057	1046	1039
F_x[N]	766	723	684	652	620	587	552
F [N]	349	378	402	419	436	459	487

Table 5. Drag force on the aircraft and the rate of increase of that drag force

v[km/h]	10	20	30	40	50	60
F_x[N]	28.9	52.2	69.4	87.1	110	137
Rate of increase [%]	0.0	80.6	140	201	281	374

The results of each trajectory are overlaid for comparison. Figure 11 shows that the turn diameter decreases as the constant wind speed given as a disturbance is increased. Furthermore, it can be seen that the distance traveled by the aircraft is shortening after the turning flight transition. The results show that the plane moved 120[m] along the negative z-axis after turning in the no-wind condition and 72.4[m] along the negative z-axis when a disturbance of 60 km/h was applied. When a disturbance of 60 km/h was applied, the distance traveled after the turn was 39.7% less than that for a turning flight in no wind conditions. The reason for this is considered to be that the Yaw angle of 90°, which is the condition for the end of the turning flight, was reached early because the disturbance wind was set in the direction that agitated the turning flight.

Figures 12(a) and (b) show the results for no wind and a constant wind speed of 60 km/h for the negative x-axis, respectively, and the visualization results during turning flight with the velocity isosurface of 25 m/s displayed for each result. Figure 12 is presented in dimensionless quantities for consistency with other papers. The dimensioned quantities can be calculated using the characteristic values in Table 1. After the turning flight transition, the wind is set to agitate the turning, and it can be confirmed that the results for a disturbance of 60 km/h have an effect on the flow toward the side of the aircraft compared to the no-wind condition. In addition, the continued acceleration of the aircraft's forward speed after the turning flight transition also confirms the turbulence of the flow in front of the aircraft.

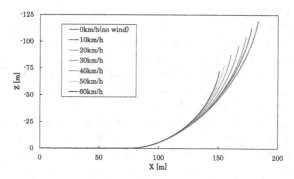

Fig. 11. Flight trajectory during acceleration turn flight

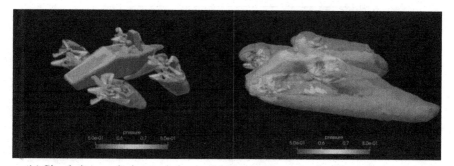

(a) Simulation results in no wind   (b) Simulation results for disturbance at 60 km/h

Fig. 12. Visualization of velocity isosurfaces during sharp turn flight

4.3 Constant Velocity Turning Flight

This section shows the results of a constant velocity turning flight simulation in which the aircraft was controlled to maintain its speed in the direction of travel at 60 km/h, which was given as the target speed during forward flight, even after the turning flight. Figure 13 shows a graph of the trajectory of a turning aircraft viewed from above with the x-axis on the horizontal axis and the z-axis on the vertical axis. As in Fig. 11, Fig. 13 shows the simulation results when no wind or a constant wind speed of 10 km/h to 60 km/h is applied as a disturbance in the negative x-axis direction, respectively. In addition, each trajectory result is overlaid for comparison. When constant velocity turning flight was used, there was no difference in the distance traveled in the negative direction of the z-axis after turning flight compared to acceleration turning flight.

Therefore, to quantify the effect of disturbance during constant-speed turning flight, differences in rotation speed are compared. Figure 14 is a zoomed-in plot of only the rotation speed results for each simulation result during constant-speed turning flight. From Fig. 14, it was observed that a large rotation speed amplitude was drawn when the flight transitioned to a turning flight. In particular, the amplitude of the results for a constant wind speed of 20 km/h to 60 km/h is smaller than that for the no-wind condition. The reason for this is thought to be that the wind was inserted as a disturbance in the direction that agitated the circling flight, which facilitated the transition to the circling flight. The amplitude of the rotation speed when 10 km/h was given as a disturbance was larger than that in the no-wind condition, and the amplitude of the rotation speed did not decrease proportionally as the disturbance increased. The reason for this is considered to be the influence of the control. In this study, the rotation speed is calculated by the proportional – derivative (PD) controller using thrust and angular velocities for roll, pitch, and yaw for the three axes as control inputs. The control gains in the PD control are considered to be inappropriate. Furthermore, the control gains may also be the reason why the forward speed of the aircraft in Fig. 8(b) did not completely follow the target speed.

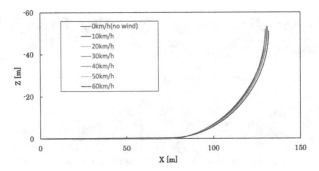

Fig. 13. Flight trajectory at constant speed turning

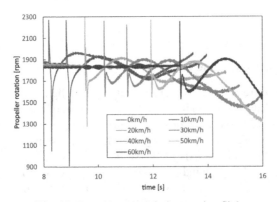

Fig. 14. Rotation speed during turning flight

5   Conclusions

The MCD method and the multi-axis sliding mesh method were used together to not only reproduce the motion of the propeller, but also to reproduce forward and turning flight by solving the flow field around the flying car's fuselage. The challenges of complex attitude change, acceleration, and long-distance travel, which had been considered difficult, were solved, and disturbances, which had not been considered previously, were successfully inserted. These results confirm that the time required to reach the target forward speed increases when a headwind disturbance is applied to the aircraft during forward flight. The simulation results also confirmed the reproduction of a headwind by calculating the force in the x-axis direction during forward flight, the thrust by the propeller, and the drag force exerted by the wind on the fuselage. In addition, two patterns of circling flight were simulated: acceleration turning flight, in which acceleration is continued after the circling flight transition, and constant velocity turning flight, in which turning is performed at a constant speed. In the acceleration turning simulation, the crosswind was set to agitate the turning motion, resulting in a turning trajectory that achieves a yaw angle of 90deg faster as the disturbance is increased. In constant velocity turning flight, the turning trajectory was compared to the RPM during the turning flight, suggesting that the control gain used for PD control was inappropriate for constant velocity turning flight. Future prospects include the development of a control design and simulation environment that can perfectly follow the set forward and turning flight trajectories.

Acknowledgments. This paper is based on results obtained from a project, JPNP14004, subsidized by the New Energy and Industrial Technology Development Organization (NEDO).

References

1. Pan, G., Alouini, M.-S.: Flying car transportation system: advances, techniques, and challenges. IEEE Access **9**, 24586–24603 (2021)
2. Rostami, M., Bardin, J., Neufeld, D., Chung, J.: EVTOL tilt-wing aircraft design under uncertainty using a multidisciplinary possibilistic approach. Aerospace **10**, 718 (2023)
3. Ugwueze, O., Statheros, T., Bromfield, M.A., Horri, N.: Trends in eVTOL aircraft development: the concepts, enablers and challenges. In: Proceedings of the AIAA SCITECH 2023 Forum, National Harbor, MD, USA, 23–27 January 2023
4. Simmons, B.M., Hatke, D.B.: Investigation of high incidence angle propeller aerodynamics for subscale eVTOL aircraft. NASA TM–20210014010, May 2021
5. Pavel, M.D.: Understanding the control characteristics of electric vertical take-off and landing (eVTOL) aircraft for urban air mobility. Aerosp. Sci. Technol., 107143 (2021)
6. Healy, R., Misiorowski, M., Gandhi, F.: A systematic CFD-based examination of rotor-rotor separation effects on interactional aerodynamics for large eVTOL aircraft. In: Proceedings of the 75th Annual Forum, (Philadelphia), VFS International, May 2019
7. Zanotti, A., Piccinini, R., Tugnoli, M.: Numerical investigation of the rotor-rotor aerodynamic interaction of evtol configurations by a mid-fidelity approach. In: Proceedings of the 47th European Rotorcraft Forum, Virtual Event, 7–9 September 2021
8. Takii, A., et al.: Turning flight simulation with fluid-rigid body interaction for flying car with contra-rotating propellers. In: ICCS, pp. 566–577 (2023)

9. Naoya, T., et al.: Numerical simulation of the octorotor flying car in sudden rotor stop. In: ICCS, pp. 33–46 (2023)
10. Danapalasingam, K.A., et al.: Robust helicopter stabilization in the face of wind disturbance. In: Proceedings of the 49th IEEE Conference on Decision and Control, pp. 3832–3837 (2010)
11. Yamakawa, M., et al.: Numerical simulation for a flow around body ejection using an axisymmetric unstructured moving grid method. Comput. Therm. Sci. **4**(3), 217–223 (2012)
12. Yamakawa, M., et al.: Numerical simulation of rotation of intermeshing rotors using added and eliminated mesh method. Procedia Comput. Sci. **108**, 1883–1892 (2017)
13. Takii, A., Yamakawa, M., Asao, S., Tajiri, K.: Six degrees of freedom flight simulation of tilt-rotor aircraft with nacelle conversion. J. Comput. Sci. **44**, 101164 (2020)
14. Yamakawa, M., et al.: Domain decomposition method for unstructured meshes in an OpenMP computing environment. Comput. Fluids **45**, 168–171 (2011)
15. SkyDrive Inc Homepage. https://en.skydrive2020.com/. Accessed 23 Jan 2023
16. Ito, Y., Nakahashi, K.: Surface triangulation for polygonal models based on CAD data. Int. J. Numer. Meth. Fluidsnt. J. Numer. Meth. Fluids **39**(1), 75–96 (2002)
17. Ito, Y.: Challenges in unstructured mesh generation for practical and efficient computational fluid dynamics simulations. Comput. Fluids **85**(1), 47–52 (2013)
18. Gomi, R., Takii, A., Yamakawa, M., Asao, S., Takeuchi, S., Nishimura, M.: Flight simulation from takeoff to yawing of eVTOL airplane with coaxial propellers by fluid-rigid body interaction. Adv. Aerodyn. **5**(1), 1–15 (2023)

DP-PINN: A Dual-Phase Training Scheme for Improving the Performance of Physics-Informed Neural Networks

Da Yan and Ligang He[✉]

Department of Computer Science, University of Warwick, Coventry, UK
{da.yan,ligang.he}@warwick.ac.uk

Abstract. Physics-Informed Neural Networks (PINNs) are a promising application of deep neural networks for the numerical solution of nonlinear partial differential equations (PDEs). However, it has been observed that standard PINNs may not be able to accurately fit all types of PDEs, leading to poor predictions for specific regions in the domain. A common solution is to partition the domain by time and train each time interval separately. However, this approach leads to the prediction errors being accumulated over time, which is especially the case when solving "stiff" PDEs. To address these issues, we propose a new PINN training scheme, called DP-PINN (Dual-Phase PINN). DP-PINN divides the training into two phases based on a carefully chosen time point t_s. The phase-1 training aims to generate the accurate solution at t_s, which will serve as the additional intermediate condition for the phase-2 training. New sampling strategies are also proposed to enhance the training process. These design considerations improve the prediction accuracy significantly. We have conducted the experiments to evaluate DP-PINN with both "stiff" and non-stiff PDEs. The results show that the solutions predicted by DP-PINN exhibit significantly higher accuracy compared to those obtained by the state-of-the-art PINNs in literature.

Keywords: Physics-Informed Neural Networks · Partial Differential Equations · Model Training · Data Sampling

1 Introduction

Traditional physics-based numerical methods [1,6,15] have had great success in solving partial differential equations (PDEs) for a variety of scientific and engineering problems. While these methods are accurate, they are computationally intensive for complex problems such as nonlinear partial differential equations and often require problem-specific techniques. Over the past decade, data-driven methods have gained significant attention in various areas of science and engineering. These methods can identify highly nonlinear mappings between inputs and outputs, potentially replacing or augmenting expensive physical simulations.

© The Author(s), under exclusive license to Springer Nature Switzerland AG 2024
L. Franco et al. (Eds.): ICCS 2024, LNCS 14832, pp. 19–32, 2024.
https://doi.org/10.1007/978-3-031-63749-0_2

However, typical data-driven deep learning methods tend to ignore a physical understanding of the problem domain [5].

In order to incorporate physical priors into the model training, a new deep learning technique, Physics-Informed Neural Networks (PINNs) [10], has been proposed. Propelled by vast advances in computational capabilities and training algorithms, including the availability of automatic differentiation methods, PINNs combine the idea of using neural networks as generalized function approximator for solving PDEs [4] and the idea of using the system of PDEs as physical priors to constrain the output of the neural networks, which makes neural networks a new and effective approach to solving PDEs.

The original PINN algorithm proposed in [10], hereafter referred to as the "standard PINN", is effective in estimating solutions that are reasonably smooth with simple boundary conditions (e.g., the specific boundary values are given), such as the viscous Burger's equation, Poisson's equation, Schrödinger's equation and the wave equation. On the other hand, it has been observed that the standard PINN has the convergence and accuracy problems when solving "stiff" PDEs [3] such as the nonlinear Allen-Cahn equation, where solutions contain sharp space transitions or fast time evolution.

To solve these problems, numerous methods have been proposed recently, including bc-PINN [7], the time adaptive approach [14] and SA-PINN [8]. Among them, bc-PINN is especially noteworthy for its simple and intuitive philosophy. Since the stiff PDE has sharp, fast space/time transitions [13], it is hard to predict a domain as a whole. The key idea of bc-PINN is to retrain the same neural network for solving the PDE over successive time segments while satisfying the already obtained solutions for all previous time segments.

However, our analysis reveals that the training scheme in bc-PINN may lead to a progressive accumulation of prediction errors across successive time segments. Our explanation for this phenomenon is that after bc-PINN trains the solutions in a time segment, the trained results are used as the ground truth to train the solutions in subsequent segments. Consequently, since the predictions in initial segments inevitably contain inaccuracies, these errors propagate and amplify throughout the training process for later segments.

To address this issue, we propose a new training method called DP-PINN. In DP-PINN, the training is divided into two distinct phases. The division is informed by our observations in the benchmark experiments with existing PINN methods. Our benchmark experiments revealed that the prediction errors became notably more severe after a certain point (denoted as t_s) within the time domain $[t_{start}, t_{end}]$, where t_{start} is usually 0. In DP-PINN, we use t_s as the pivotal time point in the dual-phase training scheme.

In phase 1, DP-PINN focuses on training the network to predict solutions from t_{start} (i.e., 0) to t_s. In phase 2, we diverge from the bc-PINN's practice of using the entire solutions predicted within $[0, t_s]$ as the ground truth for subsequent training during $(t_s, t_{end}]$, Rather, we found that what is more important is the accuracy of the solutions predicted at t_s. In phase 1 of DP-PINN, we will obtain the predicted solutions on t_s.

In phase 2, we extend the training across the entire time domain $[0, t_{end}]$ using the same neural network architecture. This phase not only trains the network for $(t_s, t_{end}]$, but also continue to refine the solutions in $[0, t_s]$. Crucially, the solutions predicted at t_s serve as "intermediate" conditions (augmenting the original boundary and initial conditions of the PDE) to guide the training in phase 2.

To improve the accuracy of predictions at t_s and overall model accuracy, we propose the new sampling strategies in DP-PINN. With the use of intermediate conditions and the new sampling strategies, DP-PINN is able to achieve much higher accuracy than the state-of-the-art methods.

In summary, DP-PINN incorporates three optimization strategies to enhance its prediction accuracy, particularly in solving complex PDEs such as Allen-Cahn equation. These strategies include: i) strategically dividing the network training into two phases around a specifically identified point t_s, ii) leveraging the predictions at t_s obtained in phase 1 as the extra "intermediate" conditions to improve accuracy in subsequent predictions, and iii) incorporating the new sampling strategies to improve the accuracy of solutions at t_s obtained in phase 1. Together, these strategies enpower DP-PINN to effectively solve a variety of PDEs, including the stiff PDE - the Allen-Cahn PDE, with significantly higher accuracy than other state-of-the-art PINN algorithms. We have conducted the experiments to validate the effectiveness of DP-PINN.

2   Related Work

The standard PINN algorithm can be unstable during training and produce inaccurate approximations around sharp space and time transitions or fail to converge entirely in the solution of "stiff" PDEs, such as the Allen-Cahn equation. Much of the recent studies on PINNs has been devoted to mitigating these issues by introducing modifications to the standard PINN algorithm that can increase training stability and accuracy of the approximation, mostly via splitting the solution domain evenly into several smaller time segments, or by using a weighted loss function during training. We discuss the main approaches of those below.

Non-adaptive Weighting. The work in [14] points out that the neural network should be forced to satisfy the initial condition closely. Accordingly, a loss function with the form, $\mathcal{L}(\theta) = \mathcal{L}_b(\theta) + C\mathcal{L}_I(\theta) + \mathcal{L}_r(\theta)$, was proposed, where C ($C \gg 1$) is a hyper-parameter.

Adaptive Weighting. In [8], PINNs are trained adaptively, using the fully-trainable weights that force the neural network to focus on the difficult regions of the solution, which is an approach reminiscent of soft multiplicative attention masks used in Computer vision [9,12]. The key idea is to increase the weights as the corresponding losses increase, which is accomplished by training the network to minimize the losses while maximizing the weights.

Backward Compatible PINN. In [7], the proposed method, termed backward compatible PINN (bc-PINN), addresses the limitation of retraining a single neural network over successive time segments by ensuring that the network satisfies the solutions obtained in all previous time segments (i.e., treating the solutions in previous time segments as the ground truth for the training in subsequent time segments) during the progressive solution of a PDE system. The bc-PINN divides the time axis into several even time intervals, ensuring a comprehensive and backward-compatible solution across the entire temporal range.

Time-Adaptive Approaches. In [8], the time axis is divided into several time intervals, then PINNs are trained on them, either separately or sequentially. The initial condition for each time interval relies on the predictions in the preceding time step. This approach is time consuming due to the dependency between time steps and the need for training multiple PINNs.

3   Method

In this section, we first give a brief overview of PINN. Next, we present DP-PINN. Finally, we describe the sampling method used in DP-PINN.

3.1   Overview of Standard PINN and Motivation of DP-PINN

Overview of Standard PINN. The general form of the PDE solved by PINN can be defined as follows:

$$u_t + \mathcal{N}[u] = 0, \quad \boldsymbol{x} \in \Omega, t \in [0, t_{end}], \tag{1}$$

$$u(x, t) = g(x, t), \quad \boldsymbol{x} \in \partial\Omega, t \in [0, t_{end}], \tag{2}$$

$$u(x, 0) = h(x), \quad \boldsymbol{x} \in \Omega \tag{3}$$

where x is a spatial vector variable, which includes a vector of spatial points on which we need to find solutions of u, t is time, u_t is the partial derivative of u over t, Ω is a subset of R^d (d is the dimension of the space), $\partial\Omega$ denotes the set of all boundary spatial vector variables where the boundary conditions of the PDE are enforced, and $\mathcal{N}[\cdot]$ is non-linear differential operator. Note that Eqs. (2) and (3) represent the boundary conditions and initial conditions of the PDE, respectively.

the solution $u(\boldsymbol{x}, t)$ is approximated by the output $u_\theta(\boldsymbol{x}, t)$ of the deep neural network (θ denotes the network parameters) with inputs $\boldsymbol{x}$ and t. The residual, $r_\theta(x, t)$, is defined as:

$$r_\theta(\boldsymbol{x}, t) = \frac{\partial}{\partial t} u_\theta(\boldsymbol{x}, t) + \mathcal{N}[u_\theta(\boldsymbol{x}, t)], \tag{4}$$

where all partial derivatives can be computed by automatic differentiation methods [2]. With the use of back-propagation [11] during training, the parameters θ can be obtained by minimizing the following loss function:

$$\mathcal{L}(\theta) = \mathcal{L}_b(\theta) + \mathcal{L}_I(\theta) + \mathcal{L}_r(\theta) \tag{5}$$

where $\mathcal{L}_b$ is the loss corresponding to the boundary condition, $\mathcal{L}_I$ is the loss due to the initial condition, and $\mathcal{L}_r$ is the loss corresponding to the residual. $\mathcal{L}_b, \mathcal{L}_I, \mathcal{L}_r$ are essentially penalties for outputs that do not satisfy (2) (3) and (4) respectively. $\mathcal{L}_b$, $\mathcal{L}_I$ and $\mathcal{L}_r$ can be defined by Eqs. (6)–(8), respectively.

$$\mathcal{L}_b(\theta) = \frac{1}{N_b} \sum_{i=1}^{N_b} ||[u_\theta(x_b^i, t_b^i) - g(x_b^i, t_b^i)]||^2, \tag{6}$$

$$\mathcal{L}_I(\theta) = \frac{1}{N_I} \sum_{i=1}^{N_I} |u_\theta(x_I^i, 0) - h(x_I^i)|^2, \tag{7}$$

$$\mathcal{L}_r(\theta) = \frac{1}{N_r} \sum_{i=1}^{N_r} |r_\theta(x_r^i, t_r^i)|^2 \tag{8}$$

where $\{x_I^i\}_{i=1}^{N_I}$ and $\{h(x_I^i)\}_{i=1}^{N_I}$ denote the initial points of the PDE and their corresponding values; $\{x_b^i, t_b^i\}_{i=1}^{N_b}$ and $\{g(x_b^i, t_b^i)\}_{i=1}^{N_b}$ are the points on the boundary of the PDE and their values; $\{x_r^i, t_r^i\}_{i=1}^{N_r}$ is the set of collocation points randomly sampled from the domain Ω; N_b, N_I and N_r denote the number of boundary points, initial points and the collocation points, respectively. To tune the parameters θ of the neural network, the training is done for 10k iterations of Adam, followed by 10k iterations of L-BFGS, consistent with the related work for a fair comparison in the experiments. L-BFGS uses Hessian matrix (second derivative) to identify the direction of steepest descent.

Motivation of DP-PINN. We conducted the experiment to use bc-PINN to train the model for solving Allen-Cahn (AC) equation, known for its stiff nature. Figure 1 shows the distribution of its prediction errors (L2-error) across the domain. It can be observed that prediction errors escalate as the training progresses over time. This outcome can be attributed to bc-PINN's unique training approach. bc-PINN divided the entire time domain into four discrete segments, and train the model sequentially across these segments. This scheme allows bc-PINN to focus on smaller, more manageable portions of the time domain at any given moment, which effectively circumvents the challenges faced by the standard PINN that train across the full domain simultaneously. Initially, prediction errors within the early segments may appear harmless. Nonetheless, as the training adopts the outcomes of preceding segments as the groundtruth for training in subsequent ones, prediction errors will accumulate as more segments are processed.

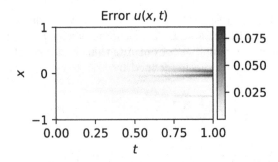

Fig. 1. bc-PINN's absolute prediction error on Allen-Cahn equation.

The benchmark experiments with bc-PINN provided insights that led us to propose the ideas in our DP-PINN. A key observation is that bc-PINN tends to generate more accurate predictions in the initial segments of the time domain. This can be attributed to two main factors. Firstly, by focusing on a smaller time segment, as opposed to tackling the entire time domain in a single sweep like standard PINN, the model is better positioned to learn the features and relationships among points. This reduced time span simplifies the learning process, enabling more precise predictions. Secondly, the proximity of the early segment to the initial conditions, which are the true ground truth, allows the model to more accurately capture the genuine relationships between points. As the training progresses to points further from the initial conditions, the reliance on previous predictions introduces a compounding effect of inaccuracies. This observation underpins the rationale of the dual-phase training adopted in DP-PINN, aiming to mitigate the propagation of errors.

The core principle behind our DP-PINN involves identifying a critical time point, t_s, beyond which prediction errors notably intensify. We divide the model training into two phases. In phase 1, DP-PINN aims to generate the predictions at t_s as accurate as possible. Subsequently, in the phase-2 training, the predictions at t_s act as the initial condition for training the model over the subsequent time segment $(t_s, t_{end}]$. We anticipate an enhancement in prediction accuracy for $(t_s, t_{end}]$ according to the understanding gleaned from bc-PINN, where predictions are more accurate in segments closer to the initial conditions. We will present the dual-phase training scheme in Sect. 3.2.

In light of the core principle underpinning DP-PINN, an important objective is to achieve utmost accuracy in predictions specifically at t_s. The accuracy of these predictions at t_s first depends critically on the selection of t_s within the time domain. Identifying a optimal t_s presents a challenge: it must be neither too close to the initial condition at time 0 nor excessively distant. The rationale behind this balance is that because the solutions predicted at t_s will be used as the initial condition for the training over $(t_s, t_{end}]$, smaller t_s means that some points in $(t_s, t_{end}]$ are further away from their initial conditions and may consequently yield more inaccurate predictions.

In order to obtain an appropriate t_s and generate accurate predictions at t_s, we first conducted benchmark experiments designed to map out the trend of prediction errors across the time domain, thereby establishing an empirical foundation for selecting t_s. Second, in the phase-2 training of DP-PINN, we incorporate a trainable parameter η in the model to fine-tune the prediction values at t_s. The approach of empirical determination of t_s followed by fine-tuning with η is instrumental in optimizing the accuracy of DP-PINN's predictions and enhancing the model's overall effectiveness.

Table 1. Training accuracy of the standard PINN on the 1D viscous Burger's equation with different numbers of sampling collocation points. The numbers of initial points and boundary points were kept unchanged, which are $N_i = 100$ and $N_b = 50$ respectively, in the three sets of sampling points.

	Number of Sampling Points		
	5k	10k	20k
Relative L2 error	3.391e-3	**1.375e-3**	2.297e-3

Another strategy of generating accurate predictions at t_s stems from another observation we made in the benchmark experiments. We applied the standard PINN to train the model with different numbers of sampling points, as shown in Table 1. These experiments revealed a relationship between the number of sampling points and model accuracy: increasing the sampling density generally led to more accurate training results. However, too many sampling points resulted in a decline in model performance, likely due to overfitting. The reason for this trend may be because while a sparse distribution of sampling points challenges the model's ability to learn underlying relationships, excessively dense sampling may cause the model to memorize the training data too closely, losing its generalization capability.

Building on this understanding, we tailor the sampling strategy for DP-PINN to circumvent these issues. Unlike bc-PINN, which samples the same number of points in each time segment, DP-PINN adopts a differentiated approach. Specifically, we allocate a higher density of sampling points to the phase-1 training. This targeted increase in sampling density for phase-1 training is strategic, aimed at obtaining highly accurate predictions at t_s. We will present our sampling strategies in Sect. 3.3.

3.2   DP-PINN

Based on the discussions in Sect. 3.1, we propose a dual-phase training scheme for PINNs. In phase 1, the model is trained on the sampled points in the time duration $[0, t_s]$, and predicts the solutions at t_s. In phase 2, the same model undergoes training across the entire domain, integrating the predictions at t_s

as the intermediate condition, which also acts as the initial condition for the training in $(t_s, t_{end}])$.

The phase-1 training in DP-PINN takes as input the initial conditions of the PDE, the boundary conditions of the points that fall in $[0, t_s]$, and trains the model on the collocation points in $[0, t_s]$. The loss function used by the phase-1 training is defined in Eq. 9, where $\mathcal{L}_I(\theta)$ is the one defined in Eq. 7, representing the initial conditions; $\mathcal{L}_{b1}(\theta)$ represents the compliance to the boundary conditions of the points in the phase-1 domain, which is defined in Eq. 10; $\mathcal{L}_{r1}(\theta)$ represents the residual of the predicted solution on the collocation points in the phase-1 domain, which is defined in Eq. 11; C is a hyper-parameter described in [14], which acts as a weight of the $\mathcal{L}_I(\theta)$ term in the overall loss function.

$$\mathcal{L}_1(\theta) = \mathcal{L}_{b1}(\theta) + C\mathcal{L}_I(\theta) + \mathcal{L}_{r1}(\theta) \tag{9}$$

$$\mathcal{L}_{b1}(\theta) = \frac{1}{N_{b1}} \sum_{i=1}^{N_{b1}} |[u_\theta(x_{b1}^i, t_{b1}^i) - g(x_{b1}^i, t_{b1}^i)]|^2, \tag{10}$$

$$\mathcal{L}_{r1}(\theta) = \frac{1}{N_{r1}} \sum_{i=1}^{N_{r1}} |r_\theta(x_{r1}^i, t_{r1}^i)|^2, \tag{11}$$

In the phase-2 training, the model is trained across the entire domain $[0, t_{end}]$. Therefore, the original boundary conditions, initial conditions of the PDE are used as the input of the phase-2 training. This phase extends the evaluation of the model's residual to include all sampled collocation points within the entire domain. Moreover, the solutions at time t_s predicted in the phase-1 training are used as an additional condition in the phase-2 training. Based on these, the loss function for the phase-2 training is defined in Eq. 12, where $\mathcal{L}_b(\theta)$ is the one defined in Eq. 6, representing the original boundary conditions of the PDE; $\mathcal{L}_{t_s}(\theta, \eta)$ represents the additional intermediate condition established based on the solutions predicted in the phase-1 training at time t_s.

The term $\mathcal{L}_{t_s}(\theta, \eta)$ is defined in Eq. 13, where $u'(x_{t_s}, t_{t_s})$ are the values predicted in phase 1 for the collocation points at t_s, and $\eta = (\eta^1, ..., \eta^{N_{t_s}})$ is a set of trainable parameters aimed at fine-tuning the predicted values for the N_{t_s} collocation points (i.e., $x_{t_s}^i$) at t_s. $\mathcal{L}_{t_s}(\theta, \eta)$ essentially calculates the residual of the predictions at t_s in phase 2 by treating the phase-1 predictions at t_s as "prior". The parameter η is updated using Eq. 14 with k indicating the learning step and μ_k the step-specific learning rate.

The base value of t_s is set to be 30% of the entire time domain. This is an empirical value gathered through our benchmark experiments. For example, in the experiments shown in Fig. 1, prediction errors were not significantly pronounced at $t = 0.3$ (given a total time span of 1) according to our experimental records.

$$\mathcal{L}_2(\theta, \eta) = \mathcal{L}_b(\theta) + C\mathcal{L}_I(\theta) + \mathcal{L}_r(\theta) + \mathcal{L}_{t_s}(\theta, \eta) \tag{12}$$

$$\mathcal{L}_{t_s}(\theta, \eta) = \frac{1}{N_{t_s}} \sum_{i=1}^{N_{t_s}} |u_\theta(x_{t_s}^i, t_s) - \eta^i * u'(x_{t_s}^i, t_s)|^2 \tag{13}$$

$$\boldsymbol{\eta}(k+1) = \boldsymbol{\eta}(k) - \mu_k \nabla_{\boldsymbol{\eta}} \mathcal{L}_2(\theta, \boldsymbol{\eta}) \tag{14}$$

3.3   Sampling Methods

We propose two sampling strategies to provide the training points for DP-PINN. N_r and N_b denote the set of collocation and boundary points sampled in the entire domain, respectively. N_{r1} and N_{b1} denote the set of collocation and boundary points for the phase-1 training, respectively. $N_{r1} \subset N_r$ and $N_{b1} \subset N_b$. N_I denotes the set of initial points. N_{t_s} denotes the set of collocation points sampled at time t_s. In both sampling strategies proposed in this work, we comply with the constraint that the total numbers of sampled collocation points and boundary points that are input for training are no more than $|N_r|$ and $|N_b|$, respectively.

Fixed Sampling Strategy. Unlike [7,14], where the time axis is evenly split into multiple time intervals (e.g., bc-PINN uses 4 intervals), our approach contains two time intervals $[0, t_s]$ and $(t_s, t_{end}]$. In the phase-1 training, i.e., the training in $[0, t_s]$, the number of sampled collocation points is determined by Eq. 15, where $\frac{t_s}{t_{end}} \times |N_r|$ is the number of collocation points that is proportional to the ratio of t_s to t_{end}, and therefore α can be regarded as an amplification factor that increases the sampling density in phase-1. α is a hyper-parameter in training.

$$|N_{r1}| = \alpha \times \frac{t_s}{t_{end}} \times |N_r| \tag{15}$$

Similarly, the number of sampled boundary points is determined by Eq. 16.

$$|N_{b1}| = \alpha \times \frac{t_s}{t_{end}} \times |N_b| \tag{16}$$

In the phase-2 training, the model is trained using the points sampled from both the initial intervals $[0, t_s]$ and the subsequent interval $(t_s, t_{end}]$. In the fixed sampling strategy, the set of sampled points in $[0, t_s]$ in the phase-2 training remains the same as that in the phase-1 training (i.e., N_{r1} and N_{b1}). To adhere to the constraint that the total number of collocation points and boundary points used for training does not exceed $|N_r|$ and $|N_b|$, respectively, we sample $|N_r| - |N_{r1}|$ collocation points and $|N_b| - |N_{b1}|$ boundary points within $(t_s, t_{end}]$ for phase-2 training. Compared to proportional sampling strategy to be presented next, this sampling strategy is deliberately designed to prioritize the accuracy of predictions at the initial stages of the timeline. This focus is particularly important for effectively addressing "stiff" PDEs, such as Allen Cahn equation, where inaccuracies in early predictions can quickly magnify as the model extends to later time points.

Proportional Sampling Strategy. In the former sampling strategy, we preserve the sampling density within $[0, t_s]$ for phase-2 training but sacrifice the sampling density for the subsequent interval $(t_s, t_{end}]$ due to the constraint on the total number of sampling points. In the proportional sampling strategy, after completing phase-1 training, we randomly select a proportion of the total collocation points, $\frac{t_s}{t_{end}} \times |N_r|$, and boundary points, $\frac{t_s}{t_{end}} \times |N_b|$, from the initial sets N_r1 and N_b1 to use during the $[0, t_s]$ interval of phase-2 training. For the remaining time span $(t_s, t_{end}]$, we then allocate the rest of the points, $(1 - \frac{t_s}{t_{end}}) \times |N_r|$ for collocation and $(1 - \frac{t_s}{t_{end}}) \times |N_b|$ for boundary points.

This strategy adjusts the distribution of sampling points in phase-2 training, reducing the number during $[0, t_s]$ so that the number of points in each interval is proportional to their respective durations. Contrary to fixed sampling, which allocates more sampling points to the early time points, this proportional sampling strategy increases the focus on the later stages (after t_s) in the phase-2 training. This strategy is suited for simpler, non-stiff PDEs.

4   Results

We conducted the experiments to compare our DP-PINN with the state-of-the-art PINN algorithms in literature, SA-PINN [8], BC-PINN [7] and TA (Time-Adaptive) approach [14]. We also used the standard PINN to solve the tested PDEs, whose performance is reported as a baseline. We applied the above PINNs to solve the Allen-Cahn PDE and the Burger's equation.

We carried out the experiments on these two PDEs because the Allen-Cahn equation, as a "stiff" PDE, is regarded as a most challenging benchmark and used in the experiments of all the three state-of-the-art PINN algorithms. The Burger's equation is relatively easier to solve (non-stiff) and is used as the benchmark in the experiments of SA-PINN. The Burger's equation is solved in our experiments in order to demonstrate the generalization of DP-PINN in solving non-stiff PDEs.

Same as in the literature, we use relative L2-error as the metric to measure the performance of the PINN algorithms. Relative L2-error is defined by Eq. 17, where N_U is the set of sampled points in the entire domain; $u(x, t)$ and $U(x, t)$ represent the predictions and the ground truth at the point x and time t, respectively.

$$Error_{L2} = \frac{\sqrt{\sum_{i=1}^{N_U} |u(x_i, t_i) - U(x_i, t_i)|^2}}{\sqrt{\sum_{i=1}^{N_U} |U(x_i, t_i)|^2}} \tag{17}$$

4.1   Allen-Cahn Equation

The Allen-Cahn reaction-diffusion equation is commonly used in the phase-field models, which are often used to model solidification and melting processes, providing insights into the behavior of phase boundaries during these transformations. In this experiment, the Allen-Cahn PDE considered is specified as follows:

$$u_t - 0.0001 u_{xx} + 5u^3 - 5u = 0, \quad x \in [-1, 1], t \in [0, 1], \tag{18}$$

$$u(0, x) = x^2 cos(\pi x), \tag{19}$$

$$u(t, -1) = u(t, 1), \tag{20}$$

$$u_x(t, -1) = u_x(t, 1) \tag{21}$$

Unlike other PDEs solved by PINNs, Allen-Cahn equation is a nonlinear parabolic PDE that challenges PINNs to approximate solutions with sharp space and time transitions, and also introduces periodic boundary conditions (Eqs. 20–21).

In order to deal with this periodic boundary conditions, the loss function $\mathcal{L}_b(\theta)$ and $\mathcal{L}_{b1}(\theta)$ defined in Eqs. 6 and 10 are replaced by:

$$\mathcal{L}_b(\theta) = \frac{1}{N_b} \sum_{i=1}^{N_b} |u_\theta(-1, t_b^i) - u_\theta(1, t_b^i)|^2 + |u_{x\theta}(-1, t_b^i) - u_{x\theta}(1, t_b^i)|^2, \qquad (22)$$

$$\mathcal{L}_{b1}(\theta) = \frac{1}{N_{b1}} \sum_{i=1}^{N_{b1}} |u_\theta(-1, t_{b1}^i) - u_\theta(1, t_{b1}^i)|^2 + |u_{x\theta}(-1, t_{b1}^i) - u_{x\theta}(1, t_{b1}^i)|^2, \quad (23)$$

The neural network architecture is fully connected with 4 hidden layers, each with 128 neurons. The input layer takes two inputs (two neurons) - x and t. The output is the prediction of $u_\theta(x, t)$. This architecture is identical to that used in [8,14], which allows a fair comparison. We set the number of collocation, initial and boundary points to $|N_r| = 20000$, $|N_I| = 100$ and $|N_b| = 100$. The amplification factor α as a hyper-parameter is set to 1.6. The number of the points sampled at time t_s for the phase-1 training is set to $|N_{t_s}| = 100$, excluding the points at the boundary. t_s is set to 0.3, and the trainable co-efficient η in Eq. 13 are initialized following a uniform distribution in the range of $[0,1)$. Mini-batches are used in training, with the batch size of 32. All experiments underwent 10 independent runs with random starts and the performance reported is the average over the 10 runs.

Table 2 shows the performance of different PINNs in solving the Allen-Cahn equation. Comparing to other state-of-the-art algorithms, DP-PINN can achieve a much lower relative L2 error, which is 0.84% ± 0.29%. Note that standard PINN cannot solve the Allen-Cahn equation.

Table 2. Comparing the performance of different PINN algorithms in solving the Allen-Cahn equation. When testing the PINN algorithms in literature, the settings are exactly same as those reported in the literature whenever applicable.

Methods	Relative L2-Error
standard PINN	99.18% ± 0.54%
Time-adaptive approach	7.55% ± 1.03%
SA-PINN	2.06% ± 1.33%
bc-PINN	1.67% ± 0.89%
DP-PINN (fixed sampling)	**0.84% ± 0.29%**

Figure 2 visualizes the numerical solutions and prediction errors obtained by DP-PINN in Table 2. Comparatively, Fig. 1 visualizes the prediction errors obtained by bc-PINN in this experiment. It can be seen from the figures that the predictions made by DP-PINN are very accurate.

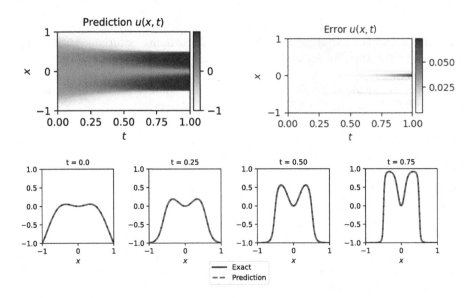

Fig. 2. Visualization of the solutions of Allen Cahn equation and their prediction errors obtained by DP-PINN. The two plots on the top are the predicted solution (left) and the prediction errors (right) across the domain. The four plots at the bottom are the predicted solution at 4 different time point.

4.2   1D Viscous Burger's Equation

The 1D viscous Burger's equations with Dirichlet boundary conditions are formalized as follows:

$$u_t + uu_x - (0.01/\pi)u_{xx} = 0, \quad x \in [-1,1], t \in [0,1], \tag{24}$$

$$u(0,x) = -sin(\pi x), \tag{25}$$

$$u(t,-1) = u(t,1) = 0. \tag{26}$$

In this set of experiments, the neural network architecture is fully connected with 8 hidden layers, each with 20 neurons. This architecture is identical to [8,10]. We set $|N_r| = 10000$, $|N_b| = 50$ and $|N_I| = 100$ for all PINNs, which is the common setting in [8,10]. All training is done by 10k Adam iterations, followed by 10k L-BFGS iterations, which is the same as those used in literature. For DP-PINN, $|N_{t_s}| = 256$. η is initialized in the same way as in the previous experiment. In this experiment, we only compared our DP-PINN with SA-PINN since other two state-of-the-art PINN algorithms (bc-PINN and Time Adaptive) are not tested on this simpler PDE.

The performance of different PINN methods are listed in Table 3. From this table, we can see that DP-PINN achieves slightly better performance than SA-PINN. The improvement is not as prominent as with Allen Cahn equation because Burger's equation is easy to solve. The state-of-the-art PINN method (SA-PINN) can already generate very accurate solutions.

Table 3. Performance of different PINNs algorithms in solving Burger's equation

Methods	Relative L2 error
Standard-PINN	1.375e-3 ± 1.191e-3
SA-PINN	4.685e-4 ± 1.211e-4
DP-PINN	**4.595e-4 ± 1.381e-4**

Figure 3 visualizes the solutions of the Burger's equation and their prediction errors obtained by DP-PINN in Table 3. Once again, the solutions generated by DP-PINN are very accurate.

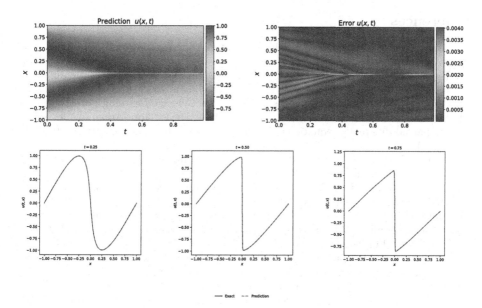

Fig. 3. Visualization of the solution of the 1D viscous Burger's equation and the prediction errors obtained by DP-PINN. The two plots on the top show the predicted solution (left) and prediction error (right) across the entire domain. The plots at the bottom show the predicted solutions at three different time points.

5   Conclusion and Future Works

In this paper, we develop a new training method for PINN, called DP-PINN, to address the limitations of the existing PINNs in solving "stiff" PDEs. By dividing the training into two phases at a carefully chosen time point t_s, DP-PINN significantly reduces prediction errors that accumulate over time in the existing methods like bc-PINN. The first phase of DP-PINN focuses on training the network and predicting accurate solutions at t_s, while the second phase trains the entire time domain, using the solutions at t_s as an intermediate condition. The experiments were conducted to compare DP-PINN with the state-of-the-art PINNs in solving Allen Cahn equation (stiff PDE) and Burger's equation (non-stiff PDE). The results show that DP-PINN achieve much higher performance in terms of relative L2 error in solving Allen Cahn equation. DP-PINN only achieves slightly lower related L2 error in solving Burger's equation because it is easier to solve and the existing methods already performs very well. Despite the aforementioned achievements, some further work can be conducted. On the one hand, there have not been theoretical analyses about the impact of network size on the approximation accuracy of PDE solutions, which will be part of our future work. On the other hand, we would like to explore the techniques for solving PDEs from other application domains such as chemistry and biology.

References

1. Ames, W.F.: Numerical Methods for Partial Differential Equations. Academic Press (2014)
2. Baydin, A.G., Pearlmutter, B.A., Radul, A.A., Siskind, J.M.: Automatic differentiation in machine learning: a survey. J. Mach. Learn. Res. **18**, 1–43 (2018)
3. Burden, R.L.: Numerical Analysis. Brooks/Cole Cengage Learning (2011)
4. Dissanayake, M., Phan-Thien, N.: Neural-network-based approximations for solving partial differential equations. Commun. Numer. Methods Eng. **10**(3), 195–201 (1994)
5. Huang, S., Feng, W., Tang, C., Lv, J.: Partial differential equations meet deep neural networks: a survey. arXiv preprint arXiv:2211.05567 (2022)
6. Lee, J.Y., Ko, S., Hong, Y.: Finite element operator network for solving parametric PDES. arXiv preprint arXiv:2308.04690 (2023)
7. Mattey, R., Ghosh, S.: A novel sequential method to train physics informed neural networks for allen cahn and cahn hilliard equations. Comput. Methods Appl. Mech. Eng. **390**, 114474 (2022)
8. McClenny, L., Braga-Neto, U.: Self-adaptive physics-informed neural networks using a soft attention mechanism. arXiv preprint arXiv:2009.04544 (2020)
9. Pang, Y., Xie, J., Khan, M.H., Anwer, R.M., Khan, F.S., Shao, L.: Mask-guided attention network for occluded pedestrian detection. In: Proceedings of the IEEE/CVF International Conference on Computer Vision, pp. 4967–4975 (2019)
10. Raissi, M., Perdikaris, P., Karniadakis, G.E.: Physics-informed neural networks: a deep learning framework for solving forward and inverse problems involving nonlinear partial differential equations. J. Comput. Phys. **378**, 686–707 (2019)
11. Rumelhart, D.E., Durbin, R., Golden, R., Chauvin, Y.: Backpropagation: the basic theory. In: Backpropagation, pp. 1–34. Psychology Press (2013)
12. Wang, F., et al.: Residual attention network for image classification. In: Proceedings of the IEEE Conference on Computer Vision and Pattern Recognition, pp. 3156–3164 (2017)
13. Wang, S., Teng, Y., Perdikaris, P.: Understanding and mitigating gradient flow pathologies in physics-informed neural networks. SIAM J. Sci. Comput. **43**(5), A3055–A3081 (2021)
14. Wight, C.L., Zhao, J.: Solving Allen-Cahn and Cahn-Hilliard equations using the adaptive physics informed neural networks. arXiv preprint arXiv:2007.04542 (2020)
15. Zawawi, M.H., et al.: A review: fundamentals of computational fluid dynamics (CFD). In: AIP Conference Proceedings, vol. 2030. AIP Publishing (2018)

Operator Entanglement Growth Quantifies Complexity of Cellular Automata

Calvin Bakker[1,2] and Wout Merbis[2,3(✉)] (iD)

[1] Instituut-Lorentz, Leiden Institute of Physics, Universiteit Leiden,
Leiden, Netherlands
`bakker@lorentz.leidenuniv.nl`
[2] Institute for Theoretical Physics (ITFA), University of Amsterdam,
Amsterdam, Netherlands
`w.merbis@uva.nl`
[3] Dutch Institute for Emergent Phenomena (DIEP), University of Amsterdam,
Amsterdam, Netherlands

Abstract. Cellular automata (CA) exemplify systems where simple local interaction rules can lead to intricate and complex emergent phenomena at large scales. The various types of dynamical behavior of CA are usually categorized empirically into Wolfram's complexity classes. Here, we propose a quantitative measure, rooted in quantum information theory, to categorize the complexity of classical deterministic cellular automata. Specifically, we construct a Matrix Product Operator (MPO) of the transition matrix on the space of all possible CA configurations. We find that the growth of entropy of the singular value spectrum of the MPO reveals the complexity of the CA and can be used to characterize its dynamical behavior. This measure defines the concept of *operator entanglement entropy* for CA, demonstrating that quantum information measures can be meaningfully applied to classical deterministic systems.

Keywords: Complexity · Complex Systems · Cellular Automata · Tensor Networks · Entanglement Entropy · Quantum Information Theory

1 Introduction

Cellular automata (CA) are models of dynamical complex systems where a large number simple components (cells) are subject to locally defined interaction rules [22]. Despite the simple nature of the cells and their interaction rules, surprisingly rich dynamical behavior can emerge at larger scales [23,24]. Known and well-studied emergent properties of CA contain complex pattern formations such as fractals [19], localized excitations (called 'lifeforms') and self-reproducing structures [11], deterministic chaos [6] and in some cases the emergence of universal Turing machines allowing for universal computation [7]. Due to their relatively simple rules at small scales and the resulting complex behavior on large scales,

© The Author(s), under exclusive license to Springer Nature Switzerland AG 2024
L. Franco et al. (Eds.): ICCS 2024, LNCS 14832, pp. 33–47, 2024.
https://doi.org/10.1007/978-3-031-63749-0_3

CA are frequently used as a testing bed for studying emergence in complex systems [10].

The simplest CA with emergent properties are the elementary cellular automata (ECA). These are defined in terms of a one-dimensional array of cells with a binary state space, together with a transition rule depending on the state of the three cell neighborhood that surrounds each cell. Wolfram empirically classified the dynamical behavior of the ECA to fall into four distinct classes [23]. Class I CA converge quickly to the uniform state; Class II CA converge quickly to a periodic state; Class III CA show chaotic behavior that does not seem to converge to any regularly repeating pattern; and finally, class IV CA show complex behavior that is *at the edge of chaos* [10]; in between the chaotic and periodic states. In this class, localized excitations are found which are able to carry information through the CA and interact with each other over a background periodic structure. While the Wolfram complexity classes can (and have) been further refined (see [21] for a recent review), the classification into four globally different types of dynamical behavior has persisted in the literature. Ultimately, these classifications are done empirically by running the CA from many different initial configurations and observing the resulting patterns. A systematic way to deduce and quantify the complexity of a CA directly from first principles is currently lacking.

In this work, we use the growth in *operator space entanglement entropy* [15,25], or simply: the operator entanglement (OE), of the ECA rule under time evolution to quantify its complexity. The OE physically represents the (lossless) compressibility of the transition matrix that evolves CA configurations into the future. To compute the OE, we use methods based on tensor networks [13,17,20]. Tensor networks decompose large dimensional vectors or matrices as a network of smaller tensors, contracted over an internal (bond) dimension. The bond dimension reflects the compressibility of the large dimensional object. Here, we represent the transition matrix as a *Matrix Product Operator* (MPO), which is a tensor network composed out of a one-dimensional array of tensors. The OE is then defined as the Shannon entropy of the distribution formed from the squares of the singular values across the middle bond of the MPO. This gives a measure of the amount of information flowing between the two halves of the CA.

The OE provides a new, time-dependent measure for the CA's complexity which does not depend on initial conditions, nor on the empirical classification of individual trajectories. Instead, it directly relates the complexity to the ability to compress the high-dimensional transition matrix which contains all possible trajectories of length t. We find that for the simplest rules, this matrix can be compressed to contain only a single relevant term, while for the most chaotic rules the transition matrix cannot be compressed at all. We find that the evolution of the OE in time serves as a powerful indicator of the CA's phenomenological behavior and allows us to classify the CA complexity into four distinct global types. We propose a further refinement within these four types based solely on the characteristics of the OE.

Previous work using tensor networks has mainly focused on quantum cellular automata (QCA) [2]. In [1], the OE growth of a quantum mechanical version of rule 54 was found to grow logarithmic in time. Reference [14] showed the amount of entanglement which can be created by the action of a QCA is limited by an area law. A quantum version of Game of Life was created in [3], which shows the emergence of complexity in the quantum domain, and the entanglement entropy growth for GCA was studied in [9,12]. In this work, we provide for the first time a tensor network analysis for the classical and deterministic Wolfram rules. We show that the operator entanglement can be used to quantify its complexity, demonstrating the usefulness of quantum information inspired measures in the classical domain.

This work is organized as follows: we start with explaining our construction of the MPO that implements the classical ECA transition rules in Sect. 2. In Sect. 3, we introduce the operator entanglement for ECA and propose a categorizations based on its dynamical behavior. We conclude in Sect. 4 and give perspective on future work along the lines of the work presented here.

2   A Matrix Product Operator for Elementary Cellular Automata

The one-dimensional, discrete state, discrete time cellular automata are defined on an array of cells, where $x_i^{(t)}$ denotes the state of cell i at time t. The state of each cell is given by an integer $x_i \in \{0, \ldots, k-1\}$. The evolution of the CA is determined by iterating the mapping:

$$x_i^{(t)} = f\left[x_{i-r}^{(t-1)}, \ldots, x_i^{(t-1)}, \ldots, x_{i+r}^{(t-1)}\right], \tag{1}$$

where r is the 'range' of the CA. In this work, we will primarily be concerned with the elementary cellular automata (ECA), that have $k = 2$ and $r = 1$. For the ECA, the transition function (1) of each cell only depends on a neighborhood containing itself and its immediate neighbors. There are then $2^3 = 8$ possible neighborhood states, which results in $2^8 = 256$ possible transition rules. These transition rules follow a conventional numbering system, where the 8-bit binary representation of the rule number determines the cells future state for each of the 8 neighbourhood configurations. An important detail of the time evolution is that all cells are updated simultaneously, which we will call 'parallel updates'.

The ECA with transition rules given by (1) are non-linear dynamical systems, where the state of a cell depends on a multi-linear function of its neighborhood. However, any finite CA of length L can also be represented as a bit string $\mathbf{x}^{(t)} = \{x_1^{(t)} x_2^{(t)} \ldots x_L^{(t)}\}$, which is vectorized into a vector of dimension 2^L. In this case, the dynamics of the CA is a map from bit string to bit string, that can be implemented as a *linear* operation on the vector space representing all possible bit strings (or: of all *state configurations* of the CA). Hence, we can represent any CA update rule as a matrix $\hat{T}$ that maps the input CA configuration to an output CA configuration. This transition matrix is $2^L \times 2^L$ dimensional, making its explicit computation for large system sizes intractable.

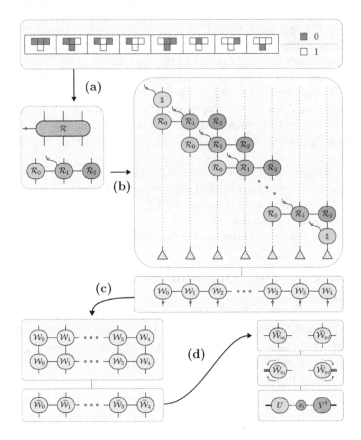

Fig. 1. Overview of our methodology: **(a)** the construction of a tensor $\mathcal{R}$ from the ECA rule, **(b)** formation of the Matrix Product Operator of the transition matrix $\hat{T}$, implementing a single time step **(c)** contraction of MPOs to build $\hat{T} \cdot \hat{T}$, implementing two time steps at once and **(d)** extraction of singular values s_i from the bonds between the MPO tensors.

To overcome this 'curse of dimensionality' for the single time-step transition matrix, we construct a Matrix Product Operator (MPO) representation for the transition matrix. An MPO can give a lower dimensional (compressed) representation for high dimensional matrices, by decomposing it into an array of smaller rank-4 tensors, that, when contracted over the internal *bond dimension*, will reproduce the high-dimensional matrix. The bond dimension represents physically the amount of compression which can be achieved, as it captures the influence neighboring sites exert on each other. The transition matrix MPO is obtained as outlined in Fig. 1 and described in the steps below. Here tensors are essentially multi-dimensional arrays, and they are indicated in the figure as boxes with their indices shown as edges. Connected (or: contracted) edges indicate inner products over the corresponding array dimensions.

(a) From the transition rule (1), we construct a rank-7 tensor $\mathcal{R}$ with components:

$$\mathcal{R}^{r_{i-1}r_i r_{i+1}}_{o_{i+1} o_i o_{i+1} w_i} = \delta^{r_{i-1}}_{o_{i-1}} \delta^{r_i}_{o_i} \delta^{r_{i+1}}_{o_{i+1}} \delta_{w_i f(r_{i-1}, r_i, r_{i+1})} \tag{2}$$

Here, the three indices indicated by (r_{i-1}, r_i, r_{i+1}) represent the state of the neighborhood of cell i. The three bottom indices $(o_{i-1} o_i o_{i+1})$ are copies of this state, which we need to implement the update rule on the neighboring cells (see (b) below). The index w_i is labeled by the green circle in Fig. 1 and indicates the new state of cell i, according to the transition rule (1). Using two consecutive *singular value decompositions* (SVDs), the tensor $\mathcal{R}$ is brought into an matrix product form:

$$\mathcal{R}^{r_{i-1}r_i r_{i+1}}_{o_{i+1} o_i o_{i+1} w_i} = \sum_{\alpha_1, \alpha_2} (\mathcal{R}_0)^{r_{i-1}}_{o_{i-1}\alpha_1} (\mathcal{R}_1)^{r_i \alpha_1}_{o_i w_i \alpha_2} (\mathcal{R}_2)^{r_{i+1}\alpha_2}_{o_{i+1}} . \tag{3}$$

Here the $\alpha_{1,2}$ indices indicate the *bond dimension* of the MPO.

(b) The MPO $\hat{T}$ is constructed by sequentially updating the sites with $\mathcal{R}$, starting from the left to the right, followed by summing over the output indices $o_1, \ldots o_L$ when all neighbours are updated. This is indicated by the triangles in Fig. 1. In this work, we consider open boundary conditions where the left and right boundary cells are not updated. The final MPO is:

$$\hat{T} = \sum_{\{\gamma_1 \ldots \gamma_{L-1}\}} (\mathcal{W}_0)^{r_1}_{o_1\gamma_1} (\mathcal{W}_1)^{r_2\gamma_1}_{w_2\gamma_2} (\mathcal{W}_2)^{r_3\gamma_2}_{w_3\gamma_3} \ldots (\mathcal{W}_3)^{r_{L-1}\gamma_{L-2}}_{w_{L-1}\gamma_{L-1}} (\mathcal{W}_4)^{r_L\gamma_{L-1}}_{w_L} . \tag{4}$$

Here the γ indices are composed out of direct products of the α indices. After the combination of α bond dimensions into the γ bonds, another SVD of the $\mathcal{W}$ tensors is performed. At this stage the resulting MPO has a bond dimension that does not exceed $D = 4$.[1]

(c) The MPO $\hat{T}$ is a compressed representation of the $2^L \times 2^L$ dimensional transition matrix that maps all CA configurations one time step into the future. Time evolution of the CA for several time steps can now be implemented by an MPO $\hat{T}^t$, where t represents the (integer) number of time steps. This is obtained by contracting the MPO $\hat{T}$ with itself repeatedly. During this process, the bond dimension of $\hat{T}^t$ increases.

(d) After each contraction with $\hat{T}$, the MPO representing the time evolution is compressed using SVDs. This way the bond dimension is reduced such that only the relevant linear combinations of configurations contributing to the systems time evolution are taken into account and the operator $\hat{T}^t$ is optimally compressed without loss of information.

We have verified that all ECA rules are implemented correctly using our MPO representation. Furthermore, it is possible to evolve probability distributions over CA configurations in time, without having to perform ensemble averages over

---

[1] Here the value of 4 is a result of the $\log_2 4 = 2$ bits of information which is exerted on each cell by its two direct neighbors..

simulated trajectories. We have checked the ECA rules are implemented correctly by contracting the MPO (4) with a Matrix Product State representing a uniform distribution over all configurations. The resulting state was matched with the empirical distribution obtained by evolving all possible initial configurations one time step into the future.

3   Operator Space Entanglement Entropy Growth

Now that we have obtained an MPO representation for any ECA transition rule, we can investigate the complexity of the CA under time evolution. We do so by investigating the *compressibility* of the MPO representation of $\hat{T}^t$ as time increases. At each time step, we perform an SVD over each bond in the MPO, indicated by the γ indices in (4). Since the SVD decomposes the bond into a product of unitary matrices (which implement an orthogonal basis transformation) and the singular values, it reveals which linear combinations of configurations are most relevant in the time evolution of the CA. If all singular values are equally large, the MPO is incompressible and the bond dimension of the MPO is multiplied by 4 at each timestep (see footnote 1), resulting in exponential growth as 4^t. In this situation, all configurations are relevant to predict the future state. If only one singular value is non-zero, the CA's evolution can be described as a map into a single configuration and the MPO can compressed into a direct (Kronecker) product of matrices with bond dimension one. Most rules, however, have a singular value spectrum somewhere in between these two extremes.

To quantify the complexity of ECA, we investigate the Shannon entropy of the distribution formed from the squares of the singular values s_i across the middle bond of the MPO $\hat{T}^t$:

$$S_{L/2}(t) = -\sum_i s_i^2 \log_2(s_i^2).$$ (5)

This we call the *operator entanglement* (OE), in analogy with the corresponding measure for operators in quantum many-body systems [15,25].[2] The OE gives a measure of the information transfer between the two halves of the CA as a function of time. The maximal growth of this information transfer is linear, since at each timestep a cell is only influenced by its direct neighbors. This theoretically maximal growth occurs when the singular value spectrum is uniform and the bond dimension grows as 4^t, such that $S(t) = 2t$. For the simplest CAs, $S(t)$ drops to zero.

We have computed the growth of OE for all 88 distinct ECA rules[3]. As the bond dimension grows rapidly for some rules, we halt the time evolution

---

[2] The analogy with quantum systems is more profound, as this measure is exactly the *entanglement entropy* $S_{L/2}(t) = -\text{Tr}\left[\hat{\rho}_{L/2}(t) \log_2 \hat{\rho}_{L/2}(t)\right]$ of a reduced density matrix $\hat{\rho}_{L/2}(t)$, constructed from the partial trace over half the CA cells of the Gram matrix of the time evolution operator: $\hat{\rho}(t) = (\hat{T}^t)^T \hat{T}^t$.

[3] Out of the 256 possible rules, many are related to each other by symmetry (left-right inversion, bit inversion or both), such that there are only 88 unique rules [5].

Table 1. ECA Wolfram rule numbers corresponding to each type of operator entanglement growth.

Type	Wolfram rule number	Wolfram classification
I.A.	0, 8, 32, 40, 128, 136, 160, 168 , 51, and 204	Class I & II
I.B.	4, 12, 13, 19, 23, 36, 44, 50, 72, 76, 77 , 78, 104, 132, 140, 164, 172, 178, 200, and 232	Class II
I.C.	1, 5, 28, 29, 33, 108, and 156	Class II
II.A.	2, 3, 7, 10, 15, 24, 27, 34, 35, 38, 42, 46, 56 , 58, 130, 138, 152, 162, 170, and 184	Class II
II.B.	6, 11, 14, 43, 57, 74, 134, 142, 154 , 30 , and 106	Class II, III & IV
III.A.	18, 22, 45, 60, 122, 126, and 146	Class III
III.B.	90, 105, and 150	Class III
IV.	9, 25, 26, 37, 41, 62, 73, 94 , 54, and 110	Class II & IV

when a maximal bond dimension ($D \sim 1000$) is reached. Still, the behavior of the OE for these rules for small system sizes gives a good indicator of the rule's phenomenological behavior. In general, we categorize the behavior of the OE into four distinct types, which may have further subdivisions. Within the theoretical lower and upper bounds, we have found the following four types of characteristic OE behaviour:

I. **Constant** and independent of system size, either immediately or after an initial peak due to transients. The peak time does not depend on L.
   A. The OE converges to zero.
   B. The OE converges to a finite value.
   C. The OE converges to an oscillating value.
II. Growing, followed by a **peak** at a time $t \sim L$, and then a decrease.
   A. The OE drops to a constant value independent of L.
   B. The OE drops to a constant value dependent on L.
III. Growing **linearly** and reaching a **plateau** that increases with L.
   A. The OE grows sub-maximally.
   B. The OE grows maximally as $S_{L/2}(t) = 2t$.
IV. Growing **sub-linearly** and reaching a L-dependent plateau at late times, without dropping significantly.

We will now discuss each of these possibilities with examples, and show which type of phenomenological behavior of the CA corresponds to which type of behavior of the operator entanglement. The Wolfram rule numbers corresponding to each type of behavior are summarized in Table 1, where one can also compare with the empirical Wolfram complexity classes.

3.1　Type I Rules: Quick Convergence to Constant Values

Type I.A: For type I behavior, the OE quickly converges to a constant. For the simplest rules, this constant is zero. A typical example of this type is shown in the top panel of Fig. 2 for rule 136. The OE may grow slightly due to transients,

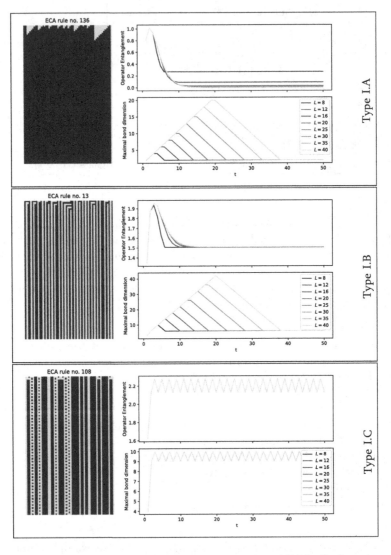

Fig. 2. Examples for typical OE growth of type I rules of ECA. Each trajectory on the left is an example of the rules evolution from a random initial configuration. The plots on the right show the operator entanglement and maximal bond dimension of the MPO. These do not depend on the initial configuration and hence say something on the complexity of the dynamical process itself.

but afterwards it quickly drops to zero or a very small value decreasing with L. The time of the peak is independent of system size, which signifies that the transients do not lead to long-range information transfer within the CA.

The ECA rules showing this OE growth all converge to uniform configurations, except for two notable examples: rule 51 and 204. Both rules have vanishing OE, but rule 51 shows oscillatory behavior as it maps each cell to the opposite state regardless of the neighbors state. Rule 204 shows striping patterns, as this rule maps each cells state into itself in the next time step. In both cases, there is no information transfer between neighboring cells. In terms of OE these rules are as complex as rule 0, which maps all configurations to the homogeneous state.

Type I.B: In the second category of the first type the OE quickly converges to a non-vanishing constant value, as exemplified by rule 13 in the middle panel of Fig. 2. By 'quickly converging' we mean that if the OE does not immediately reach a constant (and L independent) value, it does so after a peak due to transients that occurs after a time t_{peak} independent of L. This implies that the transients do not give rise to long-range information transfers in the CA and the CA quickly settles into either a constant striping pattern, or an oscillating pattern where cells are mapped to its opposing states. In both cases, there are no patterns propagating through the CA, and hence based on the OE these striped patterns and oscillating patterns are equally complex.

Type I.C: In this category, the OE oscillates between two non-zero values, which do not depend on L. The example is illustrated in the bottom panel of Fig. 2 by rule 108. All rules in this category show the coexistence of striped and oscillation patterns, but there is no long range lateral information transfer of information in the form of propagating shapes through the CA.

3.2   Type II Rules: Long Transients and Information Transfer

Type II.A: For all rules of the second type, we see the OE initially increase (at most) linearly, however, it reaches a peak after a time t of the order of the system size L, after which it decreases again. The distinction we make between type II.A and II.B is whether the OE decreases to a L independent or L dependent constant value, respectively. All rules of type II.A show common phenomenological behavior (see the top panel of Fig. 3 for an example with rule 34). We observe simple structures propagating linearly through the CA. After a transient time comparable to the CAs size L, the system reaches either a homogeneous state, a constant striped state, or an oscillating state with a short period. At this point all propagating structures have reached the boundaries, such that there is no more information being transferred within the CA, leading to a constant OE.

Type II.B: Here the final value of the OE does depend on system size L. Phenomenologically, most of these rules are similar to the above type, where initially

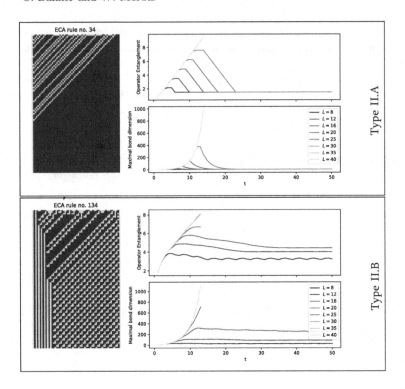

Fig. 3. Two typical representatives of the second type of OE behavior: rule 34 and rule 134. The computation is performed until a maximum bond dimension is found to be larger than $D \approx 2000$, or numerical convergence issues are encountered.

simple structures propagate through the CA until they reach the boundaries. Only now, the dynamics does not necessarily settle into a simple background pattern, but there may be coexisting domains of different background behaviors (such as stripped and with a part oscillating or a part homogeneous, or a separation between two different oscillating patterns).

There are also two notable exceptions within this type. Rule 30 is a chaotic rule with periodic boundary conditions, but it shows OE belonging to this type. This is because the open boundary conditions enforce a striping pattern after many time steps. Another exception is rule 106, which has complex phenomenology (class IV), but does show a significant drop in operator entanglement after the initial linear increase. Here, the open boundary conditions enforce a homogeneous state at late times whenever the rightmost cell is empty.

3.3   Type III Rules: Chaotic CAs

Type III.A: This type of CA produces an exponential increase in bond dimension, and therefore it is computationally hard to propagate these rules over long time scales. For these rules we observe an initial linear growth of OE,

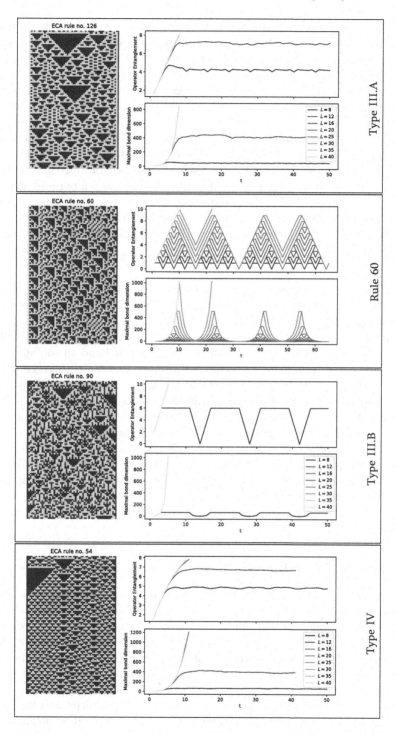

Fig. 4. Representatives of the third and fourth type of OE behavior: rule 126, 60, 90 and 54.

which reaches a plateau when t becomes of the order of L and barely decreases afterwards. This type of behavior is exemplified by rule 126 in the top panel of Fig. 4. All of these rules are characterized by chaotic behavior starting from random initial conditions. This type of behavior is exactly the same as the OE growth in quantum chaotic systems [8,16,26], demonstrating that linear OE growth indicates chaos not only quantum mechanically, but also in classical, deterministic systems.

An interesting and unique behavior of OE growth is seen for rule 60 (Fig. 4, second panel). Here the initial growth is linear, as a sign of chaotic behavior, but then the OE decreases linearly to a minima at multiples of four time steps. In fact, rule 60 with open boundary conditions is periodic with a period τ depending on L as $\tau(L) = 2^{\lceil \log_2(L-1) \rceil}$, *regardless of the initial conditions*. The CA returns to a configuration close to the initial configuration whenever the OE has a minimum, and it returns to the exact starting configuration at times when the OE drops to zero.

Type III.B: For this type the initial growth in OE is linear, but distinctive for these rules is that the growth rate is maximal, such that $S_{L/2} = \log_2(4^t) = 2t$ (see the third panel in Fig. 4 for an example). This implies that in these rules, all singular values in the MPO representation are relevant and equally large. In other words, the time evolution operator cannot be compressed and all configurations are relevant for determining the systems future evolution. In these cases, the transition matrix becomes a permutation matrix, where each configuration is mapped one-to-one to another configuration. Just as is observed for rule 60, for these rules the OE drops to a value of zero periodically. For instance, for $N = 8$, all rules of this type return to the initial configuration after 14 time steps, regardless of initial configuration. This hints at a form of synchronized behavior, even for the apparently most chaotic rules. For finite system sizes, the system does not explore all possible configurations before returning to the initial configuration, but rather gets trapped in cycles of length $t \sim L$. So, rules of this type have a transition matrix composed out of cyclic permutations with uniform cycle length.

3.4   Type IV Rules: Domain Walls, Lifeforms and Complexity

For the final type, we see an operator entanglement that increases sub-linearly (either as t^α with $\alpha < 1$ or as $\log(t)$) and then reach a plateau. The plateau height increases with system size. Phenomenologically, these rules are characterized by complex pattern formations. There may be areas of repeating patterns with local excitations (lifeforms) on top of these patterns. The excitations can also function as domain walls between areas with different background patterns, and they may interact in non-trivial ways. The famous rule 110, which is Turing complete, falls into this type. It is worth noticing that many of these rules ultimately settle into a periodically repeating pattern, implying that strictly taken they belong to the Wolfram class 2 cellular automata. The classification we are considering here

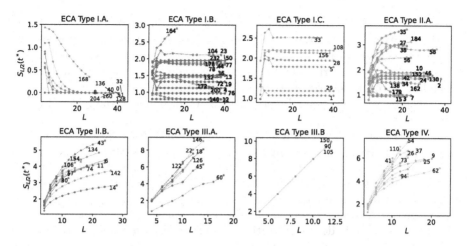

Fig. 5. The operator entanglement after converging to a steady value plotted against systems size L for all unique ECA rules. Each subfigure shows the ECA rules of the corresponding type. Curves are labelled by their rule number.

concerns the initial growth of the operator entanglement, which is sub-linear for all of these rules, and the phenomenology shows there is complex behavior in the transients for all of these rules.

3.5   Parallel with Quantum Systems

Figure 5 shows the value where the OE saturates for all different ECA rules as a function of L. There is a clear parallel with the area law scaling of entanglement entropy [18] in quantum systems. The top panels in Fig. 5 satisfy an 'area law', which in our case implies the OE is (ultimately) constant in system size. For chaotic systems, the OE grows with the volume of the system, in analogy with entanglement growth in quantum chaotic systems [8,16,26]. The CA rules that show complex behavior of type IV have OE growing sub-linearly in both t and L. This is reminiscent of quantum mechanical systems at criticality [8] or integrable quantum systems [1,16]. We wish to explore this parallel in more detail in future work.

4   Conclusion

We have investigated the *operator space entanglement entropy* growth of classical, deterministic cellular automata by mapping the transition rule to a matrix product operator (MPO). This provides an indicator of the complexity of the ECA rule, as it quantifies information transfer within the ECA, regardless of initial conditions. We find that the operator entanglement growth curves can be used to classify ECA with open boundary conditions. We distinguish four main types of behavior of the OE under time evolution, with further refinements for some types, and relate this to the phenomenology of the ECA. We find that the OE either settles to a constant value relatively quickly (type I), or after an initial growth period, displaying a peak at times increasing with system size (type II). Type III is characterized by linear growth leading to a plateau with height depending on L, in analogy with quantum chaotic systems. Type IV rules also have OE growing towards an L dependent plateau, but here the initial growth is sub-linear.

Our work enables new insight into the information transfer within the CA and thus the inner structure and complexity of a CA rule. It furthermore opens the door to apply the rich toolbox of tensor networks [13] to classical CA and other non-linear discrete dynamical system, introducing new computational methods into the study of complex dynamical systems. Possible extensions are the use of density matrix renormalization group algorithms for stochastic (noisy) CA and a study into the large deviation statistics of classical cellular automata [4]. We wish to explore these topics in future work.

References

1. Alba, V., Dubail, J., Medenjak, M.: Operator entanglement in interacting integrable quantum systems: the case of the rule 54 chain. Phys. Rev. Lett. **122**(25), 250603 (2019). https://doi.org/10.1103/PhysRevLett.122.250603
2. Arrighi, P.: An overview of quantum cellular automata. Nat. Comput. **18**(4), 885–899 (2019). https://doi.org/10.1007/s11047-019-09762-6
3. Bleh, D., Calarco, T., Montangero, S.: Quantum game of life. Europhys. Lett. **97**(2), 20012 (2012). https://doi.org/10.1209/0295-5075/97/20012
4. Buča, B., Garrahan, J.P., Prosen, T., Vanicat, M.: Exact large deviation statistics and trajectory phase transition of a deterministic boundary driven cellular automaton. Phys. Rev. E **100**(2), 020103 (2019)
5. Cattaneo, G., Formenti, E., Margara, L., Mauri, G.: Transformations of the one-dimensional cellular automata rule space. Parallel Comput. **23**(11), 1593–1611 (1997). https://doi.org/10.1016/S0167-8191(97)00076-8
6. Cattaneo, G., Formenti, E., Margara, L., Mauri, G.: On the dynamical behavior of chaotic cellular automata. Theor. Comput. Sci. **217**(1), 31–51 (1999). https://doi.org/10.1016/S0304-3975(98)00149-2
7. Cook, M., et al.: Universality in elementary cellular automata. Compl. Syst. **15**(1), 1–40 (2004)
8. Dubail, J.: Entanglement scaling of operators: a conformal field theory approach, with a glimpse of simulability of long-time dynamics in 1+ 1d. J. Phys. A: Math. Theor. **50**(23), 234001 (2017). https://doi.org/10.1088/1751-8121/aa6f38
9. Hillberry, L.E., et al.: Entangled quantum cellular automata, physical complexity, and goldilocks rules. Quant. Sci. Technol. **6**(4), 045017 (2021). https://doi.org/10.1088/2058-9565/ac1c41
10. Langton, C.G.: Computation at the edge of chaos: phase transitions and emergent computation. Physica D: Nonl. Phenom. **42**(1–3), 12–37 (1990). https://doi.org/10.1016/0167-2789(90)90064-V
11. Langton, C.G.: Self-reproduction in cellular automata. Physica D **10**(1–2), 135–144 (1984)

12. Ney, P.M., Notarnicola, S., Montangero, S., Morigi, G.: Entanglement in the quantum game of life. Phys. Rev. A **105**(1), 012416 (2022). https://doi.org/10.1103/PhysRevA.105.012416
13. Orús, R.: A practical introduction to tensor networks: matrix product states and projected entangled pair states. Annal. Phys. **349**, 117–158 (2014). https://doi.org/10.1016/j.aop.2014.06.013
14. Piroli, L., Cirac, J.I.: Quantum cellular automata, tensor networks, and area laws. Phys. Rev. Lett. **125**(19), 190402 (2020). https://doi.org/10.1103/PhysRevLett.125.190402
15. Prosen, T., Pižorn, I.: Operator space entanglement entropy in a transverse ising chain. Phys. Rev. A **76**(3), 032316 (2007). https://doi.org/10.1103/PhysRevA.76.032316
16. Prosen, T., Žnidarič, M.: Is the efficiency of classical simulations of quantum dynamics related to integrability? Phys. Rev. E **75**(1), 015202 (2007). https://doi.org/10.1103/PhysRevE.75.015202
17. Schollwöck, U.: The density-matrix renormalization group in the age of matrix product states. Annal. Phys. **326**(1), 96–192 (2011). https://doi.org/10.1016/j.aop.2010.09.012
18. Schuch, N., Wolf, M.M., Verstraete, F., Cirac, J.I.: Entropy scaling and simulability by matrix product states. Phys. Rev. Lett. **100**(3), 030504 (2008). https://doi.org/10.1103/PhysRevLett.100.030504
19. Takahashi, S.: Cellular automata and multifractals: dimension spectra of linear cellular automata. Physica D: Nonl. Phenom. **45**(1–3), 36–48 (1990). https://doi.org/10.1016/0167-2789(90)90172-L
20. Verstraete, F., Murg, V., Cirac, J.I.: Matrix product states, projected entangled pair states, and variational renormalization group methods for quantum spin systems. Adv. Phys. **57**(2), 143–224 (2008). https://doi.org/10.1080/14789940801912366
21. Vispoel, M., Daly, A.J., Baetens, J.M.: Progress, gaps and obstacles in the classification of cellular automata. Physica D: Nonl. Phenom. **432**, 133074 (2022). https://doi.org/10.1016/j.physd.2021.133074
22. Wolfram, S.: Statistical mechanics of cellular automata. Rev. Mod. Phys. **55**(3), 601 (1983). https://doi.org/10.1103/RevModPhys.55.601
23. Wolfram, S.: Cellular automata as models of complexity. Nature **311**(5985), 419–424 (1984). https://doi.org/10.1038/311419a0
24. Wolfram, S.: Universality and complexity in cellular automata. Physica D: Nonl. Phenom. **10**(1–2), 1–35 (1984). https://doi.org/10.1016/0167-2789(84)90245-8
25. Zanardi, P.: Entanglement of quantum evolutions. Phys. Rev. A **63**(4), 040304 (2001). https://doi.org/10.1103/PhysRevA.63.040304
26. Zhou, T., Luitz, D.J.: Operator entanglement entropy of the time evolution operator in chaotic systems. Phys. Rev. B **95**(9), 094206 (2017)

HarVI: Real-Time Intervention Planning for Coronary Artery Disease Using Machine Learning

Cyrus Tanade$^{(\boxtimes)}$ (iD) and Amanda Randles$^{(\boxtimes)}$ (iD)

Department of Biomedical Engineering, Duke University, Durham, NC 27705, USA
{cyrus.tanade,amanda.randles}@duke.edu

Abstract. Virtual planning tools that provide intuitive user interaction and immediate hemodynamic feedback are crucial for cardiologists to effectively treat coronary artery disease. Current FDA-approved tools for coronary intervention planning require days of preliminary processing and rely on conventional 2D displays for hemodynamic evaluation. Immersion offered by extended reality (XR) has been found to benefit intervention planning over traditional 2D displays. To bridge these gaps, we introduce HarVI, a coronary intervention planner that leverages machine learning for real-time hemodynamic analysis and extended reality for intuitive 3D user interaction. The framework uses a predefined set of 1D computational fluid dynamics (CFD) simulations to perform one-shot training for our machine learning-based blood flow model. In a cohort of 50 patients, we calculated fractional flow reserve (FFR), the gold standard biomarker of ischemia in coronary disease, using HarVI (FFR_{HarVI}) and 1D CFD models (FFR_{1D}). HarVI was shown to almost perfectly recapitulate the results of 1D CFD simulations through continuous and categorical validation scores. In this study, we establish a machine learning-based process for virtual coronary treatment planning with an average turnaround time of just 74 min, thus reducing the required time for one-shot training to less than a working day.

Keywords: coronary artery disease · machine learning · virtual reality

1 Introduction

Enabling intervention planning can give physicians an intuitive approach to performing virtual interventions and receiving real-time hemodynamic feedback to guide clinical decision making. By integrating patient-specific computational fluid dynamics (CFD) models, our planning tool offers a seamless and intuitive platform to perform virtual interventions in extended reality (XR). The role of immersion beyond traditional 2D displays has been explored and shown to improve user interaction for intervention planning [1–7]. More importantly, our tool also provides real-time hemodynamic feedback, which is crucial for informed clinical decision making. Advances in personalized CFD have already demonstrated the ability to non-invasively and accurately determine key hemodynamic

L. Franco et al. (Eds.): ICCS 2024, LNCS 14832, pp. 48–62, 2024.
https://doi.org/10.1007/978-3-031-63749-0_4

metrics vital for decision-making processes. For example, fractional flow reserve (FFR) [8] is a pressure-based metric that indicates coronary ischemia [9]. There is even FDA-approved coronary intervention planning software that predicts FFR in response to treatment [10,11]. However, these state-of-the-art models face limitations in turnaround time and user interaction.

In coronary artery disease (CAD), stent implantation is the leading percutaneous coronary intervention (PCI) to treat functionally significant lesions (i.e., FFR $\leq$ 0.80), with more than 600,000 stents implanted annually in the United States alone [12]. However, 25% of the patients evaluated for a successful PCI procedure still have residual ischemia associated with long-term adverse outcomes [13]. This issue of incomplete functional revascularization could be mitigated by allowing interventional cardiologists to interactively experiment with a variety of possible PCI strategies and their impact on the hemodynamics of that patient to determine how best to relieve ischemia. Another use case is in complex coronary lesions. Bifurcation lesions represent 20% of cases and occur in arterial bifurcations or branch points [14]. Determining how best to treat these lesions remains a particular challenge, as they can affect the main branch (MB), the side branch (SB), or both vessels, and bifurcation stenting is associated with a higher risk of adverse cardiac events [15]. Using intervention planning, we could rapidly test strategies in MB, SB, or both lesions to help identify options to achieve complete revascularization and improve outcomes.

Machine learning has emerged as a promising option to predict FFR without explicitly running CFD simulations [10]. The only FDA-approved coronary intervention planning tool uses an interpolation model, which requires running a series of 3D CFD simulations as part of model training [10]. Although accurate, these models require 24–48 h of processing time before a clinician can apply the planning tool. 3D [8] and 1D CFD models of FFR [16] have both been shown to recover invasive FFR accurately, and we hypothesized that real-time predictors of post-intervention FFR derived from reduced-order models would accurately recapitulate virtual PCI hemodynamics with short turnaround times. Enabling a faster turnaround time is helpful for some patient subgroups with unstable coronary disease, such as patients with ST-elevated myocardial infarction, where door-to-balloon time—the time from hospital admission to intervention—is recommended to be less than 90 min [17].

In addition to efficiently training machine learning models, another essential part of intervention planning is enabling intuitive interaction with 3D geometries to simulate intervention. The utility of intervention planning tools can benefit from the immersion offered by XR devices. The role of immersion in intervention planning and evaluating hemodynamic feedback has been extensively explored through user studies. Visualizing complex anatomy in traditional 2D displays could be challenging with vessel overlap and foreshortening effects [18]. These user studies [1–4] have demonstrated that immersive displays are beneficial over 2D displays in the analysis of hemodynamic maps and performing some intervention planning tasks. There is well-established geometry editing software in the literature [5,19], but all of these frameworks do not provide instantaneous hemodynamic updates for geometric modification. The only FDA-approved coronary intervention planning tool that provides instantaneous hemodynamic feedback uses 2D displays. To our knowledge, no intervention planning tool exists with a turnaround time of one working day and is compatible with commonly available XR headsets.

To enable intervention planning with short turnaround times that leverage XR, we present HarVI (**HAR**VEY [20] **V**irtual **I**ntervention). HarVI is the back-end that predicts post-PCI FFR. On the user interaction side, we used Harvis [2]—an established computational platform to modify geometries, deploy massively parallel simulations, and visualize hemodynamic results [2]. We introduce an innovative tool for intuitive clinical decision support by leveraging the real-time flow prediction we establish here with HarVI alongside the immersive

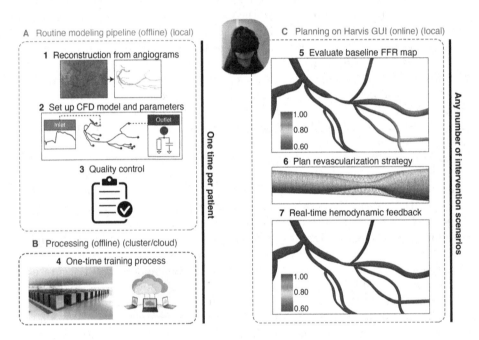

Fig. 1. Overview of the HarVI pipeline. (A) Clinically validated patient-specific modeling pipeline for coronary arteries. The computational domain was derived from coronary angiogram reconstructions. Boundary conditions were informed by clinical measurements. Quality control was performed to ensure reconstruction accuracy compared to coronary angiogram analysis. **(B)** A set of modifications was made to the reconstructed geometries to sample probable intervention scenarios. This step is a one-shot learning process and was invoked once per patient. The results of this training process were used to train a machine learning model to enable real-time prediction of post-intervention FFR. **(C)** After training the HarVI model, users could plan a revascularization strategy using extended reality headsets and receive real-time hemodynamic feedback in the Harvis GUI. HarVI refers to the machine learning model used for intervention planning. We coupled HarVI with Harvis—a virtual reality platform for patient-specific modeling.

anatomical editing capabilities from Harvis (Fig. 1). HarVI uses a clinically validated 1D FFR modeling pipeline [16] to train the machine learning model for real-time flow prediction (Fig. 1A–B). This process is invoked only once per patient as part of a training routine. The results of the training process are used to create a machine learning model to predict post-PCI FFR in real-time. Finally, Harvis allows users to perform a virtual intervention in an immersive environment and visualize updated hemodynamic results, and HarVI provides instantaneous hemodynamic feedback (Fig. 1C). This paper takes a foundational step in the establishment of immersive, intuitive, and real-time techniques for clinical decision support for CAD interventions.

2   Methods

2.1   Personalized Hemodynamic Analysis Using Reduced-Order Models

Image-derived data provided the complex 3D geometries needed for this study. Using anonymized and deidentified data from previous research [16], our analysis incorporated clinical measurements from 50 patients who had confirmed coronary artery disease by angiography. These data were originally collected during invasive coronary angiography procedures, to create patient-specific blood flow models. The key clinical measurements used for this purpose included cardiac output, cuff pressure, heart rate, and hematocrit.

We applied our in-house 1D blood flow simulator to create patient-specific 1D blood flow models of the coronary arteries [16,21]. Our approach differs from FDA-approved software [16]. For flow simulations, image-derived 3D anatomies were required. In this study, we use anonymized 3D meshes from our previous work [16]. These 3D anatomies were reconstructed from coronary angiograms using the SnakeTree3D software described in [18]. The 1D computational domain was defined by computing centerlines and corresponding hydraulic diameters from the 3D reconstructed geometry using Mimics (Materialise, Leuven, BE).

Reduced-order blood flow simulations to compute pre-intervention FFR were performed using a well-established 1D blood flow simulator [16,22]. Pre-intervention FFR calculated from the 1D simulator has been clinically validated against invasively measured FFR in all 50 patients in this study [16]. We maintain the same model assumptions as in [16] where blood was modeled as an incompressible Newtonian fluid with a density of 1060 kg/m^3 and dynamic viscosity evaluated per patient. Pulsatile flow rate waveforms were incorporated at the inlet, and 2-element Windkessel models (resistance and compliance) were employed at the outlets. All simulations were tuned to simulate hyperemic conditions.

2.2   Integrating with Harvis and Editing 3D Stenosis Geometry

The first part of establishing a new virtual intervention tool was to enable users to easily edit 3D coronary geometries to emulate intervention. Within Harvis,

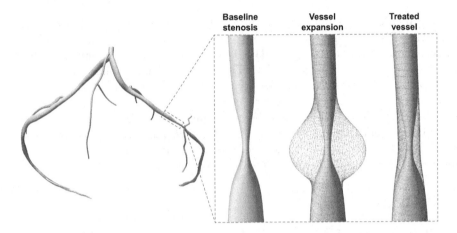

Fig. 2. Geometry modification process. Solid 3D geometry represents baseline anatomy and wireframes show mesh transformations used to treat the vessel. Vertices belonging to the stenosis were selected by the user and then translated following a sinusoidal function to simulate intervention. For the vessel expansion case, we show an extreme case displaying the sinusoidal nature of the modification process.

the user would place two dots to delineate the endpoints of the stenosis and intervention to be simulated. The user-defined dots act as landmarks to identify all vertices in the mesh to be edited. The distance between the two dots was taken as the length of the stenosis for the intervention. Once the two dots are placed, a slider pops up that controls the radius of the stenosis. The mesh vertices between the dots would be displaced radially outward, normal to the vector between the two dots. The displacement of the vertices follows a sinusoidal function such that the vertices at the minimal luminal diameter, typically at the center, expand radially outward the most and the vertices closer to the endpoints do not expand much [2]. To illustrate this point, Fig. 2 demonstrates the vessel expansion process for one patient in our study cohort. As the geometry modification process involves only displacing vertices, no further post-processing was required for blood flow simulations.

From the user's perspective, stenosis modification occurs within the Harvis GUI (Fig. 3). Users would first load the baseline 3D geometry of the coronary arteries before intervention in the format of .off into Harvis. At this point, the one-shot training process shown in Fig. 1A–B would be completed. The pre-intervention FFR map could then be displayed. If the FFR map shows locations with FFR < 0.80, an intervention would be required, as shown in Fig. 3. To perform a virtual intervention, two dots would be placed at the endpoints of a stenosis. A slider would appear to allow the user to control the diameter of the stenosis, and the distance between the two dots would control the length of the stenosis. The corresponding stenosis length and radius specified by the user would be relayed from the Harvis GUI to the HarVI machine learning backend to query the resulting post-intervention FFR map. This planning stage corresponds to Fig. 1C and can be performed any number of times without running any new 1D computational fluid dynamics simulations.

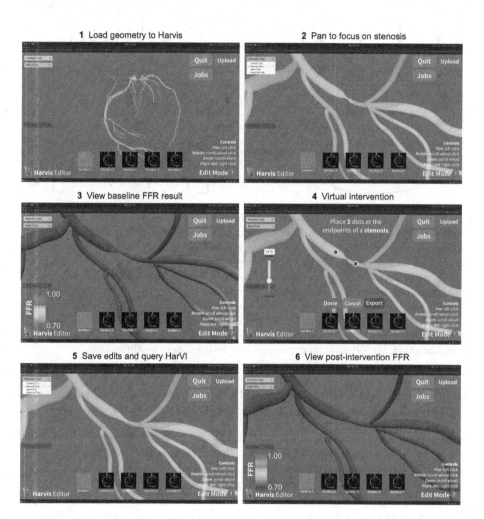

Fig. 3. Intervention planning workflow for the end user. Geometry modification was enabled using Harvis. The steps for virtual coronary intervention include: **1** loading baseline geometry into Harvis, **2** panning around the geometry to focus on the stenosis of interest, **3** evaluating baseline FFR results, **4** modifying stenosis to plan for intervention, **5** saving geometry modification and querying HarVI for hemodynamic feedback, and **6** evaluating and visualizing the resulting post-intervention FFR.

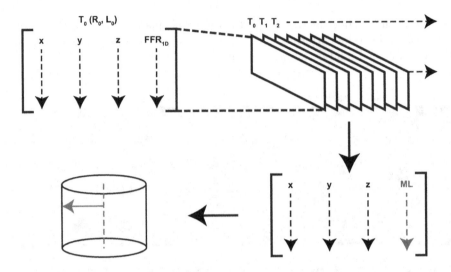

Fig. 4. Machine learning architecture for HarVI. Each predefined intervention scenario (T_N) for training consisted of creating a 2D matrix with 3D centerline data and corresponding FFR results from 1D simulations. Each of these intervention scenarios ran on 1 CPU, and all the 1D simulations were deployed simultaneously on a compute cluster in parallel. Each centerline point was fitted with a bivariate cubic-spline interpolation model. When queried a test radius and length, the resulting predictors would return a post-intervention FFR prediction per centerline point. The centerline points (interrupted green line) with post-intervention FFR predictions were then projected to the triangles of the modified geometry mesh (solid green arrow) to create an FFR map. Note that the cylinder shown is just an idealized case for illustration.

2.3 Establishing a Real-Time Prediction Model of Post-intervention Hemodynamics Using Machine Learning

To improve clinical decision support, it is critical to quickly evaluate how various intervention strategies can influence the hemodynamics of that particular patient. In this work, we developed a machine learning model to quickly and accurately predict post-intervention FFR maps from a set of 1D CFD simulations used for training, directly addressing this need. The training set was constructed from a series of randomly sampled geometric modifications, in terms of stenosis length and radius, and then 1D CFD simulations were performed to obtain post-intervention FFR maps. Intuitively, the larger the set of 1D CFD simulations reserved for training, or in other words, the more samples used for training, the better the machine learning model would perform. Our objective in designing the interpolation model was to sample the parameter space of the radius and length of the stenosis more efficiently to minimize the training set while ensuring that the interpolation model would adequately capture the variance in the FFR distribution in response to intervention.

To this end, we applied Latin Hypercube Sampling (LHS) from the SALib library that implements the algorithms presented in [23,24]. To minimize sampling and achieve faster convergence, we opted for a quasirandom over pseudorandom sampling method so that samples would be more evenly distributed in the parameter space. We initialized the LHS for all patients from 90% the minimal luminal diameter to 110% the unstenosed radius (or radius if there was no stenosis). The radius not stenosed ($R_{unstenosed}$) was calculated according to the following equation:

$$\%_{stenosis} = \left(1 - \frac{R_{stenosed}}{R_{unstenosed}}\right) \times 100\% \tag{1}$$

where the degree of stenosis ($\%_{stenosis}$) and $R_{stenosed}$ were both labeled by intervention cardiologists. We added these bounds to buffer our predictions and minimize uncertainty at the extreme bounds. Similarly, lengths were sampled from 50% to 150% of the pre-intervention stenosis length labeled by interventional cardiologists. Radii and lengths were uniformly sampled between the standardized bounds we set for all patients and used as input to generate an LHS sample matrix.

The training matrix was applied to automatically modify the coronary geometries to reflect the scenarios in the training set. So far, all operations have been performed locally. To run the 1D simulations and generate the training set, all input files and coronary geometries were transferred to the Duke Compute Cluster (Fig. 4). For each training instance (T_N), each 1D simulation was allocated one CPU and all simulations were deployed simultaneously. This parallel approach allowed us to complete the training process in the time it takes to run one 1D simulation. The post-intervention FFR was computed at all centerline points for all training scenarios. Centerline-parsed FFR data was transferred back to the local machine to construct the machine learning model. Bivariate cubic splines were fitted for each centerline point as a function of the radius of the stenosis and the lengths of the LHS. The specific bivariate cubic spline interpolation we applied was from SciPy and based on implementations from FITPACK [25,26]. The centerline points were spaced at 100 μm resolution and compact enough to run efficiently on local machines. To generate the post-intervention FFR map, K-d trees were constructed to efficiently project the FFR data at the centerline to the nearest surface triangles on the 3D mesh.

2.4   Experimental Protocol to Establish and Validate HarVI

To establish HarVI, the first pertinent step was to determine the level of sampling needed to accurately capture the hemodynamic changes in response to intervention. The number of samples impacts the accuracy of the machine learning model—the more samples, the better the interpolation. The objective was to find the minimum level of sampling that would result in the same FFR 2D heatmaps predicted using HarVI as the ground-truth with high sampling. We determined the level of sampling needed for all patients through convergence studies, where we assumed a ground-truth sampling with $n = 250$ and compared it with interpolation results with progressively fewer samples. We created a $5,000 \times 5,000$ grid to uniformly evaluate the 2D parameter space between the minimum and maximum bounds of stenosis radius and length. In short, we made 25 million predictions for each sampling level. To compare with the ground-truth, we computed the maximum and average percentage error over all grid points. We set the error tolerances at 5% for maximum error and 1% for average error.

Once we determined the level of sampling needed across patients, we evaluated how closely FFR_{HarVI} compares to FFR_{1D}. We performed two test intervention cases that were held out during the training process. Specifically, we took treatment 1 (Rx 1) as $0.8R_{unstenosed}$ and $1.2L_{stenosis}$ and treatment 2 (Rx 2) as $R_{unstenosed}$ and $0.8L_{stenosis}$. All baseline geometries were modified to match these test intervention configurations. Post-intervention FFR was validated using continuous metrics and categorical metrics. The post-intervention FFR_{1D} ground-truth was generated by running 1D simulations with the corresponding test geometry modifications.

Lastly, we measured the total turnaround time for HarVI to compare with state-of-the-art FDA approved software. The turnaround time consisted of averages for all 50 patients for the patient-specific modeling pipeline (reconstruction and quality control) and the processing time needed for one-shot model training approach. Comparing the turnaround time of HarVI with established methods was important to gauge clinical translatability.

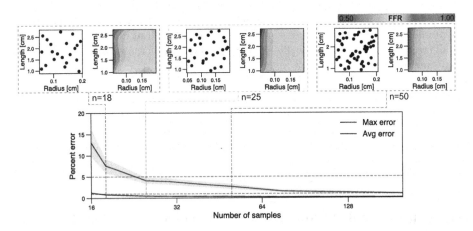

Fig. 5. Evaluating convergence in sampling stenosis radius and length. Ground-truth was assumed to have 250 samples. For all patients (n=50), we evaluated the maximum and average percentage errors for increasing number of samples as compared to the ground-truth. The area around the maximum error curve represents standard errors across all patients. For a representative patient, we showcase three insets at 18, 25, and 50 samples that show sampling on the left and resulting FFR 2D prediction heatmaps on the right.

3   Results and Discussion

3.1   Few Samples Are Needed to Capture Post-intervention Fractional Flow Reserve Accurately

To establish HarVI, it was important to determine the minimum level of sampling needed to capture changes in hemodynamics as a result of intervention. Figure 5 demonstrates that the maximum error curve and average error curve decreases to below 5% and 1%, respectively, when using 25 samples. The 2D heatmaps presented in Fig. 5 show that sampling with 18 intervention configurations was insufficient. Although we show heatmaps for one representative case, there were apparent aberrations in the 2D heatmap that correspond to sites with a paucity of samples. 2D heatmaps for the 25 and 50 sample cases show nearly identical patterns without aberrations. Inter-patient variability was small for the maximum error curve and negligible for the average error curve. This result demonstrates that the convergence in sampling was predictable between patients. All lesions investigated in this study were intermediate focal stenoses, which may explain why a similar level of sampling was found to be sufficient for all patients.

3.2   Post-intervention Fractional Flow Reserve Predicted Using HarVI Agrees with 1D Ground-Truths

After determining the level of sampling needed for convergence, we validated HarVI against 1D ground-truths for two test intervention configurations. The

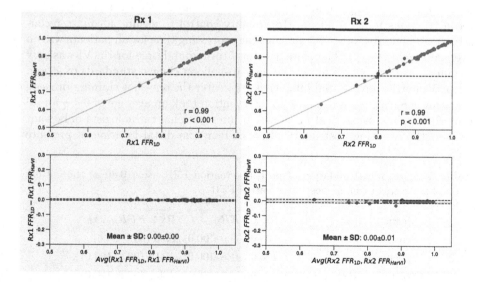

Fig. 6. Continuous validation of post-intervention FFR. For validation, two test cases (Rx 1 - left and Rx 2 - right) were used to compare FFR_{HarVI} against FFR_{1D}. The top row show scatter plots with correlation scores and the bottom row show Bland-Altman plots with bias and imprecision scores.

correlation coefficient was 0.99 ($p < 0.001$) and the bias was 0.0004 ± 0.0020 for Rx 1. The correlation coefficient was 0.99 ($p < 0.001$) and the bias was -0.0006 ± 0.0060 for Rx 2. In short, FFR_{HarVI} almost perfectly recovers FFR_{1D} for continuous metrics with negligible bias (Fig. 6) in both test treatments. For context, FDA-approved software in the literature has reported a bias of around 0.01 ± 0.05 [8, 27] when comparing FFR computed using CFD to invasively measured FFR. The percentage discrepancy between FFR_{HarVI} and FFR_{1D} was on average 0.2% and 0.3% for Rx 1 and Rx 2, respectively.

The diagnostic performance of FFR_{HarVI} to discern stenoses with ischemia is summarized in Table 1. Diagnostic performance perfectly matched FFR_{1D} for Rx 1. Although there was a drop in sensitivity for Rx 2, there was, in fact, only one case misclassified in the Rx 2 test set. One case had post-intervention FFR_{1D} of 0.799—right at the 0.80 threshold. The FFR_{HarVI} prediction was slightly above the threshold at 0.812, which resulted in a false negative. Except for the single case located right at the ischemic threshold, HarVI classified patients in full agreement with 1D CFD ground truths for both test cases.

3.3 End-to-End Turnaround Time to Enable Intervention Planning Within One Working Day

The clinical translatability of HarVI depends on the turnaround time required for the one-shot training process before real-time predictions could be made. FDA-approved software was estimated to require 24–48 h of processing time, including the entire segmentation workflow, CFD simulations, and quality control, before allowing intervention planning [10,11,28–31]. For the HarVI end-to-end pipeline, reconstructions took 15–30 min [18], quality control to ensure adequate reconstruction required another 30 min, and the processing time for all patients was on average 14 min (Fig. 7). Therefore, the total turnaround time for HarVI was only 74 min on average, which fits in the door-to-balloon clinical standards of 90 min for patients with myocardial infarction [17]. After the one-shot training process, updating FFR_{HarVI} in response to intervention took an average of 0.4 s for all centerline points using local machines. If interventional cardiologists only want to view the resulting post-intervention FFR at the distal location of pressure

Table 1. Categorical validation of post-intervention FFR using 0.80 as the cut-off. Ground-truth was taken as post-intervention FFR$_{1D}$.

Metric	Rx 1 FFR_{HarVI}	Rx 2 FFR_{HarVI}
Sensitivity	100.0 (59.0–100.0)	83.3 (35.9–99.6)
Specificity	100.0 (91.8–100.0)	100.0 (92.0–100.0)
Positive predictive value	100.0 (59.0–100.0)	100.0 (47.8–100.0)
Negative predictive value	100.0 (91.8–100.0)	97.8 (88.0–99.6)
Accuracy	100.0 (92.9–100.0)	98.0 (89.4–99.9)
Area under the curve	1.00	1.00

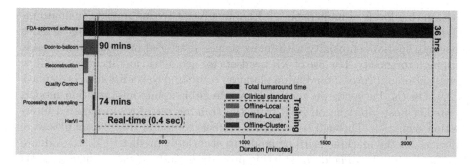

Fig. 7. Gantt chart showing timing breakdown for HarVI compared to clinical standards and FDA approved software. Top section (purple) shows estimated total turnaround time for FDA-approved intervention planning software and clinical standards for door-to-balloon time. Bottom section (pink) shows timing breakdowns (reconstruction, quality control, processing and sampling) for the one-shot training process needed for HarVI. After the training process, with an average turnaround time of 74 min, intervention planning in HarVI is near-instantaneous, requiring only 0.4 s per query (red box with interrupted lines). (Color figure online)

measurement, FFR_{HarVI} updates in only 0.3 ms. HarVI would enable clinicians the ability to test a series of intervention scenarios pre-operatively within the same working day a patient is admitted.

4 Conclusion

Real-time virtual treatment technology is a recent innovation [10,11,31] and can give interventional cardiologists the option to interactively experiment with a variety of possible strategies to determine how best to relieve ischemia. However, the current turnaround time required to train FDA-approved software per patient ranges from 24–48 h [8,29]. This long turnaround time hinders the ability of cardiologists to test intervention strategies within the same day a patient is admitted and, perhaps more pertinently, prevents planning for patients who require immediate treatment, such as those with ST-elevated myocardial infarction. Moreover, current software in the literature relies on conventional 2D displays and is not deployed on immersive devices that help with intuitive evaluation of resulting post-intervention hemodynamics. To address both unmet needs, we established HarVI, a coronary intervention planning software that minimizes turnaround time to enable virtual intervention within one working day and leverages extended reality hardware.

In this study, we developed a machine learning model to predict FFR in response to intervention. We employed LHS to efficiently capture the variance in FFR in response to any possible intervention scenario in terms of radius and length of stenosis. For intuitive user interaction, interventions were captured as geometric modifications to 3D coronary meshes in virtual reality. We coupled the HarVI backend to the Harvis GUI, which is deployable on multiple XR headsets with varying levels of immersion [2]. Enabling intuitive and immersive anatomical editing is important for more accurate treatment planning. Coronary arteries have high spatial complexity with many points of arterial branching and vessels with high tortuosity. The use of XR headsets for geometric modification and flow visualization improves intervention planning compared to traditional 2D displays [4,32]. On the backend, we applied bivariate cubic spline interpolation models to predict post-intervention FFR from the training set of pre-configured intervention scenarios. Through convergence studies, we found that 25 samples were sufficient for the machine learning model to accurately predict FFR. To evaluate HarVI, we tested two held-out stenting procedures and compared them with 1D CFD ground truths. HarVI was able to recapitulate the 1D CFD results accurately (0.2–0.3% error) as evaluated using continuous and categorical validation metrics for the 50 patients in the cohort. Finally, we recorded the time needed in critical parts of the HarVI pipeline and found that the average end-to-end turnaround time required to enable virtual intervention was 74 min.

In this study, we have taken initial steps toward integrating machine learning and extended reality for real-time planning of coronary interventions, focusing on focal lesions. The findings suggest a promising direction for expanding this approach to more complex diseases, such as serial and bifurcation lesions. HarVI offers a more streamlined method for interventional cardiologists to evaluate different treatment options within hours instead of days. The bivariate cubic spline model used here demonstrates strong accuracy. Future exploration of other machine learning-based approaches could further improve accuracy and reduce turnaround time. While the developments presented here are in their early stages, they lay the critical foundation for improving the efficiency of clinical decision making in cardiology.

Acknowledgements. The authors thank Samreen Mahmud, Aristotle Martin, and Runxin Wu for fruitful discussions and Juliet Jiang for her picture with the immersive headset. Computing support for this work came from the Duke Compute Cluster. The work of Cyrus Tanade was supported by the National Science Foundation Graduate Research Fellowship under Grant No. NSF GRFP DGE 1644868. The work of Amanda Randles was supported by the National Institute On Aging of the NIH under Award Number DP1AG082343 and the Duke/Duke-NUS Research Collaboration Pilot Project program. The content does not necessarily represent the official views of the NSF or NIH.

References

1. Vardhan, M., et al.: Investigating the role of VR in a simulation-based medical planning system for coronary interventions. In: Shen, D., et al. (eds.) MICCAI 2019. LNCS, vol. 11768, pp. 366–374. Springer, Cham (2019). https://doi.org/10.1007/978-3-030-32254-0_41
2. Shi, H., Ames, J., Randles, A.: Harvis: an interactive virtual reality tool for hemodynamic modification and simulation. J. Comput. Sci. **43**, 101091 (2020)

3. Vardhan, M., Shi, H., Urick, D., Patel, M., Leopold, J.A., Randles, A.: The role of extended reality for planning coronary artery bypass graft surgery. In: 2022 IEEE Visualization and Visual Analytics (VIS), pp. 115–119. IEEE (2022)

4. Shi, H., Vardhan, M., Randles, A.: The role of immersion for improving extended reality analysis of personalized flow simulations. Cardiovasc. Eng. Technol. 14(2), 194–203 (2023)

5. Luffel, M., et al.: A solid modeling tool for planning and optimizing pediatric heart surgeries. Comput.-Aided Des. 70, 3–12 (2016)

6. Leo, J., et al.: Interactive cardiovascular surgical planning via augmented reality. In: Asian CHI Symposium 2021, pp. 132–135 (2021)

7. Yang, H., et al.: Evaluating cardiovascular surgical planning in mobile augmented reality. arXiv preprint arXiv:2208.10639 (2022)

8. Nørgaard, B.L., et al.: Diagnostic performance of noninvasive fractional flow reserve derived from coronary computed tomography angiography in suspected coronary artery disease: the NXT trial (analysis of coronary blood flow using CT angiography: next steps). J. Am. College Cardiol. 63(12), 1145–1155 (2014)

9. Tonino, P.A.L., et al.: Fractional flow reserve versus angiography for guiding percutaneous coronary intervention. New Engl. J. Med. 360(3), 213–224 (2009)

10. Sankaran, S., et al.: Physics driven real-time blood flow simulations. Comput. Methods Appl. Mech. Eng. 364, 112963 (2020)

11. Modi, B.N., et al.: Predicting the physiological effect of revascularization in serially diseased coronary arteries: clinical validation of a novel CT coronary angiography–based technique. Circulat. Cardiovasc. Intervent. 12(2), 007577 (2019)

12. AHA Statistical Update. Heart disease and stroke statistics-2017 update. Circulation 135, e146–e603 (2017)

13. Fournier, S., et al.: Association of improvement in fractional flow reserve with outcomes, including symptomatic relief, after percutaneous coronary intervention. JAMA Cardiol. 4(4), 370–374 (2019)

14. Katritsis, D.G., Theodorakakos, A., Pantos, I., Gavaises, M., Karcanias, N., Efstathopoulos, E.P.: Flow patterns at stented coronary bifurcations: computational fluid dynamics analysis. Circulat.: Cardiovasc. Interventi. 5(4), 530–539 (2012)

15. Antoniadis, A.P., et al.: Biomechanical modeling to improve coronary artery bifurcation stenting: expert review document on techniques and clinical implementation. JACC: Cardiovasc. Intervent. 8(10), 1281–1296 (2015)

16. Tanade, C., Chen, S.J., Leopold, J.A., Randles, A.: Analysis identifying minimal governing parameters for clinically accurate in silico fractional flow reserve. Front. Med. Technol. 4, 1034801 (2022)

17. Menees, D.S., et al. Door-to-balloon time and mortality among patients undergoing primary PCI. New Engl. J. Med. 369(10), 901–909 (2013)

18. Chen, S.J., Carroll, J.D.: 3-d reconstruction of coronary arterial tree to optimize angiographic visualization. IEEE Trans. Med. Imaging 19(4), 318–336 (2000)

19. Pham, J., Wyetzner, S., Pfaller, M.R., Parker, D., James, D.L., Marsden, A.L.: svmorph: interactive geometry-editing tools for virtual patient-specific vascular anatomies. J. Biomech. Eng. 145(3), 031001 (2023)

20. Randles, A.P., Kale, V., Hammond, J., Gropp, W., Kaxiras, E.: Performance analysis of the lattice Boltzmann model beyond Navier-Stokes. In: 2013 IEEE 27th International Symposium on Parallel and Distributed Processing, pp. 1063–1074. IEEE (2013)

21. Tanade, C., Feiger, B., Vardhan, M., Chen, S.J., Leopold, J.A., Randles, A.: Global sensitivity analysis for clinically validated 1d models of fractional flow reserve. In: 2021 43rd Annual International Conference of the IEEE Engineering in Medicine and Biology Society (EMBC), pp. 4395–4398. IEEE (2021)
22. Feiger, B., Kochar, A., Gounley, J., Bonadonna, D., Daneshmand, M., Randles, A.: Determining the impacts of venoarterial extracorporeal membrane oxygenation on cerebral oxygenation using a one-dimensional blood flow simulator. J. Biomech. **104**, 109707 (2020)
23. McKay, M.D., Beckman, R.J., Conover, W.J.: A comparison of three methods for selecting values of input variables in the analysis of output from a computer code. Technometrics **42**(1), 55–61 (2000)
24. Iman, R.L., Helton, J.C., Campbell, J.E.: An approach to sensitivity analysis of computer models: Part i–Introduction, input variable selection and preliminary variable assessment. J. Quality Technol. **13**(3), 174–183 (1981)
25. Dierckx, P.: An algorithm for surface-fitting with spline functions. IMA J. Numer. Anal. **1**(3), 267–283 (1981)
26. Dierckx, P.: Curve and Surface Fitting with Splines. Oxford University Press (1995)
27. Pellicano, M., et al.: Validation study of image-based fractional flow reserve during coronary angiography. Circulat.: Cardiovasc. Intervent. **10**(9), e005259 (2017)
28. Torii, R., Yacoub, M.H.: Ct-based fractional flow reserve: development and expanded application. Glob. Cardiol. Sci. Pract. **2021**(3) (2021)
29. HeartFlow FFRCT for Estimating Fractional Flow Reserve from Coronary CT Angiography — Guidance — NICE. NICE (2017). https://www.nice.org.uk/guidance/mtg32
30. Gaur, S., et al.: Rationale and design of the heartflownxt (heartflow analysis of coronary blood flow using CT angiography: next steps) study. J. Cardiovasc. Comput. Tomogr. **7**(5), 279–288 (2013)
31. Taylor, C.A., et al.: Patient-specific modeling of blood flow in the coronary arteries. Comput. Methods Appl. Mech. Eng. **417**, 116414 (2023)
32. Pekkan, K., et al.: Patient-specific surgical planning and hemodynamic computational fluid dynamics optimization through free-form haptic anatomy editing tool (surgem). Med. Biol. Eng. Comput. **46**, 1139–1152 (2008)

Krylov Solvers for Interior Point Methods with Applications in Radiation Therapy and Support Vector Machines

Felix Liu[1,2]([✉]) [iD], Albin Fredriksson[2], and Stefano Markidis[1]

[1] KTH Royal Institute of Technology, Stockholm, Sweden
felixliu@kth.se
[2] RaySearch Laboratories, Stockholm, Sweden

Abstract. Interior point methods are widely used for different types of mathematical optimization problems. Many implementations of interior point methods in use today rely on direct linear solvers to solve systems of equations in each iteration. The need to solve ever larger optimization problems more efficiently and the rise of hardware accelerators for general purpose computing has led to a large interest in using iterative linear solvers instead, with the major issue being inevitable ill-conditioning of the linear systems arising as the optimization progresses. We investigate the use of Krylov solvers for interior point methods in solving optimization problems from radiation therapy and support vector machines. We implement a prototype interior point method using a so called doubly augmented formulation of the Karush-Kuhn-Tucker linear system of equations, originally proposed by Forsgren and Gill, and evaluate its performance on real optimization problems from radiation therapy and support vector machines. Crucially, our implementation uses a preconditioned conjugate gradient method with Jacobi preconditioning internally. Our measurements of the conditioning of the linear systems indicate that the Jacobi preconditioner improves the conditioning of the systems to a degree that they can be solved iteratively, but there is room for further improvement in that regard. Furthermore, profiling of our prototype code shows that it is suitable for GPU acceleration, which may further improve its performance in practice. Overall, our results indicate that our method can find solutions of acceptable accuracy in reasonable time, even with a simple Jacobi preconditioner.

Keywords: Interior point method · Krylov solver · Radiation therapy · Support Vector Machines

1 Introduction

Mathematical optimization is used in many areas of science and industry, with applications in fields like precision medicine, operations research and many others. In this paper, we focus on the solution of quadratic programs (QP), which are optimization problems with a quadratic objective function and linear constraints, using an interior point method (IPM). These arise in many applications naturally, for instance in training support vector machine classifiers, but can also be used as part of a sequential quadratic programming solver [3] to solve more general nonlinear optimization problems.

Computationally, IPMs for optimization rely on Newton's method to find search directions. This involves the solution of a large, often sparse and structured linear system of equations, which we will refer to henceforth as the Karush-Kuhn-Tucker (KKT) system, at each iteration. Traditionally, this system is often solved using direct linear solvers, such as LDL^T-factorization for indefinite matrices or Cholesky factorization for positive definite methods, but a topic of interest for much research in the field is the use of iterative linear solvers [23] instead. Indeed, the move to iterative linear algebra for interior point methods has been identified by some authors as a key step in enabling interior point methods to handle very large optimization problems [15].

Another advantage of iterative algorithms for solving linear systems is their suitability for modern computing hardware, such as GPUs. With their rising dominance in High-Performance Computing (HPC), extracting the maximum performance from modern computing hardware all but requires the use of some type of accelerator. Direct linear solvers can suffer performance wise on these types of hardware for a variety of reasons, e.g., unstructured memory accesses, which has been seen in previous studies [25] in the context of interior point methods. The major challenge of using iterative solvers lies in a structured form of ill-conditioning that inevitably exists in the linear systems, which can be severely detrimental to the convergence of the linear solver. Still, given potential performance gains from the use of significantly more powerful computing resources, we believe the trade-off between numerical stability and parallel performance is worthy of further investigation.

One type of problems we consider arise from treatment planning for radiation therapy, which loosely speaking is the process of optimizing treatment plans (treatment machine parameters) for each individual patient case to deliver as accurate a dose as possible to the tumor volume. This inverse problem is often solved by formulating it as a constrained optimization problem, for which finding a solution can be both computationally expensive and present an important bottleneck in the clinical workflow. Computational speed and efficiency is thus of crucial importance, and in the ideal case the optimization would be performed in real time, with the patient present at the clinic. To demonstrate the applicability of our proposed method to other problems as well, we also consider problems from the training of support vector machine classifiers [19], an important method from classical machine learning. All in all, we are interested in studying and evaluating the potential of using IPMs with iterative linear solvers as an avenue to enable us to utilize accelerators and powerful computing resources for solving these problems more efficiently.

In this paper, we propose a complete IPM solver prototype based on the work by Forsgren and Gill in [11], where a special formulation of the KKT systems from interior point methods is considered, which guarantees that the KKT system is positive-definite throughout the optimization for convex problems. Our contribution a complete IPM solver prototype using the doubly augmented formulation and iterative linear solvers, which is capable of solving real world optimization problems. We demonstrate its effectiveness for the solution of quadratic optimization problems arising from real world applications in both radiation therapy treatment planning as well as support vector machines.

2   Background

2.1   Interior Point Methods

We are concerned with the solution of convex, continuous quadratic programs using interior point methods. The following section introduces the relevant aspects for our proposed method. Readers interested in a more thorough overview of interior point methods for optimization are referred to e.g. [13, 26]. In general, we will be dealing with a problem on the form:

$$\begin{aligned} \text{min.} \quad & \frac{1}{2}x^T H x + p^T x \\ \text{s.t.} \quad & l \leq Ax \leq u \\ & Cx = b \end{aligned} \qquad (1)$$

where H is the (positive definite) Hessian of the objective function, $p \in \mathbb{R}^n$ are the linear coefficients of the objective, A, C are the Jacobians of the (linear) inequality- and equality constraints, respectively. In general, we allow components of the constraints to be unbounded, but we do not account for this explicitly to simplify the exposition. Inequality constraints are often treated by the introduction of auxiliary *slack variables*. We convert the problem to having only lower bounds, introduce slack variables s_l, s_u (for lower and upper bounds respectively), use a log-barrier term for the inequality constraints and finally, we use a penalty barrier method [10, 13] to handle equality constraints, giving:

$$\begin{aligned} \text{min.} \quad & \frac{1}{2}x^T H x + p^T x- \\ & - \mu \sum \log((s_l)_i) - \mu \sum \log((s_u)_i)+ \\ & + \frac{1}{2\mu}\|Cx - b\|^2 \\ \text{s.t.} \quad & Ax - s_l - l = 0 \\ & -Ax - s_u + u = 0 \end{aligned} \qquad (2)$$

Here μ is the so called barrier parameter. Intuitively, the barrier terms diverge towards $+\infty$ as the boundary of the feasible set is approached, thus encouraging feasibility throughout the iterations. As is common in primal-dual interior point methods, we work with the *perturbed* optimality conditions. These state that an optimal solution to the equality constrained subproblem in (2) satisfies the following:

$$r_H := Hx + p - A^T\lambda_l + A^T\lambda_u - C^T\lambda_e = 0$$
$$r_{A_l} := Ax - s_l - l = 0$$
$$r_{A_u} := -Ax + s_u + u = 0$$
$$r_e := Cx - b + \mu\lambda_e = 0 \tag{3}$$
$$r_{c_1} := (\lambda_l)_i(s_l)_i - \mu = 0, \quad i = 1, ..., m_l$$
$$r_{c_2} := (\lambda_l)_i(s_u)_i - \mu = 0, \quad i = 1, ..., m_l.$$

These conditions are very similar to the first order Karush-Kuhn-Tucker conditions for optimality [20, Ch. 12.3], except that the final two conditions are perturbed by μ, and the inclusion of the penalty barrier method for equality constraints. Primal-dual interior point methods generally seek points satisfying the perturbed optimality conditions above using, e.g., Newton's method, while successively decreasing the barrier parameter $\mu \to 0$. As μ approaches 0, we expect our solution to approach a point satisfying the KKT conditions for optimality.

2.2   Optimization Problems in Radiation Therapy and SVMs

The first type of problem we consider are from radiation therapy, and are all exported from the treatment planning system RayStation, developed by the Stockholm-based company RaySearch Laboratories. The problems all arise from *treatment planning* for radiation therapy, where an optimization problem is solved to determine a *treatment plan* for each individual patient. A view of treatment planning from an optimization perspective can be found in e.g. [8,9].

The optimization problems from RayStation are QP-subproblems in a Sequential Quadratic Programming solver used for nonlinear optimization, and have the form:

$$\text{min.} \quad \frac{1}{2}p^T\nabla_{xx}\mathcal{L}(x,\lambda)p + (\nabla f)^T(x)p \tag{4}$$
$$\text{s.t.} \quad \nabla g(x)^T p + g(x) \geq 0.$$

Here, $\mathcal{L}(x,\lambda) = f(x) - g(x)\lambda^T$ is the Lagrangian, and λ are the Lagrange multipliers. In practice, the Hessian of the Lagrangian $\nabla_{xx}\mathcal{L}(x,\lambda)$ can be expensive to form, and it is common to use a quasi-Newton type approximation of the Hessian instead. The SQP solver uses Broyden-Fletcher-Goldfarb-Shanno (BFGS) updates [4] to estimate the Hessian of the non-linear problem, which means that the Hessian for each of our QP-subproblems can be written on matrix form as:

$$H = H_0 + UWU^T, \tag{5}$$

where H_0 is the initial approximation to the Hessian, the dense $n \times k$ matrix U consists of the update vectors on the columns, and W is a diagonal matrix with the scalar update weights on the diagonal. With a suitable line-search method, the updates to the Hessian can be ensured to preserve positive definiteness, thus making the QP-subproblem we need to solve at each SQP iteration convex.

The second type of problem we consider are QP dual problems from support vector machine training for classification. These problems are of the form:

$$\min. \quad \frac{1}{2}\alpha^T H\alpha - \alpha^T e$$

$$\text{s.t.} \quad \alpha^T y = 0 \tag{6}$$

$$0 \le \alpha \le c,$$

where the Hessian H is of the form $H = yy^T Q$, and the entries of Q are $K(x_i, x_j)$, for some "kernel" K used to map the data into a (more) separable space. For our experiments, we use a radial basis function kernel

$$K(x_i, x_j) = \exp(-||x_i - x_j||^2/2\sigma).$$

3   Prototype Interior Point Method

We now describe the key components of our prototype interior point method implementation used in this paper.

3.1   KKT System Formulation

Many optimization problems from real applications include bounds on the (primal) variables themselves. Such bounds can be included by the introduction of appropriate rows in the A matrix. For an efficient formulation, we will consider the bounds on variables separately from general linear constraints. Separating the handling of the variable bounds (i.e. lower and upper bounds on the values of the variables x) and then using Newton's method to solve the perturbed optimality conditions (3) gives a linear system of the form:

$$\begin{pmatrix} H & -A^T & A^T & -I & I & C^T & & & & & \\ C & & & & & M & & & & & \\ A & & & & & & -I & & & & \\ -A & & & & & & & -I & & & \\ I & & & & & & & & -I & & \\ -I & & & & & & & & & -I & \\ S_{l_A} & & & & & \Lambda_{l_A} & & & & & \\ & S_{u_A} & & & & & \Lambda_{u_A} & & & & \\ & & S_{l_x} & & & & & \Lambda_{l_x} & & & \\ & & & S_{u_x} & & & & & \Lambda_{u_x} & & \end{pmatrix} \begin{pmatrix} \Delta x \\ \Delta\lambda_{l_A} \\ \Delta\lambda_{u_A} \\ \Delta\lambda_{l_x} \\ \Delta\lambda_{u_x} \\ \Delta\lambda_e \\ \Delta s_{l_A} \\ \Delta s_{u_A} \\ \Delta s_{l_x} \\ \Delta s_{u_x} \end{pmatrix} = \begin{pmatrix} -r_H \\ -r_e \\ -r_{l_A} \\ -r_{u_A} \\ -r_{l_x} \\ -r_{u_x} \\ -r_{c1} \\ -r_{c2} \\ -r_{c3} \\ -r_{c4} \end{pmatrix}, \tag{7}$$

where $\lambda_e, \lambda_{l_A}, \lambda_{u_A}, \lambda_{l_x}, \lambda_{u_x}$ are the Lagrange multipliers for the equality constraints, lower and upper bounds on the (general) linear constraints, and lower and upper bounds for the variable bounds respectively. The slack variables s are subscripted in the same way. The residuals on the RHS are similar to the ones shown in (3). Λ, S, M denote diagonal matrices with the corresponding Lagrange multipliers, slack variables or barrier parameter μ on the diagonal respectively, and e is an appropriately sized column-vector with a value of 1 in all coefficients.

The above system can be reduced in size by block-row elimination. Multiplying the sixth and seventh block row by $\Lambda_{l_A}^{-1}$ and the fifth block row by $\Lambda_{u_A}^{-1}$ and adding them to the second and third rows, followed by multiplying the sixth and seventh block rows by $S_{l_x}^{-1}$ and $-S_{u_x}^{-1}$ respectively and adding to the top row, as well as multiplying the fourth and fifth block rows by $S_{l_x}^{-1}\Lambda_{l_x}$ and $-S_{u_x}^{-1}\Lambda_{u_x}$ and adding to the top row. The final reduced linear system of equations (with the same row operations on the RHS) can be written as:

$$\begin{pmatrix} Q & -B^T \\ B & D \end{pmatrix} \begin{pmatrix} \Delta x \\ \Delta\lambda_A \end{pmatrix} = \begin{pmatrix} r_1 \\ r_2 \end{pmatrix}, \tag{8}$$

where:

$$Q = H + S_{l_x}^{-1}\Lambda_{l_x} + S_{u_x}^{-1}\Lambda_{u_x}, \quad B = \begin{pmatrix} C \\ A \\ -A \end{pmatrix}, \quad D = \begin{pmatrix} M & & \\ & \Lambda_{l_A}^{-1}S_{l_A} & \\ & & \Lambda_{u_A}^{-1}S_{u_A} \end{pmatrix}.$$

$$\Delta\lambda_A = \begin{pmatrix} \Delta\lambda_e \\ \Delta\lambda_{l_A} \\ \Delta\lambda_{u_A} \end{pmatrix}, \quad r_1 = -r_H - S_{l_x}^{-1}r_{c3} + S_{u_x}^{-1}r_{c4} - S_{l_x}^{-1}\Lambda_{l_x}r_{l_x} + S_{u_x}^{-1}\Lambda_{u_x}r_{u_x}$$

$$r_2 = \begin{pmatrix} -r_e \\ -r_{A_l} - \Lambda_{l_A}^{-1}r_{c1} \\ -r_{A_u} - \Lambda_{u_A}^{-1}r_{c2} \end{pmatrix}$$

At this point the purpose of handling the variable bounds separately becomes clear, since they now only contribute a diagonal term $S_{l_x}^{-1}\Lambda_{l_x} + S_{u_x}^{-1}\Lambda_{u_x}$ in the Hessian block. A challenge with the system (8) is that it becomes inevitably ill-conditioned as the optimization approaches a solution. Intuitively this can be seen by noting that as $\mu \to 0$, some elements of the diagonal block D become very small and some become unbounded, since for active constraints the slack variables tend to zero, while for inactive constraints the Lagrange multipliers do. For more details on this ill-conditioning see, e.g., [12,27]. The system (8) is unsymmetric, and it is common to consider many equivalent but symmetric systems instead. Our implementation uses the *doubly augmented* formulation, proposed in [11]. This formulation can be derived through block-row operations on the system in (8), by multiplying the second block row by $2B^T D^{-1}$ and adding it to the first block row:

$$\begin{pmatrix} Q + 2B^T D^{-1}B & B^T \\ B & D \end{pmatrix} \begin{pmatrix} \Delta x \\ \Delta\lambda_A \end{pmatrix} = \begin{pmatrix} r_1 + 2B^T D^{-1}r_2 \\ r_2 \end{pmatrix}. \tag{9}$$

The major advantage is that the matrix is symmetric and positive definite for convex problems [11]. This enables us to use a preconditioned conjugate gradient method to solve the system efficiently. To precondition the system we use Jacobi (diagonal scaling) preconditioning, which is motivated by the ill-conditioning arising primarily from the diagonal D block in the matrix.

3.2   IPM Implementation

We implement a prototype interior point method to assess the performance and accuracy of the method when using iterative linear algebra. The following is a brief description of the design of our implementation. Our implementation is a primal-dual interior point method based on the KKT system formulation in (9). We use a ratio test to determine the maximum step length to take each iteration to maintain the positivity of the slack variables and Lagrange multipliers:

$$
\begin{aligned}
\alpha_x &= \min\left\{1.0, \gamma\left(\min\left\{-\frac{s_i}{\Delta s_i} : \Delta s_i < 0\right\}\right)\right\} \\
\alpha_\lambda &= \min\left\{1.0, \gamma\left(\min\left\{-\frac{\lambda_i}{\Delta \lambda_i} : \Delta \lambda_i < 0\right\}\right)\right\}.
\end{aligned}
\tag{10}
$$

The scalar $\gamma < 1$ is to ensure strict positivity of the slacks and Lagrange multipliers throughout the optimization, and we use a value of $\gamma = 0.99$ in our implementation. The step lengths from the line search are used to scale the search direction.

Finally, we decrease the barrier parameter μ based value of the residuals (shown in the right-hand side of (7)). Namely, when the 2-norm of the residuals is smaller than the current value of μ, we divide μ by 10 and continue the optimization. The optimizer terminates when the barrier parameter decreases below a tolerance threshold, which by default is set to 10^{-6}.

Algorithm 1. Interior Point Method

1: **for** $i \leftarrow 1$ to N **do**
2:     Find search direction by solving (8) using PCG
3:     Line search for α_x, α_λ from (10)
4:     $x \leftarrow x + \alpha_x \Delta x$
5:     $\lambda \leftarrow \lambda + \alpha_\lambda \Delta \lambda$
6:     $s \leftarrow s + \alpha_x \Delta s$
7:     Update diagonal D in KKT system
8:     Compute residuals (RHS of (7))
9:     **if** $\|r\| < \mu$ **then**   ▷ r is the RHS of (7)
10:         **if** $\mu < \mu_{tol}$ **then**
11:             Return solution
12:         **end if**
13:         $\mu \leftarrow \mu/10$
14:     **end if**
15: **end for**

We summarize the main components of our implementation in Algorithm 1. The use of a Krylov solver for our implementation provides some practical benefits in how the computation is structured, as it allows us to work in a matrix-free manner. Concretely, we do not explicitly form the $2B^T D^{-1} B$ term in (8) (nor the entirety of the matrix), but always work with the different components separately. This is also especially advantageous for the quasi-Newton structure of the Hessian. as discussed in Sect. 2.2. Recall that the BFGS-Hessian can be written in matrix form as $H = H_0 + UWU^T$, which is a dense $n \times n$ matrix, where n is the number of variables in the QP. Similarly to before, this matrix is also not explicitly formed in our solver, saving significant computational effort when the number of variables is large.

We have implemented the method described in C++ (using BLAS for many computational kernels), which is also the implementation we use for the experiments conducted in this work.

4   Experimental Setup

Table 1. Problem dimensions for the considered QPs

Problem	Vars	Lin. cons	Bound cons
Proton H&N	55770	0	90657
Proton Liver	90657	15	90657
Photon H&N	13425	42273	13425
SVM a1a	1605	1	3210

The radiation therapy optimization problems evaluated in this work are exported directly from the RayStation optimizer. We export the QP subproblems to files directly from RayStation which permits us to use them for our experiments, without relying on RayStation itself. In particular, we consider three problem cases, two cases treated using proton therapy, one for the head-and-neck region and a liver case and one case treated using photons. The dimensions of the corresponding optimization problems are shown in Table 1. In all considered problems, the problems are exported from the later stages of the SQP iterations, which are typically the most challenging.

For the SVM training problem, we use the a1a problem available from the LIBSVM dataset [5]. We pre-compute the (dense) Hessian H using the radial basis function kernel as described in Sect. 2.2.

The performance measurements were all carried out on a local workstation equipped with an AMD Ryzen 7900x CPU with 64 GB of DDR5 DRAM. The BLAS library used was OpenBLAS 0.3.21 with OpenMP threading. The measurements of the condition number were run on a node on the Dardel supercomputer in PDC at the KTH Royal Institute of Technology, for memory reasons.

5   Results

In the following section, we present some experimental evaluation of our prototype method in terms of convergence of the conjugate gradient solver and interior point solver itself, as well as the conditioning of the KKT-systems and the computational performance.

5.1 Krylov Solver Convergence

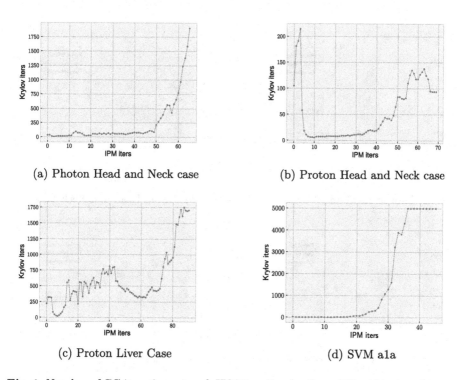

(a) Photon Head and Neck case

(b) Proton Head and Neck case

(c) Proton Liver Case

(d) SVM a1a

Fig. 1. Number of CG iterations at each IPM iteration for three different test problems.

Figure 1 shows the number of CG iterations required for the linear solver to converge within each IPM iteration for our test problems. Note that the maximum number of CG iterations was set to 5000 for each of the test cases, with a convergence tolerance of 10^{-7} for the unpreconditioned residual. As a trend, we see that all problems show a sharp increase in the number of CG iterations towards the later IPM iterations, which is consistent with the observation that the ill-conditioning of the systems arises when the barrier parameter μ gets close to zero. The proton head and neck case stands out in that it has a spike in CG at the beginning of the optimization as well, which is also seen in the estimated condition numbers being large in the beginning in Fig. 3. To note is that the proton head and neck case is the only one considered without general linear constraints (it has only variable bound constraints). For some of the considered cases, especially the proton liver case and SVM cases, the number of CG iterations for convergence is very large, indicating that improved preconditioning is an interesting prospect for future work.

5.2 IPM Solver Convergence

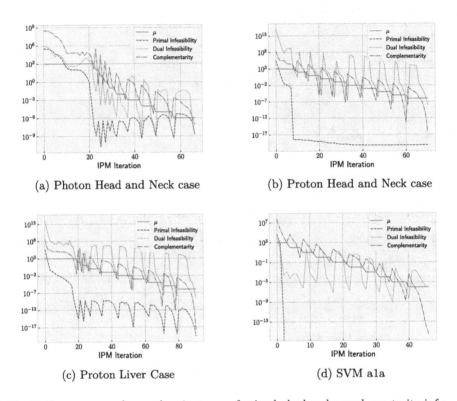

(a) Photon Head and Neck case

(b) Proton Head and Neck case

(c) Proton Liver Case

(d) SVM a1a

Fig. 2. Convergence of our solver in terms of primal, dual and complementarity infeasibility. The value of the barrier parameter μ is shown as well, which is successively decreased as optimization progresses.

Another interesting aspect to consider is the convergence of the interior point method as a whole. In optimization solvers, the convergence is often measured with respect to the *primal, dual* and *complementarity* infeasibility (among others). The primal infeasibility is the (Euclidean) norm of $\left(r_{l_A} \; r_{u_A} \; r_{l_x} \; r_{u_x} \right)^T$, the dual infeasibility is the norm of r_H and the complementarity infeasibility the norm of $\left(r_{c_1} \; r_{c_2} \; r_{c_3} \; r_{c_4} \right)^T$. In other words, the primal infeasibility measures the error with respect to satisfying the constraints, the dual infeasibility measures the error in stationarity of the Lagrangian, and the complementarity infeasibility the error with respect to the perturbed complementary slackness condition.

In Fig. 2, we show the convergence in terms of the primal, dual and complementarity infeasibility over IPM iterations for our test-problems. From the figures, we see that the convergence towards optimality is far from monotonous, with the spikes in the infeasibility norms coinciding with the points when the barrier parameter μ is decreased in the solver. This could indicate that the update of μ is too aggressive. The reason could be that we use a relatively crude update rule for μ in our prototype implementation, and more sophisticated methods may give better performance. Another interesting observation from the infeasibility plots is that the dual infeasibility exhibits slower convergence compared to the complementarity and primal infeasiblities, especially for the proton cases.

Speculatively, one can observe that the dual infeasibility r_H, appears only once in the right-hand side of (8). The block in the RHS in which the dual infeasibility appears is also contains multiple terms scaled by the inverse of the slack variables, which may be large for *active* constraints. The doubly augmented system (9) introduces an additional term in the top block of the RHS. In view of Krylov methods as algorithms seeking least-norm solutions in a given Krylov subspace for a set of linear system of equations, this may give a partial explanation why the dual infeasibility lags behind in our case, as the contribution to the RHS in the linear system from the corresponding term can be relatively small.

5.3   Numerical Stability and Conditioning of KKT System

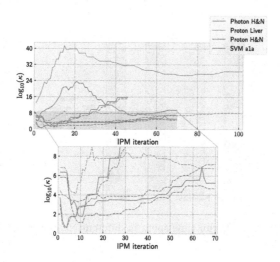

Fig. 3. Condition number estimates using Matlab's `condest` throughout the optimization iterations. The bottom plot is zoomed in on the red region (with the Photon H&N line removed for clarity). The dashed lines are condition numbers after Jacobi preconditioning.

As discussed previously, one of the main challenges in using iterative linear solvers for interior point methods is the structured ill-conditioning in the linear systems. To study how this conditioning affects our problem, we evaluate how the condition number κ of the doubly augmented KKT system (9) that we solve in each iteration changes throughout the optimization. The condition numbers of the resulting matrices are estimated using Matlab's `condest` function, which we modify slightly by using Cholesky instead of LU-factorization internally, since our matrices are symmetric positive definite. `condest` gives an estimate of the condition number in the $L1$-norm and is based on an algorithm proposed by Higham in [16].

Figure 3 shows the results of the condition number analysis. The solid lines show the un-preconditioned condition numbers, with the dashed lines showing the condition numbers with Jacobi preconditioning. The un-preconditioned KKT systems for the Photon H&N and Proton Liver cases show extreme ill-conditioning, especially in the middle of the optimization, with estimated condition numbers up to the order 10^{41} for the Proton Liver case and 10^{23} for the Photon H&N case. While the accuracy of Matlab's estimation using `condest` at such extreme ill-conditioning may be questioned, suffice it to say that the un-preconditioned matrices are close to singular. However, we see that the Jacobi preconditioning does manage to improve the conditioning of those matrices significantly, reducing the condition number to around 10^8 (or less) for both cases.

5.4   Performance Analysis

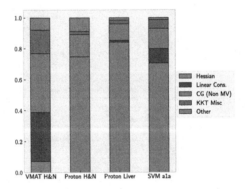

Fig. 4. Relative (normalized) run-time spent in different computational kernels for our solver.

Figure 4 shows the relative time spent in different parts of the code for the three test problems considered in this work. The absolute run-times for the different parts is given in Table 2. The parts we included in the profiling are: Matrix-vector multiplication with the quasi-Newton Hessian, matrix-vector multiplication with the constraint matrix, remaining time in the CG solver (excluding matrix vector multiplication) remaining time spent in the matrix-vector multiplication with

Table 2. Total run times in seconds spent in different parts of the solver for the different problems. CG (Not MV) is the time spent in the conjugate gradients solver excluding matrix-vector products. The relative times in different parts are also visualized in Fig. 4.

Case	Total	Hessian	Lin. Cons	CG (Not MV)	KKT Misc	Other
Photon H&N	13.6	0.997	4.30	5.20	2.06	1.08
Proton H&N	6.28	4.70	0.002	0.907	0.127	0.551
Proton Liver	324	222	62.8	28.4	6.13	4.29
SVM a1a	2.78	1.96	0.256	0.367	0.157	0.035

the doubly augmented KKT system in (9), and finally the "Other" category comprising the remaining time spent in other parts of the solver.

From Fig. 4, we see that for both proton cases and the SVM problem, the solver spends the majority of the time in computing (dense) matrix-vector products with the Hessian, while for the photon radiation therapy problem, a significant amount of time is spent in the CG solver itself, as well as for multiplication with the linear constraints. Overall, we believe the performance analysis shows that there is potential for improved performance when moving to GPU, especially for the proton radiation therapy problems and SVM problems.

6 Related Work

The topic of iterative linear solvers in interior point methods has attracted much research, which we briefly summarize in the following section. Preconditioners for KKT systems in interior point method have been studied extensively previously, for instance in [2,7,14,17,21,28] among many others. A general overview of HPC in the space of optimization and optimization software can be found in [18]. The topic of parallel computing in optimization and operations research in general has also been surveyed previously [24].

Practical studies where iterative linear solvers are used for different kinds of optimization problems can be found in [22], where a type of hybrid direct-iterative solution method is evaluated on very large problems in optimal power flow. In the context of interior point methods for linear programming (LP), preconditioned Krylov methods were studied in e.g. [6,7].

7 Conclusions and Future Work

In this work, we have presented our prototype interior point method for quadratic programming that uses an iterative linear solver for the KKT systems arising in each iteration. We demonstrate that the method can solve real optimization problems from radiation therapy to acceptable levels of accuracy and within reasonable time. From analyzing the performance of our implementation using tracing and profiling, we believe that our method is suitable for GPU acceleration, which we will investigate further in future work. Overall, we believe that interior point methods using Krylov solvers give a promising path forward for GPU accelerated interior point methods, which hold great promise for e.g. computational efficiency in treatment plan optimization for radiation therapy.

There are many interesting questions and problems remaining for future research. Among those is the porting of the code to be able to run on GPU accelerators and looking at improved preconditioners for the KKT systems. One concrete possiblity for improved preconditioning would be to consider a method similar to the one proposed in [1], based on low-ranks updates of a factorized Schur-complement. For some of the problems considered in this paper with only a few linear constraints, and thus a correspondingly small Schur complement, explicit re-factorization of the Schur complement may be so cheap that the low-rank update scheme is not required at all. Finally, another interesting possibility for future work is to investigate the suitability of the proposed method for optimization problems from other domains.

Acknowledgment. The computations were enabled by resources provided by the National Academic Infrastructure for Supercomputing in Sweden (NAISS) at PDC, partially funded by the Swedish Research Council through grant agreement no. 2022-06725.

References

1. Bellavia, S., De Simone, V., di Serafino, D., Morini, B.: Updating constraint preconditioners for kkt systems in quadratic programming via low-rank corrections. SIAM J. Optim. **25**(3), 1787–1808 (2015)
2. Bergamaschi, L., Gondzio, J., Zilli, G.: Preconditioning indefinite systems in interior point methods for optimization. Comput. Optim. Appl. **28**, 149–171 (2004)
3. Boggs, P.T., Tolle, J.W.: Sequential quadratic programming. Acta Numer. **4**, 1–51 (1995)
4. Broyden, C.G.: The convergence of a class of double-rank minimization algorithms 1. general considerations. IMA J. Appl. Math. **6**(1), 76–90 (1970)
5. Chang, C.C., Lin, C.J.: Libsvm: a library for support vector machines. ACM Trans. Intell. Syst. Technol. **2**(3), 1–27 (2011)
6. Chowdhury, A., Dexter, G., London, P., Avron, H., Drineas, P.: Faster randomized interior point methods for tall/wide linear programs. J. Mach. Learn. Res. **23**(336), 1–48 (2022)
7. Cui, Y., Morikuni, K., Tsuchiya, T., Hayami, K.: Implementation of interior-point methods for lp based on krylov subspace iterative solvers with inner-iteration preconditioning. Comput. Optim. Appl. **74**, 143–176 (2019)
8. Ehrgott, M., Güler, Ç., Hamacher, H.W., Shao, L.: Mathematical optimization in intensity modulated radiation therapy. Ann. Oper. Res. **175**(1), 309–365 (2010)
9. Engberg, L.: Automated radiation therapy treatment planning by increased accuracy of optimization tools. Ph.D. thesis, KTH Royal Institute of Technology (2018)
10. Fiacco, A.V., McCormick, G.P.: Nonlinear programming: sequential unconstrained minimization techniques. SIAM (1990)
11. Forsgren, A., Gill, P.E., Griffin, J.D.: Iterative solution of augmented systems arising in interior methods. SIAM J. Optim. **18**(2), 666–690 (2007)

12. Forsgren, A., Gill, P.E., Shinnerl, J.R.: Stability of symmetric ill-conditioned systems arising in interior methods for constrained optimization. SIAM J. Matrix Anal. Appl. **17**(1), 187–211 (1996)
13. Forsgren, A., Gill, P.E., Wright, M.H.: Interior methods for nonlinear optimization. SIAM Rev. **44**(4), 525–597 (2002)
14. Gill, P.E., Murray, W., Ponceleón, D.B., Saunders, M.A.: Preconditioners for indefinite systems arising in optimization. SIAM J. Matrix Anal. Appl. **13**(1), 292–311 (1992)
15. Gondzio, J.: Interior point methods 25 years later. Eur. J. Oper. Res. **218**(3), 587–601 (2012)
16. Higham, N.J.: Fortran codes for estimating the one-norm of a real or complex matrix, with applications to condition estimation. ACM Trans. Math. Softw. **14**(4), 381–396 (1988)
17. Karim, S., Solomonik, E.: Efficient preconditioners for interior point methods via a new schur complement-based strategy. SIAM J. Matrix Anal. Appl. **43**(4), 1680–1711 (2022)
18. Liu, F., Fredriksson, A., Markidis, S.: A survey of hpc algorithms and frameworks for large-scale gradient-based nonlinear optimization. J. Supercomput. **78**(16), 17513–17542 (2022)
19. Noble, W.S.: What is a support vector machine? Nat. Biotechnol. **24**(12), 1565–1567 (2006)
20. Nocedal, J., Wright, S.: Numerical Optimization, 2nd edn. Springer Series in Operations Research and Financial Engineering. Springer, New York (2006)
21. Rees, T., Greif, C.: A preconditioner for linear systems arising from interior point optimization methods. SIAM J. Sci. Comput. **29**(5), 1992–2007 (2007)
22. Regev, S., et al.: Hykkt: a hybrid direct-iterative method for solving KKT linear systems. In: Optimization Methods and Software, pp. 1–24 (2022)
23. Saad, Y.: Iterative methods for sparse linear systems. SIAM (2003)
24. Schryen, G.: Parallel computational optimization in operations research: a new integrative framework, literature review and research directions. Eur. J. Oper. Res. **287**(1), 1–18 (2020)
25. Świrydowicz, K., et al.: Linear solvers for power grid optimization problems: a review of GPU-accelerated linear solvers. Parallel Comput. **111**, 102870 (2022)
26. Wright, M.H.: Interior methods for constrained optimization. Acta Numer. **1**, 341–407 (1992)
27. Wright, M.H.: Ill-conditioning and computational error in interior methods for nonlinear programming. SIAM J. Optim. **9**(1), 84–111 (1998)
28. Zilli, G., Bergamaschi, L.: Block preconditioners for linear systems in interior point methods for convex constrained optimization. Annali Dell'Universita'di Ferrara **68**(2), 337–368 (2022)

From Fine-Grained to Refined: APT Malware Knowledge Graph Construction and Attribution Analysis Driven by Multi-stage Graph Computation

Rongqi Jing[1,2], Zhengwei Jiang[1,2], Qiuyun Wang[1,2(✉)], Shuwei Wang[1,2], Hao Li[1,2], and Xiao Chen[1,2]

[1] Institute of Information Engineering, Chinese Academy of Sciences, Beijing 100093, China
{jingrongqi,jiangzhengwei,wangqiuyun}@iie.ac.cn
[2] School of Cyber Security, University of Chinese Academy of Sciences, Beijing 100049, China

Abstract. In response to the growing threat of Advanced Persistent Threat (APT) in network security, our research introduces an innovative APT malware attribution tool, the APTMalKG knowledge graph. This knowledge graph is constructed from comprehensive APT malware data and refined through a multi-stage graph clustering process. We have incorporated domain-specific meta-paths into the GraphSAGE graph embedding algorithm to enhance its effectiveness. Our approach includes an ontology model capturing complex APT malware characteristics and behaviors, extracted from sandbox analysis reports and expanded intelligence. To manage the graph's granularity and scale, we categorize nodes based on domain knowledge, form a correlation subgraph, and progressively adjust similarity thresholds and edge weights. The refined graph maintains crucial attribution data while reducing complexity. By integrating domain-specific meta-paths into GraphSAGE, we achieve improved APT attribution accuracy with an average accuracy of 91.16%, an F1 score of 89.82%, and an average AUC of 98.99%, enhancing performance significantly. This study benefits network security analysts with an intuitive knowledge graph and explores large-scale graph computing methods for practical scenarios, offering a multi-dimensional perspective on APT malware analysis and attribution research, highlighting the value of knowledge graphs in network security.

Keywords: APT malware · attribution analysis · graph clustering · graph embedding · ensemble machine learning

1 Introduction

In the realm of cybersecurity, Advanced Persistent Threats (APTs) are a major concern. APTs are intricate and covert threats, designed to infiltrate networks

Supported by Youth Innovation Promotion Association, CAS (No. 2023170).

for extended periods with the goal of stealing sensitive information or executing destructive attacks. These threats often employ sophisticated malware to bypass traditional security defenses [10,32]. Understanding and mitigating APT attacks, especially APT malware, are crucial.

Researchers have employed various techniques like machine learning, behavioral analysis, code signature recognition, and malware sandbox testing to enhance APT malware detection and attribution [13,15]. However, cybersecurity faces challenges due to the evolving nature of APT attacks and the complexity of large-scale networks and datasets [5].

Knowledge graphs have proven effective in network security analysis. They provide a comprehensive way to understand malware, attacker behavior, and attack chains by integrating data from various sources [11,16,25]. However, constructing and analyzing complex knowledge graphs pose challenges, such as the scale of nodes and edges in large networks [19]. Traditional knowledge graphs derived from abstract data sources struggle to support multidimensional attribution analysis.

Therefore, we propose a refined representation of an APT malware knowledge graph: APTMalKG. Based on in-depth analysis of malware, covering static and dynamic features, rule-based TTPs, and communication behavior, among other information. Through the construction of the APTMalKG knowledge graph, we capture multidimensional features and behaviors in a structured manner, providing a more comprehensive view for network security analysts. However, due to the fine-grained nature of knowledge graphs, their vast scale and complexity pose challenges for efficient graph computation. To address this issue, we introduce a multi-stage graph clustering method, which organizes nodes effectively into related heterogeneous subgraphs based on node attributes and structural similarity, thereby achieving efficient node fusion and graph refinement. Additionally, we consider the optimization of graph representation, defining malware aggregation features and abstract features as two dimensions of graph representation, gradually reducing graph complexity while preserving critical attribution information. Finally, we apply domain-defined meta-path-enhanced GraphSAGE graph embedding methods, guided by domain knowledge-defined malware behavior paths, to provide a streamlined and efficient APT attribution classification model. In summary, our research offers several key advantages:

– Firstly, the construction of knowledge graphs enables us to capture various features and behaviors of malware in a structured manner, providing a more comprehensive view.
– Secondly, the multi-stage graph clustering method refines knowledge graphs and optimizes graph representations, reducing graph complexity while retaining important information.
– Lastly, domain-defined meta-path-enhanced GraphSAGE graph embedding enhances attribution classification accuracy and efficiency, enabling us to better understand and respond to APT threats.

Our study aims to delve into how to construct and optimize knowledge graphs from the perspective of APT malware, providing robust support for attribution analysis. Simultaneously, we provide a new multidimensional visualization graph for APT malware analysis and attribution research, aiding security teams in better understanding and responding to APT threats, and showcasing the potential value of knowledge graphs in the field of network security.

2   Related Work

In recent years, APT malware has emerged as a major concern in cybersecurity. The in-depth analysis and accurate attribution of such attacks are crucial. Researchers have been focusing on leveraging machine learning and deep learning to enhance the identification and attribution of APT attacks. Methods like deep learning algorithms and transfer learning have been instrumental in classifying APT malware, particularly in understanding API behaviors [30]. Techniques such as sequence pattern feature mining and combining BERT with LSTM have also been explored for improved detection [2,4].

APT attribution research confronts challenges like data scarcity and attacker deception. To combat these, various methodologies, including attack pattern analysis and cross-domain data fusion, are being utilized to gain more precise insights [22–24]. Different feature dimensions like static, dynamic, and network features each provide unique perspectives for malware analysis.

The construction of knowledge graphs, representing malware attributes and relationships, has become a vital tool in APT malware analysis [8,16,21]. These graphs use standardized malware descriptions covering functionality, behavior, and targets to support analysis and detection tasks [1].

Graph representation learning, converting data into vector representations, has advanced in malicious software classification and attribution. Techniques like graph clustering, graph embedding, and graph convolution are being applied to improve the precision of malware classification and attribution [3,14,18,29,31].

In summary, current research highlights the importance of a multidimensional approach, knowledge graph construction, and attribution research in addressing APT threats. However, challenges in constructing high-quality APT malware attribution graph data and selecting appropriate models persist. Addressing these issues is crucial for the development of graph representation learning in malware classification and attribution. This study aims to construct a refined APT malware knowledge graph through multi-stage graph clustering, enhancing the effectiveness of attribution research.

3   Proposed Method

Figure 1 illustrates the construction of APTMalKG, which involves the analysis of malware in both static and dynamic dimensions, incorporating Tactics, Techniques, and Procedures (TTPs) as well as location data. The process employs a multi-stage graph clustering approach to refine the graph, reduce complexity, and enhance the GraphSAGE algorithm with domain-specific metapaths to enable efficient APT attribution.

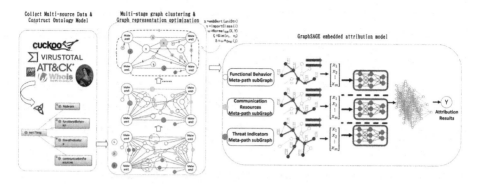

Fig. 1. System architecture of APTMalKG attribution model.

Initially, the research extracts multidimensional malware data and constructs a comprehensive knowledge graph. Subsequently, the multi-stage graph clustering method creates node-associated subgraphs based on attribute and structural similarity, optimizing the graph's representation while simplifying its complexity. The article distinguishes between two graph dimensions: aggregate and abstract malware features, further enhancing information processing efficiency. Lastly, the incorporation of malware metapaths enhances GraphSAGE, resulting in a more effective APT attribution model tailored for network security analysis.

3.1   APT Malware Ontology Mode

In the evolving field of network security, constructing and applying APT knowledge graphs has become vital. This approach organizes complex information about APT malware and its activities to enhance network security analysis. We introduce the APT malware ontology model, refining existing APT knowledge graphs. Our new framework categorizes entities into static properties, functional behaviors, communication resources, and threat indicators. We select 20 feature dimensions from malware analysis, including static, dynamic, and intelligence aspects. These dimensions help form a comprehensive ontology model, as depicted in Fig. 2, providing insights into APT attacks and improving defense strategies.

In this model, we define four core classes: APT Malware, Functional Behaviors, Communication Resources, and Threat Indicators. APT Malware captures static analysis information, Functional Behaviors describe system-level behaviors, Communication Resources cover network-related information, and Threat Indicators highlight signatures and classifications. Associations between entities are categorized into interactions with causality and process-centered calls, outputs, and mapping edges. This ontology model enhances our understanding of APT attacks and aids network security analysts.

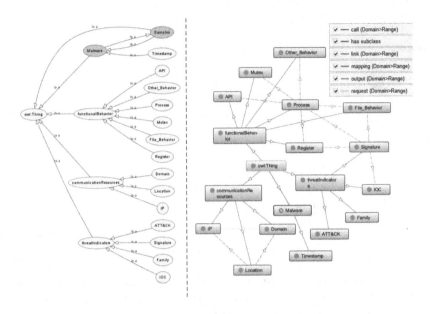

Fig. 2. APT Malware Ontology Model.

3.2   Multistage Graph Clustering and Graph Representation Optimization

In APT malware attribution analysis, the typical method involves examining shared code functionality, attack resources, or TTPs families to uncover the groups and intentions behind APT attacks. Our study constructs a knowledge graph using various behaviors and resource associations from malware analysis, aiming to identify key association dimensions to pinpoint the APT malware group. To handle the complexity of APT malware hiding among other behaviors, we use a multi-stage graph clustering method to reduce graph volume and complexity, improving computational efficiency while maintaining analysis quality. We streamline the graph volume and optimize representation post-refinement to enhance the accuracy and efficiency of APT malware attribution analysis.

Since there is no dynamic interaction between static features, this paper considers static properties as APT Malware node attributes in the graph calculation. In addition to the Malware node, we subdivided sample information into three categories: functional behaviors, communication resources, and threat indicators, assigning a unified unique identifier (uniStr) to each node type for efficient data preprocessing. Then, we introduced a multi-stage graph clustering method and defined two graph representation dimensions: aggregation features and abstract features. The aggregation dimension simplifies the graph by merging similar nodes, revealing malware macrostructure and behavioral patterns. The abstract dimension focuses on extracting abstract characteristics, improving graph representation efficiency. By dividing malware feature dimensions into

---

Algorithm 1. Multi-stage Graph Clustering for Aggregation Dimension

---

Require: Original fine-grained graph $G_{\text{fine-grained}}$, number of stages S, similarity thresholds $\{\theta_s\}_{s=1}^{S}$

Ensure: Refined graph G_{new}

1: **Preprocessing**: Initialize $G_{\text{clustering\_dimension}}$ with node embeddings from $G_{\text{fine-grained}}$: $\chi \leftarrow \text{embBert(uniStr)}$

2: **Stage 1 - Pairwise Similarity Calculation**:

3: Get pair (v_i, v_j) based on shared neighbors.

4: **for** each node pair (v_i, v_j) in $G_{\text{fine-grained}}$ **do**

5:     Calculate $Sim(v_i, v_j) = \mu_1\text{Pro}_{\text{sim}}(v_i, v_j) + \mu_2\text{Stru}_{\text{sim}}(v_i, v_j), \mu_1 + \mu_2 = 1$

6:     Form edges in $G_{\text{clustering\_dimension}}$ using these similarities: $\zeta \leftarrow Sim(v_i, v_j)$

7: **end for**

8: **Stage 2 - Multi-stage Community Clustering**:

9: **for** stage $s = 1$ to S **do**

10:     Use θ_s to filter edges ζ in $G_{\text{clustering\_dimension}}$.

11:     Apply graph community detection algorithms on $G_{\text{clustering\_dimension}}$.

12:     Merge nodes and edges based on communities to update $G_{\text{clustering\_dimension}}$.

13: **end for**

14: **Stage 3 - Final Normalization**:

15: Normalize weights from nodes X to Y based on the statistical frequency of records pointing to the aggregated node using the formula: $Normal_{sum}(X, Y) = \frac{\sum w_{X \rightarrow Y}}{\sum w_{X*}}$

16: Update the edge weight $\omega \leftarrow Normal_{sum}(X, Y)$ on $G_{\text{clustering\_dimension}}$

17: **return** $G_{\text{new}} \leftarrow \text{RefineGraph}(G_{\text{clustering\_dimension}})$

---

aggregate and abstract dimensions, we reduce computational resource requirements and enhance information processing efficiency.

Aggregate Dimension Representation Optimization. In the aggregation dimension, our focus is on handling nodes characterized by random strings and large quantities, which are often challenging to obtain through simple traversal. To reduce the graph's complexity, we cluster and merge nodes that are similar or strongly related, thus preserving essential structural and informational features. We calculate the similarity between node pairs, considering both structure and attributes, and introduce new edges and weights to construct a refined heterogeneous graph. We perform multiple stages of graph clustering calculations to enhance the graph's structure. To ensure the unsupervised graph clustering's credibility, we employ various graph community detection algorithms. During these stages, we carry out node fusion and replacement within the same community. We also define the attributes and edge weights for the merged nodes, creating a detailed and enriched APT malware knowledge graph, as illustrated in Algorithm 1. The optimized graph representations are stored in the graph structure, with χ and γ denoting node feature representations in different aggregation and abstract dimensions. The weight representations ω and δ are used to quantify the strength and relevance of connections between nodes.

Where, embBert(uniStr) represents the embedding vector of a node obtained by processing uniStr using the BERT model. Prosim(v_i, v_j) indicates the attribute similarity determined by the node's embedding vector, while Strusim(v_i, v_j) signifies the structural similarity influenced by the graph's topology. The term $\sum w_{X \to Y}$ represents the cumulative original weights of edges originating from node X and pointing towards the merged node Y, whereas $\sum w_{X*}$ denotes the total weight of all edges originating from node X and directed towards any other node. In this approach, the community clustering algorithm integrates various clustering techniques [17], which are instrumental in revealing the community structure and behavioral patterns of malware in APT malware analysis.

Abstract Dimension Representation Optimization. Unlike the aggregation dimension, the abstract dimension deals with types that have clear naming conventions and a limited number of nodes. These nodes can be traversed within a certain time frame, allowing for deeper and more specific analysis. In the abstract dimension, we enhance the quality and usability of the knowledge graph by precisely capturing and expressing the key features of these limited nodes.

At the same time, for measuring the importance of terms this paper introduces the Feature Node Importance (FNI) metric. FNI is used to measure the uniqueness of node features for certain categories, reducing the influence of features that appear in almost all categories in subsequent community partitioning. The specific formula for calculating FNI is as follows:

$$\varphi_{FNI}(j) = 1 - \left(\frac{\sum (g_j) - 1}{\sum (G)} \right)^2, \quad g_j = \text{Uniq}(R_{ij}) \tag{1}$$

$\sum(G)$ refers to the total count of all cluster categories across the entire graph, and $\sum(g_j)$ refers to the count of distinct cluster categories associated with a particular feature node after deduplication.

In Eq. (1), $\sum(G)$ represents the total number of cluster categories included, and R_{ij} represents the set of relevant edges connecting samples in cluster i to feature node j. g_j represents the set of cluster associations contained in feature node j after deduplication based on feature node j. $\sum(g_j)$ represents the number of cluster categories associated in feature node j. To capture the trend of decreasing influence as features become more general, a non-linear decreasing function $\varphi_{FNI}(j)$ is fitted. Based on the importance indicator, we define edge weights in the abstract dimension as follows:

$$\delta : \omega \cdot \varphi_{FNI}(j) \tag{2}$$

Furthermore, nodes evaluate their impact on classification by computing the attribute γ for node i, as expressed by the equation (3). Where $F(i)$ represents the total frequency of feature node i in the sample set. $\sum_{C \in \text{class All}}$ represents the summation over all categories $class$ C. $f(i, C)$ represents the frequency of feature node i included in category C within the sample set.

$$\gamma : ImportClass(i) = \sum_{C \in \text{class All}} \frac{f(i, C)}{F(i)} \tag{3}$$

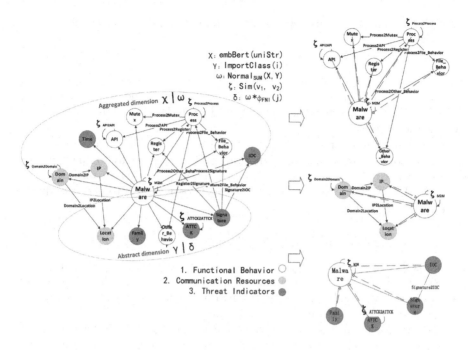

Fig. 3. Graph Embedding Representation Based on Metapath.

This formula accounts for node i's influence across different classifications, considering both its frequency within the sample set and its distribution among various categories. The resulting node attribute γ quantifies the extent of nodes' impact on different classifications.

3.3 Threat Actor Groupal Attribution Based on Meta-Path Graph Embedding

In this research, we propose a graph embedding method based on meta-paths, which improves upon the GraphSAGE algorithm to construct an accurate and efficient APT attribution classification model. This method leverages malware behavior paths defined by domain knowledge, effectively focusing on the most relevant structures in the graph, and captures critical information about nodes and their neighbors through specific aggregation functions. This guided Graph-SAGE strategy contributes to the achievement of a more precise and efficient APT attribution classification model.

As shown in Fig. 3, the representation of the symbols corresponds to the descriptions provided in the previous section. It should be noted that ζ is employed for multi-stage graph clustering and is not part of the graph embedding calculations. By utilizing domain meta-paths defined for Functional Behavior, Communication Resources, and Threat Indicators, we partition the graph into different subgraphs aligned with these meta-paths. Subsequently, we apply the

GraphSAGE algorithm to extract information surrounding Malware center nodes as described below.

$$h_v^k = \sigma\left(W^k \cdot \text{CONCAT}\left(h_v^{k-1}, \text{AGG}\left(\{h_u^{k-1}, \forall u \in N(v)\}\right)\right)\right) \tag{4}$$

Here, h_v^k is the hidden state of node v at layer k, $N(v)$ is the set of neighboring nodes of v, W^k is the weight matrix at layer k, σ is the activation function, and AGG is the aggregation function.

To combine the advantages of different aggregation strategies and comprehensively capture information from neighboring nodes, we use a hybrid aggregation function that combines direct concatenation, pooling, long short-term memory networks (LSTM), and graph convolutional networks (GCN) to generate different embeddings as follows:

$$\text{AGG}(S) = (\text{SPLICE}(S), \text{POOL}(S), \text{LSTM}(S), \text{GCN}(S)) \tag{5}$$

Here, S is the feature set of neighboring nodes. This hybrid aggregation approach preserves the integrity of the original features while enhancing the extraction of information from neighboring nodes through different strategies.

Next, we combine these graph embedding results with ensemble machine learning and use an auto-sklearn [9] for ensembled model selection and parameter tuning. We explore different machine learning models and parameter configurations to find the best combination. GraphSAGE primarily focuses on structural feature extraction in this process, while ensemble learning is responsible for model selection and parameter optimization. This ensemble method combines multiple machine learning techniques to improve prediction accuracy and generalization.

With this optimization approach, our model retains the information of the graph data structure while achieving effective model selection and optimization. This combination of meta-path-guided graph embedding and ensemble machine learning techniques provides an innovative and efficient solution for APT attribution classification.

4   Evaluation and Discussion

In this chapter, we compare our ontology-based APT malware knowledge graph with existing research on APT-related knowledge graphs. We focus on evaluating whether graph refinement negatively impacts classification performance and comparing different graph computation methods using meta-path graph embeddings. Our goal is to demonstrate the advantages of our domain graph embedding approach based on the refined knowledge graph in APT group recognition tasks.

We collected APT malware from public intelligence sources [6,12,20] using VirusTotal [28], excluding samples that were either reused pieces of malware or could not be analyzed in sandboxes. We focused on 10 groups from different countries and regions for experimental verification and obtained labeled samples that were manually verified by security experts. This study also encompasses

Table 1. APT Malware Distribution and Corresponding Threat Actor Groups.

Country	Threat Actor Groups	Number
Iran	APT-C-07	139
USA	Equation	464
Turkey	PROMETHIUM	203
Vietnam	APT32	1243
North Korea	Lazarus Group	4028
Russia	BlackEnergy	101
South Korea	Darkhotel	525
India	Patchwork	1380
India	APT-C-35	379
Russia	Carbanak	557
COUNT	10	9019

Table 2. Comparisons with APT Malware Classification Approaches.

Node Type	Finegrained	Refined Knowledge Graph		
		Abstract	Aggregation	All
API	208061		3935	3935
ATTCK	143	164		164
Location	56262	49722		49722
Domain	3605		1761	1761
Family	3142	717		717
File_Behavior	982223		12039	12039
IOC	27846		107	107
IP	3109741		291169	291169
Malware	9019			9019
Mutex	3161		1655	1655
Other_Behavior	2053	27		27
Process	15516		8633	8634
Register	82104		3850	3850
Signature	481	477		477
All Nodes	4503357			383276
Related Edge	14085227			1249510

the analysis of APT groups with imbalanced samples, ranging from just over 100 samples for the least represented group to over 4,000 samples for the most represented group. The specific distribution of APT malware and their corresponding group lists are shown in Table 1.

The fine-grained knowledge graph constructed through the ontology model in this paper contains various types of nodes, such as API, ATT&CK, location, Domain, etc., with a total of approximately 4.5 million nodes. Through multi-stage graph clustering methods, these nodes are significantly reduced in the aggregation dimension, ultimately reducing the total number of nodes in the "Refined Knowledge Graph" to approximately 380,000, as shown in Table 2.

4.1 Discussion 1: Whether the Key Information of APT Malware Knowledge Graph Attribution Is Reduced After Refining

We visualize the similarity relationship between two malware samples by considering the associated nodes around them, using a relationship-based layout as depicted in Fig. 4. The results show significant associations within the same group. This highlights the effectiveness of the ontology model we propose for constructing the APTMalKG, especially in classifying APT groups.

Furthermore, the analysis indicates that the Refined Knowledge Graph's performance is on par with that of the Fine-grained Knowledge Graph, showing a slight advantage in differentiating certain groups. For instance, the distinction between PROMETHIUM (green) and Lazarus Group (pink) is more distinct, and the separation of Darkhotel (blue) from other groups is clearer. This suggests that while we reduced the graph's complexity, essential features for groupal analysis have been effectively retained.

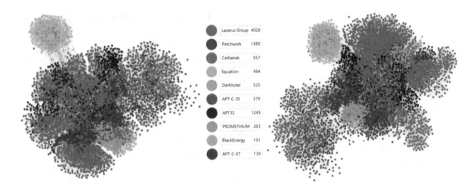

Fig. 4. Comparative Analysis of Association Effects in APT Threat Actor Groups: Fine-Grained vs. Refined Knowledge Graphs of Malware Samples. (left: the fine-grained knowledge graph, right: refined knowledge graph, nodes: individual malware samples, colors: 10 group categories.)

4.2   Discussion 2: What Is the Effect of Refined APT Malware Knowledge Graph Attribution Analysis

In this section, we delve into the attribution analysis of refined APT malware knowledge graphs. Through a comprehensive comparison of various graph computation methods and ensemble machine learning techniques, we showcase their performance in APT malware attribution analysis.

Table 3. Comparison with Other Graph Calculation Methods on F1 score.

Group	RGCN	Hetro-GNN	Graph-SAGE	SAGE-Metapath	Ensemble (our)
Lazarus Group	0.8151	0.8351	0.7984	0.834	0.9373
Patchwork	0.6909	0.705	0.6277	0.6996	0.9274
APT32	0.7	0.692	0.6186	0.6891	0.8722
Darkhotel	0.5161	0.5579	0.7089	0.7264	0.8128
Carbanak	0.6827	0.6698	0.9278	0.9326	0.8768
Equation	0.8432	0.8508	0.6712	0.7199	0.9348
APT-C-35	0.609	0.5854	0.5979	0.5998	0.8050
PROMETHIUM	0.8648	0.8266	0.8919	0.8981	0.9512
APT-C-07	0.84	0.8727	0.6455	0.8679	1
BlackEnergy	0.6444	0.5589	0.7331	0.7947	0.8648
$F1_{Macro}$	0.72062	0.71542	0.7221	0.77621	0.8982
Test ACC	0.7533	0.7572	0.7641	0.8021	0.9116
Test AUC	0.9253	0.9205	0.9506	0.9612	0.9899

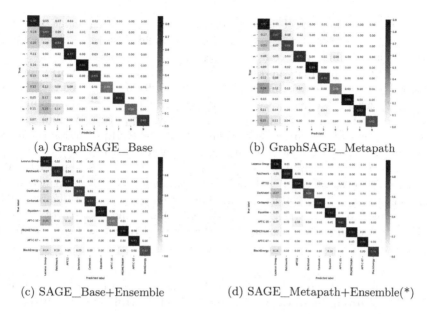

(a) GraphSAGE_Base   (b) GraphSAGE_Metapath

(c) SAGE_Base+Ensemble   (d) SAGE_Metapath+Ensemble(*)

Fig. 5. Confusion Matrices for Test Data Set with Model Selection (* indicates the final chosen model)

First, Table 3 data illustrates the significant performance of our specific method, which combines GraphSAGE embeddings based on metapath with an ensemble of machine learning models (denoted as SAGE_Metapath+Ensemble), across all APT groups. Particularly, in the cases of "Lazarus Group" and "Equation," this combined method outperforms others with F1 scores of 0.9373 and 0.9348, respectively. This indicates the effectiveness of our composite approach in identifying complex and covert APT activities compared to traditional methods like RGCN, Hetero-GNN, and standalone GraphSAGE.

Table 4. Comparison with Other Integrated Machine Learning Methods.

Model	ACC	AUC	F1_Macro
GraphSAGE_XGBoost	0.8620	0.9802	**0.8286**
GraphSAGE_CatBoost	0.8242	0.9731	0.7652
GraphSAGE_LightGBM	**0.8624**	**0.9823**	0.8274
SAGEMetapath_XGBoost	**0.9082**	0.9874	**0.8927**
SAGEMetapath_CatBoost	0.9054	0.9880	0.8856
SAGEMetapath_LightGBM	0.9060	**0.9884**	0.8884
Our	**0.9116**	**0.9899**	**0.8982**

Table 5. Comparisons with APT Malware Classification Approaches.

Model	ACC	AUC	F1_Macro
Shudong et al. [15]	0.7401	0.9497	0.7610
Adem et al. [27]	0.8286	0.9414	0.8092
Hrishabh et al. [26]	0.8054	0.9038	0.7850
Do Xuan et al. [7]	**0.8398**	**0.9541**	**0.8217**
Our	0.9116	0.9899	0.8982

We further compare the performance of our ensemble machine learning method in Table 4. Our method, which is a combination of SAGE_Metapath and Ensemble learning, excels in terms of accuracy (ACC), area under the curve (AUC), and macro-average F1 score ($F1_{Macro}$), highlighting its efficiency and accuracy when dealing with complex datasets.

The confusion matrices in Fig. 5 further confirms the superiority of our method. The SAGE_Metapath+Ensemble model performs best in correctly classifying APT groups, demonstrating its highly precise classification capabilities.

Our research indicates that a comprehensive approach combining the strengths of multiple advanced algorithms enhances overall prediction accuracy and robustness. The SAGE-Metapath algorithm improves feature representation of malicious software samples, effectively capturing key features and intricate behavioral patterns of APT malware, thus enhancing classification accuracy. Moreover, our method exhibits outstanding performance across diverse APT group samples, demonstrating its broad applicability and robustness. This is particularly crucial in the context of the increasing diversity and complexity of APT malware.

4.3 Discussion 3: How Meaningful Is Our Research in Comparison with Similar Studies in the Field of APT Malware Analysis

Nowadays, existing research on related knowledge graphs primarily relies on manual intelligence sources, requiring manual processing and often lacking in-depth behavioral characteristics. We have completed the data collection on the malware side through APT malware analysis. This connects detailed low-level behavioral analysis with broader infrastructure and threat intelligence information. Despite the raw nature of the data obtained from samples, which has not undergone manual processing, our graph data encompasses a more comprehensive range of dimensions available at the sample level. After constructing an extensive dataset, we refined it to capture critical groupal classification features.

In the realm of APT group classification research, we conducted experiments on our dataset using feature processing techniques from established multi-classification models in malware studies. As shown in Table 5, this highlights the comparison models' limitations in classifying high-level APT groups, emphasizing the need for more targeted research. And our methodology proves effective in APT group classification.

Due to constraints like the experimental environment and parameters, our study focused on basic model replication without parameter tuning or addressing unknowns, potentially not reaching the reference models' optimal performance. Nevertheless, considering the highly unbalanced nature of the APT group dataset, these results remain credible, especially given their reliance on superficial malware features, possibly missing complex APT attack behavior patterns. Additionally, these methods may struggle with malware's semantic and structural information, limiting their effectiveness in complex APT scenarios. Our approach, integrating deep graph networks and ensemble learning techniques, thoroughly analyzes APT malware's multidimensional features, showcasing its potential in complex APT attack classification.

5   Conclusion

In summary, refining the knowledge graph reduces complexity while preserving attribution information. This research aims to enhance APT malware analysis and attribution, aiding network security analysts in countering threats efficiently. The APTMalKG is constructed through graph clustering and optimization, utilizing metapath-enhanced graph embedding methods. It integrates various data dimensions, enhancing attribution accuracy by capturing malware behaviors comprehensively. In the future, we plan to enhance the model's ability to generalize to new categories by combining multiple approaches such as embeddings and attribute learning. This approach empowers analysts with a more refined knowledge graph, bolstering their capabilities against APT threats and strengthening network security.

Funding Information. Supported by Youth Innovation Promotion Association, CAS (No. 2020166) and Youth Innovation Promotion Association, CAS (No. 2023170).

References

1. Malware Attribute Enumeration and Characterization (MAEC) (2023). https://maecproject.github.io/. Accessed 11 Nov 2023
2. Balan, G., Gavriluţ, D.T., Luchian, H.: Using API calls for sequence-pattern feature mining-based malware detection. In: Su, C., Gritzalis, D., Piuri, V. (eds.) ISPEC 2022, pp. 233–251. Springer, Cham (2022). https://doi.org/10.1007/978-3-031-21280-2_13
3. Busch, J., Kocheturov, A., Tresp, V., Seidl, T.: Nf-gnn: network flow graph neural networks for malware detection and classification. In: Proceedings of the 33rd International Conference on Scientific and Statistical Database Management, pp. 121–132. Association for Computing Machinery (2021)
4. Chang, H.Y., Yang, T.Y., Zhuang, C.J., Tseng, W.L.: Ransomware detection by distinguishing api call sequences through lstm and bert models. Comput. J. **13**, 5439 (2023)
5. Cremer, F., Sheehan, B., Fortmann, M., Kia, A.N., Mullins, M., Murphy, F., Materne, S.: Cyber risk and cybersecurity: a systematic review of data availability. Geneva Papers Risk Insur. Issues Pract. **47**, 698–736 (2022)
6. CyberMonitor, Robert Haist, K., et al.: APT and cybercriminals campaign collection. GitHub repository (2022). https://github.com/CyberMonitor/APT_CyberCriminal_Campagin_Collections
7. Do Xuan, C., Huong, D.: A new approach for apt malware detection based on deep graph network for endpoint systems. Appl. Intell. **52**(12), 14005–14024 (2022)
8. Dutta, S., Rastogi, N., Yee, D., Gu, C., Ma, Q.: Malware knowledge graph: a comprehensive knowledge base for malware analysis and detection. In: 2021 IEEE Network Security and Privacy Protection International Conference (NSPW) (2021)
9. Feurer, M., et al.: auto-sklearn: automated machine learning toolkit (2023). https://automl.github.io/auto-sklearn/master/. gitHub repository
10. Hasan, M.M., Islam, M.U., Uddin, J.: Advanced persistent threat identification with boosting and explainable AI. SN Comput. Sci. **4**, 271–279 (2023)

11. Kiesling, E., Ekelhart, A., Kurniawan, K., Ekaputra, F.: The SEPSES knowledge graph: an integrated resource for cybersecurity. In: Ghidini, C., et al. (eds.) ISWC 2019. LNCS, vol. 11779, pp. 198–214. Springer, Cham (2019). https://doi.org/10.1007/978-3-030-30796-7_13

12. Kiran Bandla, S.C.: Aptnotes data. GitHub repository (2021). https://github.com/aptnotes/data

13. Lee, K., Lee, J., Yim, K.: Classification and analysis of malicious code detection techniques based on the apt attack. Appl. Sci. **13**, 2894 (2023)

14. Li, S., Zhou, Q., Zhou, R., Lv, Q.: Intelligent malware detection based on graph convolutional network. J. Supercomput. **78**, 4182–4198 (2022)

15. Li, S., Zhang, Q., Wu, X., Han, W., Tian, Z.: Attribution classification method of apt malware in IoT using machine learning techniques. Secur. Commun. Netw. **2021**, 1–12 (2021)

16. Li, Z., Zeng, J., Chen, Y., Liang, Z.: AttacKG: constructing technique knowledge graph from cyber threat intelligence Reports. In: Atluri, V., Di Pietro, R., Jensen, C.D., Meng, W. (eds.) ESORICS 2022, pp. 589–609. Springer, Cham (2022). https://doi.org/10.1007/978-3-031-17140-6_29

17. MLG at Neo4j. Community detection (2022). https://neo4j.com/docs/graph-data-science/current/algorithms/community/

18. Moon, H.-J., Bu, S.-J., Cho, S.-B.: Directional graph transformer-based control flow embedding for malware classification. In: Yin, H., et al. (eds.) IDEAL 2021. LNCS, vol. 13113, pp. 426–436. Springer, Cham (2021). https://doi.org/10.1007/978-3-030-91608-4_42

19. Peng, C., Xia, F., Naseriparsa, M., Osborne, F.: Knowledge graphs: opportunities and challenges. Artif. Intell. Rev. **56**, 13071–13102 (2023)

20. RedDrip7. Apt_digital_weapon: indicators of compromise (IOCS) collected from public resources and categorized by qi-anxin. GitHub repository (2022)

21. Ren, Y., Xiao, Y., Zhou, Y., Zhang, Z., Tian, Z.: Cskg4apt: a cybersecurity knowledge graph for advanced persistent threat organization attribution. IEEE Trans. Knowl. Data Eng. **35**, 5695–5709 (2023)

22. Renz, M., Kröger, P., Koschmider, A., Landsiedel, O., de Sousa, N.T.: Cross domain fusion for spatiotemporal applications: taking interdisciplinary, holistic research to the next level. Informatik Spektrum **45**, 271–277 (2022)

23. Sahoo, D.: Cyber threat attribution with multi-view heuristic analysis. In: Choo, K.-K.R., Dehghantanha, A. (eds.) Handbook of Big Data Analytics and Forensics, pp. 53–73. Springer, Cham (2022). https://doi.org/10.1007/978-3-030-74753-4_4

24. Sharma, A., Gupta, B.B., Singh, A.K., Saraswat, V.K.: Advanced persistent threats (apt): evolution, anatomy, attribution and countermeasures. J. Ambient. Intell. Humaniz. Comput. **14**, 9355–9381 (2023)

25. Sikos, L.F.: Cybersecurity knowledge graphs. Knowl. Inf. Syst. **65**, 3511–3531 (2023)

26. Soni, H., Kishore, P., Mohapatra, D.P.: Opcode and API based machine learning framework for malware classification. In: 2022 2nd International Conference on Intelligent Technologies (CONIT), pp. 1–7 (2022)

27. Tekerek, A., Yapici, M.M.: A novel malware classification and augmentation model based on convolutional neural network. Comput. Secur. **112**, 102515 (2022)

28. VirusTotal. Virustotal: analyse suspicious files and URLs to detect malware. Website (2022). https://www.virustotal.com/

29. Wai, F.K., Thing, V.L.L.: Clustering based opcode graph generation for malware variant detection. In: 2021 18th International Conference on Privacy, Security and Trust (PST), pp. 1–11 (2021)

30. Wei, C., Li, Q., Guo, D., Meng, X.: Toward identifying apt malware through API system calls. Secur. Commun. Netw. **2021**, 8077220 (2021)
31. Wu, X.W., Wang, Y., Fang, Y., Jia, P.: Embedding vector generation based on function call graph for effective malware detection and classification. Neural Comput. Appl. **34**, 8643–8656 (2022)
32. Xuan, C.D., Dao, M.H.: A novel approach for apt attack detection based on combined deep learning model. Neural Comput. Appl. **33**, 13251–13264 (2021)

Flow Field Analysis in Vortex Ring State Using Small Diameter Rotor by Descent Simulation

Ryuki Mori[1($\boxtimes$)], Ayato Takii[1,2], Masashi Yamakawa[1], Yusei Kobayashi[1], Shinichi Asao[3], and Seiichi Takeuchi[3]

[1] Kyoto Institute of Technology, Matsugasaki, Sakyo-ku, Kyoto 606-8585, Japan
m3623032@edu.kit.ac.jp

[2] RIKEN Center for Computational Science, 7-1-26 Minatojima-minami-machi, Chuo-ku, Kobe 650-0047, Hyogo, Japan

[3] College of Industrial Technology, 1-27-1, Amagasaki 661-0047, Hyogo, Japan

Abstract. While the unstable turbulence condition known as Vortex Ring State in rotorcraft has been studied mainly in helicopters, there have not been many studies of drones, which have become more active in recent years. In particular, there are few studies using numerical simulations focusing on small diameter rotors such as quadcopters. In this paper, descent simulations are performed using a rotor model with a diameter of 8 inches and a propeller pitch of 4.5 inches, which is used for quadcopters. In this study, Moving Computational Domain method is used to reproduce the descent motion of the rotor over the entire moving computational domain. In addition, Sliding mesh method is applied to reproduce the rotor rotation within the computational grid. This method allows flow field analysis under free rigid body motion of the analytical model. The displacement of the computational domain itself is applied at each step to reproduce the descent conditions. By combining these methods, fluid flow simulations under vertical descent and conditions are performed to visualize the flow field. Flow field evaluation using the Q criterion showed that even in a small rotor, circulating vortices are generated at a velocity close to the induced velocity v_h. VRS was also observed around the rotor under the conditions of horizontal speed $V_H = 2.0v_h$, *descent speed* $V_Y = 1.0v_h$, and descent angle of 26.6°, but at the same time vortex divergence was also observed. It was inferred that the forward velocity component helps to avoid Vortex Ring State.

Keywords: VRS (Vortex Ring State) · CFD · quadcopter · small propeller

1 Introduction

Helicopters and other rotorcrafts enter an unstable state called Vortex Ring State (VRS) by descending vertically or nearly vertically at a speed close to the rotor wake velocity (induced speed). VRS is a very unstable and turbulent state, characterized by the formation of a toroidal vortex around the main rotor. At the same time, during VRS, wake dominates the inflow region of the rotor during descent, reducing the blade angle of attack and causing thrust loss and thrust oscillations at the mean thrust value [1, 2]. This

© The Author(s), under exclusive license to Springer Nature Switzerland AG 2024
L. Franco et al. (Eds.): ICCS 2024, LNCS 14832, pp. 94–106, 2024.
https://doi.org/10.1007/978-3-031-63749-0_7

instability causes the rotorcraft to oscillate and crash due to significant changes in thrust and attitude angle. Once the rotor enters the VRS, it is difficult to predict the motion of the rotorcraft, so it is important to avoid the VRS. Therefore, various studies on rotors have been conducted.

Efimov (2022) et al. investigated VRS in a single rotor in a wind tunnel test and found that thrust is regained when the angle between the axis of rotation of the propeller and the air velocity vector exceeds 40° [3]. Wind tunnel tests using Particle Image Velocimetry (PIV) revealed that the vortex core exists in the upper and lower regions of the rotor surface and that the vortex moves freely from its development to its extinction [4]. Due to experimental cost challenges and advances in numerical computation, many numerical simulations have investigated the VRS characteristics of rotors. Stalewski (2020) et al. performed unsteady calculations of the flow field in the VRS by coupling the URANS equation with the helicopter's equation of motion and found qualitative agreement for both experimental studies and flight tests [5]. There have also been many studies, both experimental and simulation, on how to avoid VRS. It was found that VRS cannot occur at all descent speeds when the angle of descent is minimal, such as below 30°, by reducing the accumulation of rotor wake by forward speed [6]. Furthermore, it was also found that the aircraft does not enter VRS when the angle of descent is less than 20° [2].

Drones are another rotary-wing aircraft that has become increasingly active in recent years. They are used for a variety of purposes, including industrial monitoring, observation of damage in disaster areas, and video production. Quadcopters, one type of drone, like other rotary-wing aircraft, are also subject to VRS during descent, so attention has been focused on safety during flight.

To study VRS in quadcopters, numerical simulations focusing on aerodynamic interference between rotors during descent were performed using the RANS equation [7]. Further numerical calculations were performed using the Open FOAM CFD package, and a method was proposed to detect VRS by differential pressure measurements when given various descent rates [8].

Actually, however, among rotary-wing aircraft, VRS has been investigated mainly for helicopters, and few studies have focused on quadcopters. Furthermore, there are only a few VRS studies that utilize CFD focusing on the small-diameter rotor portion represented by these aircraft. In addition, there are still few studies that consider the effect of the rotor model's own motion on the VRS in the flow field simulation of VRS research. The reason is that realizing this numerical simulation involves a very complex moving boundary problem, which makes it difficult to combine the interaction with fluid dynamics, which deals with the flow around the model.

And most of these studies are based on steady-state wind tunnel testing and CFD. However, as a practical phenomenon, VRS is unstable and transient. Therefore, the authors propose a method that combines the MCD (Moving Computational Domain) method [9], which is based on the unstructured mesh finite volume method [10, 11], and the sliding mesh method [12]. MCD method is good at discretizing a 4-dimensional inspection volume by applying the finite volume method to it and performing calculations that strictly satisfy geometric conservation laws. Furthermore, sliding mesh method can represent motions such as rotation by forming a computational domain divided by regions with different states of motion and transferring physical quantities at the boundaries of the computational domain. Therefore, this method is considered suitable for general-purpose calculations. Yamakawa et al. (2021) applied this method to investigate the effect of rotating screws on the free surface of water during submarine and other submerged motions [13]. Takii et al. (2020) also applied this technique to the behavior of a tilt-rotor aircraft modeled after the Osprey V-22 in VRS and the surrounding flow field [14]. These results were conducted under the unstructured parallel computational environment [15].

This study focuses on the rotor portion of a quadcopter and simulates the flow field of VRS by solving the unsteady flow during the rotor's descent. For the simulations, unsteady calculations are performed using MCD method based on the unstructured moving mesh finite volume method and the sliding mesh method with the equations of motion of the fluid and the rotational axis motion of the propeller. In order to give different descent velocities to a single rotor, the entire computational domain is subjected to displacements to create a descent condition.

2 Numerical Approach

2.1 Governing Equations

To solve for the flow field around the rotor model, the three-dimensional Euler equations, which are the fundamental equations for inviscid compressible fluids, are used as governing equations. The equations in conserved form, after non-dimensionalization, are written as follows:

$$\frac{\partial q}{\partial t} + \frac{\partial E}{\partial x} + \frac{\partial F}{\partial y} + \frac{\partial G}{\partial z} = 0, \tag{1}$$

$$q = \begin{pmatrix} \rho \\ \rho u \\ \rho v \\ \rho w \\ e \end{pmatrix}, E = \begin{pmatrix} \rho u \\ \rho u^2 + p \\ \rho u v \\ \rho u w \\ u(e+p) \end{pmatrix}, F = \begin{pmatrix} \rho v \\ \rho u v \\ \rho v^2 + p \\ \rho v w \\ v(e+p) \end{pmatrix}, G = \begin{pmatrix} \rho w \\ \rho u w \\ \rho v v \\ \rho w^2 + p \\ w(e+p) \end{pmatrix} \tag{2}$$

where q is the conserved quantity vector, and E, F, G are the flux vector. Furthermore ρ is the density of the fluid, u, v, w are the x, y, z components of velocity, respectively, and e is the total energy per unit volume. The pressure p is determined by the ideal gas equation of state shown below, assuming that the fluid to be handled is an ideal gas:

$$p = (\gamma - 1)\left[e - \tfrac{1}{2}\rho(u^2 + v^2 + w^2)\right] \tag{3}$$

where the specific heat ratio is assumed to be $\gamma = 1.4$. In a computational environment, the Reynolds number is approximately 3,400,000 and the maximum Mach number is 0.136.

2.2 Unstructured Moving-Grid Finite-Volume Method

Unstructured Moving-Grid Finite-Volume Method is used for calculations that involve moving and deforming the computational grid. In this method, the flux is evaluated on a 4-dimensional (x, y, z, t) inspection volume so that the geometric conservation law (GCL) [16] is satisfied. Applying Gauss' divergence theorem to the 4-dimensional inspection volume, the 3-dimensional Euler equations are transformed as follows:

$$
\int_\Omega \left(\frac{\partial q}{\partial t} + \frac{\partial E}{\partial x} + \frac{\partial F}{\partial y} + \frac{\partial G}{\partial z} \right) d\Omega = \int_\Omega (E, F, G, q) \cdot \tilde{n} dV
$$
$$
= \sum_{l=1}^{6} \left(E\tilde{n}_x + F\tilde{n}_y + G\tilde{n}_z + q\tilde{n}_t \right)_t = 0
$$

(4)

where Ω is the inspection volume, $\boldsymbol{\tilde{n}} = \left(\tilde{n}_x, \tilde{n}_y, \tilde{n}_z, \tilde{n}_t \right)$ is the outward unit normal vector on $\partial \Omega$, l is the surface number of the inspection volume.

2.3 Moving Computational Domain Method

This paper calculates the flow field generated during the descent of a quadcopter rotor. Conventional CFD is an alternative to wind tunnel testing, in which a stationary model is subjected to flow and the flow around it is calculated. In the MCD method, the entire computational domain moves with the objects in the domain, as shown in Fig. 1.

The fluid flow around the object is generated by the movement of the boundary surface. In order to handle flows with no known inflow or outflow, this study deals with moving boundaries by applying a far-field boundary condition in which the inflow and outflow are determined by Riemann invariants. Furthermore, the rotor model is free to move with the computational domain, regardless of the size of the computational grid, since this method eliminates the limitation of computational space. By combining the MCD method with the sliding mesh method, this method is also applicable to rotor rotation.

Flow variables are defined at the cell centers of the unstructured mesh. The flux vectors are evaluated using the Roe flux difference separation method [17]. The vanLeer-like restriction function proposed by Hishida et al. is used with respect to the physical quantity gradient of each cell [18]. This reconstructs the solution and extrapolates the cell boundary values to a higher order to achieve higher spatial accuracy. Then, a two-stage rational Runge-Kutta method is used to perform pseudo-temporal progression.

The sliding mesh method [19] is also applied to reproduce the rotation of the rotor model in the computational domain by transferring physical quantities. This method allows the mesh itself to slide at specific boundaries, deforming the mesh while maintaining its volume.

In this study, the sliding mesh method is used only for the axis of rotation because of the rotor alone.

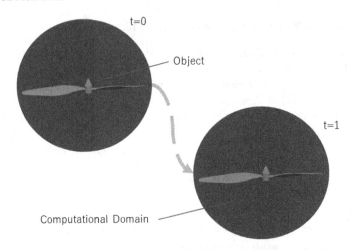

Fig. 1. Moving Computational Domain Method

2.4 Hovering Induced Velocity

VRS is considered to occur when a rotorcraft descends at a speed close to its hovering induced velocity v_h, which is given by momentum theory in the following equation [7].

$$v_h = \sqrt{\frac{T_0}{2\rho A_b}} \tag{5}$$

where T_0 is the rotor thrust [N] during hovering, ρ is the density [kg/m^3], and A_b is the rotor blade area [m^2]. In this case, a quadcopter propeller model was used. The weight of the quadcopter model is $M = 0.9721$ kg, and the total force of all rotors is obtained from the equation of the balance of force with gravity. Then, by dividing by the number of rotors, the force T_0 is derived as follows:

$$T_0 = \frac{Mg}{N_{rotor}} \tag{6}$$

where $N_{rotor} = 4$ is the number of rotors in a rotorcraft and $g = 9.80665$ m/s^2 is the gravitational acceleration constant. In this study, the conditions for VRS generation are set by giving the descent velocity based on the hovering induced velocity $v_h = 5.253$ m/s for the rotor model.

3 Descent Simulation of Rotor

3.1 Computational Mesh

The rotor model used in this study is modeled after the propeller of a commercial quadcopter drone, DJI's Flame Wheel ARF KIT F450, shown in Fig. 2. Table 1 shows the various parameters of the propeller model (Fig. 3).

Table 1. Specifications of the computational model

Rotor Size	8 inches (0.203 m)
Blade Pitch	4.5 inches (0.114 m)
Number of Rotor Blades	2
Hovering Rotor Rotation Speed	4214 rpm
Direction of Rotation	Counter Clockwise

Fig. 2. F450 quadcopter   **Fig. 3.** Computational surface grid

The computational domain consists of two regions separated by a sliding mesh. The first is a cylindrical region that encompasses the rotor model, and the second is a spherical region that encompasses the cylindrical region. The cylindrical part has a sliding mesh structure as shown in Fig. 4, and is used to propagate physical quantities. The computational grid was created by MEGG3D software [20, 21] and consists of a tetrahedral lattice. The number of lattices in the entire computational domain is 684,965. The number of lattices and minimum lattice width of the rotor model are 177,385 and 0.008 (4.8 mm), respectively.

Fig. 4. Sliding mesh domain

Figure 5 shows a cross-sectional view of the entire computational domain and a grid cross-section around the rotor model. The rotor length and the radius of the spherical computational domain are 0.2778L and 50L, respectively, for a representative length L = 0.60 m for the entire quadcopter length.

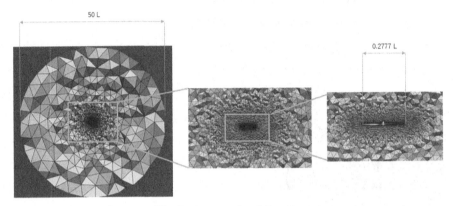

Fig. 5. Computational Domain

3.2 Computational Mesh

In this study, the descent state is reproduced by applying a displacement in the negative Y-axis direction at each step over the entire computational domain that encompasses the rotor model. Furthermore, the computational domain performs a descent motion with constant velocity. Therefore, the model motion does not depend on the translational and rotational equations of motion, but only on the following equations:

$$r(n+1) = \frac{dr}{dt}\Delta t + r(n) \tag{7}$$

where $r(n)$ is the position vector of the model center at step n and Δt is the time tick width.

Rotor rotation rises with equal acceleration from 0 to 5000 steps from the stop condition to the target speed of 4214 rpm at the same time as the descent condition. After reaching the target rpm, the rotor continues to rotate at a constant speed.

Table 2. Initial condition and Boundary condition

ρ	1
p	$1.0/\gamma$
u, v, w	0
Rotor Surface	Slip wall condition
Outer Boundary	Riemann invariant boundary condition

Table 2 shows the initial and boundary conditions for the variables. The representative velocity and density are calculated to be 340.29 m/s and 1.247 kg/m^3, respectively. The variables are also non-dimensionalized by these representative values.

4 Numerical Simulation Results

4.1 Simple Vertical Descent

This section presents the results of a vertical descent simulation for a rotor model with descent velocities $V_Y = 0.4v_h$, $V_Y = 0.9v_h$, respectively. The output results are visualized using Paraview [16]. The graph is a csv file converted from the output file using the standard Paraview method "CellDatatoPointData". A positive value of Q criterion is

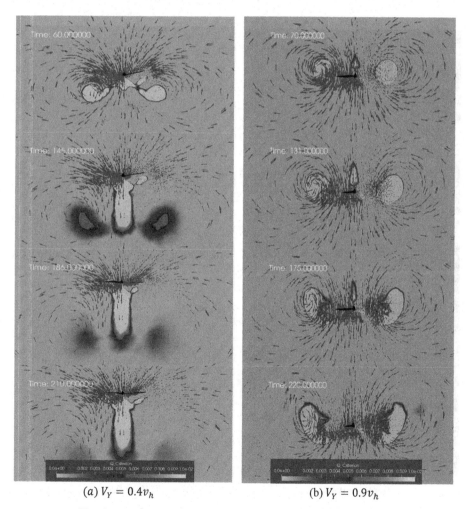

(a) $V_Y = 0.4v_h$ (b) $V_Y = 0.9v_h$

Fig. 6. Q criterion contour map with velocity vector in Y direction

used in the output figure. This is the second invariant of the velocity gradient tensor and helps identify regions in the flow field where rotation is more dominant than strain.

Figure 6 shows contour plots of the Q criterion for $V_Y = 0.4v_h$, $V_Y = 0.9v_h$, respectively, together with the velocity vector in the Y direction. When $V_Y = 0.4v_h$, most of the vortex portion stays at a position lower than the rotor surface. After a while, it is seen that the vortex region is pushed further down the rotor due to the downwash effect. In addition, focusing on the velocity vector in the Y direction, a large vortex flow can be seen below the rotor surface. On the other hand, when $V_Y = 0.9v_h$, the vortex area stays at the same or higher position than the rotor surface. This is thought to be because the rotor is descending at the same rate as the downwash. The vortex is expected to continue to stay at the rotor surface. The magnitude of the Q criterion is also large near the rotor surface for $V_Y = 0.9v_h$. The Y-directional velocity vector shows the flow circulating at a height close to the rotor surface.

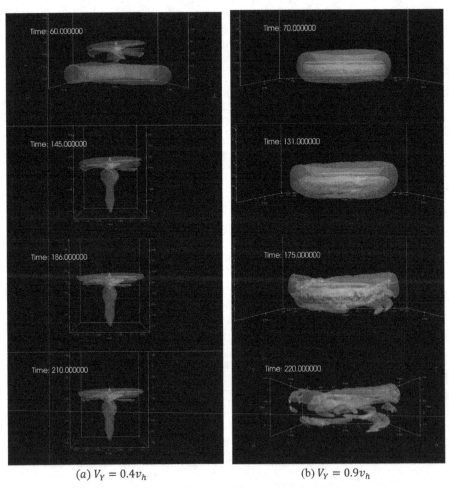

(a) $V_Y = 0.4v_h$ (b) $V_Y = 0.9v_h$

Fig. 7. Isosurfaces with Q criterion = 0.05

Figure 7 shows the isosurfaces at Q criterion = 0.05 for $V_Y = 0.4v_h$, $V_Y = 0.9v_h$ respectively. At $V_Y = 0.4v_h$, the vortex is strongly affected by downwash with time and moves away from the rotor surface. At $V_Y = 0.9v_h$, a toroidal shape of Q criterion isosurface is seen covering the rotor surface for almost the entire time period. The descending motion close to the induced velocity confirms the characteristics of the VRS, in which the wake of the rotor circulates around the rotor surface.

4.2 Oblique Descent Simulation (Angle of Descent 26.6°)

This section shows the results of a descent simulation for the rotor model with descent conditions $V_H = 2.0v_h$, $V_Y = 1.0v_h$ (angle of descent 26.6°).

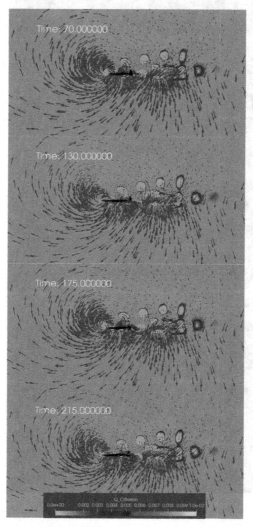

Fig. 8. Q criterion contour map with velocity vector in Y direction ($A t V_H = 2.0v_h$, $V_Y = 1.0v_h$)

Figure 8 shows the contour plot of the Q criterion at $V_H = 2.0v_h$, $V_Y = 1.0v_h$, together with the velocity vector in the Y direction. The rotor is moving toward the lower left in the figure, and it can be seen that the vortex portion is also stagnant at approximately the same position as the rotor surface, as it was at $V_Y = 0.9v_h$. This may be due to the fact that the descent velocity V_Y is about the same as the hovering induced velocity v_h. Furthermore, due to the rotor's own horizontal velocity motion, it tries to move away from the vortex relatively. This tendency is expected to continue. However, the vortex is generally dominant at the top of the rotor, and a VRS may be detected.

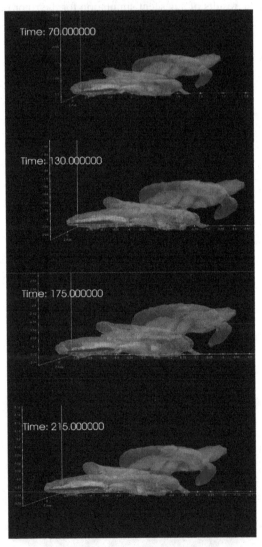

Fig. 9. Isosurfaces with Q criterion $= 0.05$ ($AtV_H = 2.0v_h$, $V_Y = 1.0v_h$)

Figure 9 shows the isosurface for Q criterion = 0.05. As with $V_Y = 0.9v_h$, there is vortex stagnation at the rotor surface over each time step. However, the vortex region is swept in the opposite direction of descent by the forward velocity, confirming that it is diverging.

5 Conclusions

Descent simulations were performed to investigate the unsteady state for the VRS of a single small diameter rotor. In this study, the MCD method based on the unstructured moving mesh finite volume method and the sliding mesh method were used. The MCD method is based on the unstructured moving mesh finite volume method, and the sliding mesh method is based on the unstructured moving mesh finite volume method. The descent simulations were performed for a simple vertical descent with $V_Y = 0.4, v_h$, $V_Y = 0.9v_h$ and a descent with $V_H = 2.0v_h$, $V_Y = 1.0v_h$ at a descent angle of 26.6°.

By evaluating the flow field around the rotor during descent, the nature of VRS as an unsteady phenomenon was reaffirmed. The flow field evaluation was visualized using Q criterion contour plots. In the simple vertical descent, when $V_Y = 0.9v_h$, the rotor model descends at the same speed as the downwash, and it was confirmed that the vortex stays around the rotor surface. This indicates that even small-diameter rotors can fall into VRS, a result that is qualitatively consistent with previous studies. Simulations were also conducted under the descent conditions of $V_H = 2.0v_h$, $V_Y = 1.0v_h$ at a descent angle of 26.6°. In this case, vortex retention was observed, but due to the forward velocity component, the flow behind the rotor was observed to flow in the opposite direction of the flight direction. This suggests that VRS may be avoided under descent conditions with a constant angle of descent by promoting the divergence of the vortex due to the forward velocity, which is in accordance with previous VRS research. In actual flight, however, the pitch angle is expected to increase as the attitude angle of the rotor. Since this point was not taken into account in this study, it needs to be reviewed.

Based on the single-rotor simulations in this study, we would like to extend the study to quadcopters equipped with the rotor model used in this study in the future and conduct VRS avoidance research. In addition, this method restricted the equation of motion of the rigid body to the rotor model, thereby providing arbitrary descent conditions. Therefore, we would like to evaluate the trajectory and attitude angle of the quadcopter using a 6-DOF coupled flight simulation that combines the equations of the fluid and rigid body.

References

1. Brand, A., Dreier, M., Kisor, R., Wood, T.: The nature of vortex ring state. In: AHS 63rd Annual Forum, Virginia Beach, USA (2007)
2. Betzina, M.D.: Tiltrotor descent aerodynamics: a small-scale experimental investigation of vortex ring state. In: AHS 57th Annual Forum, Washington, USA (2001)
3. Efimov, V.V., Chernigin, K.O.: Vortex ring state as a cause of a single-rotor helicopter unanticipated yaw. Aerosp. Syst. **5**, 413–418 (2022)

4. Surmacz, K., Ruchała, P., Stryczniewicz, W.: Wind tunnel tests of the development and demise of Vortex Ring State of the rotor. Adv. Mech. Theor. Comput. Interdiscip. Issues, 551–554 (2016)
5. Stalewski, W., Surmacz, K.: Investigations of the vortex ring state on a helicopter main rotor based on computational methodology using URANS solver. In: MATEC Web Conference, 9th EASN International Conference on Innovation in Aviation & Space, vol. 304 (2019)
6. Taamallah, S.: A qualitative introduction to the vortex-ring-state, autorotation, and optimal autorotation. In: Nationaal Lucht- en Ruimtevaartlaboratorium National Aerospace Laboratory NLR, 36th European Rotorcraft Forum, Paris, France (2010)
7. Wang, J., Chen, R., Lu, J., Zhao, Y.: Numerical simulation of the quadcopter flow field in the vertical descent state. Proc. IMechE Part G J. Aerosp. Eng., 1–13 (2022)
8. McQuaid, J., Kolaei, A., Ph.D., Bramesfeld, G., Ph.D. , Walsh, P., Ph.D.: Early on-set prediction of vortex-ring state of quadrotors. J. Aerosp. Eng. 33(6) (2020)
9. Yamakawa, M., et al.: Numerical simulation of rotation of intermeshing rotors using added and eliminated mesh method. Procedia Comput. Sci. **108**, 1883–1892 (2017)
10. Yamakawa, M., et al.: Numerical simulation for a flow around body ejection using an axisymmetric unstructured moving grid method. Comput. Thermal Sci. 4(3), 217–223 (2012)
11. Yamakawa, M., et al.: Optimization of knee joint maximum angle on dolphin kick. Phys. Fluids 32(61) (2020). Article number 067105
12. Takii, A., Yamakawa, M., Asao, S., Tajiri, K.: Six degrees of freedom numerical simulation of tilt-rotor plane. In: Rodrigues, J.M.F., Cardoso, P.J.S., Monteiro, J., Lam, R., Krzhizhanovskaya, V.V., Lees, M.H., Dongarra, J.J., Sloot, P.M.A. (eds.) ICCS 2019. LNCS, vol. 11536, pp. 506–519. Springer, Cham (2019). https://doi.org/10.1007/978-3-030-22734-0_37
13. Yamakawa, M., Yoshioka, K., Asao, S., Takeuchi, S., Kitagawa, A., Tajiri, K.: Numerical simulation of free surface affected by submarine with a rotating screw moving underwater. In: Paszynski, M., Kranzlmüller, D., Krzhizhanovskaya, V.V., Dongarra, J.J., Sloot, P.M.A. (eds.) ICCS 2021. LNCS, vol. 12747, pp. 268–281. Springer, Cham (2021)
14. Takii, A., Yamakawa, M., Asao, S.: Descending flight simulation of tiltrotor aircraft at different descent rates. In: Krzhizhanovskaya, V.V., Závodszky, G., Lees, M.H., Dongarra, J.J., Sloot, P.M.A., Brissos, S., Teixeira, J. (eds.) ICCS 2020. LNCS, vol. 12143, pp. 178–190. Springer, Cham (2020). https://doi.org/10.1007/978-3-030-50436-6_13
15. Yamakawa, M., et al.: Domain decomposition method for unstructured meshes in an OpenMP computing environment. Comput. Fluids **45**, 168–171 (2011)
16. Roe, P.L.: Approximate Riemann solvers, parameter vectors, and diference schemes. J. Comput. Phys. **43**(2), 357–372 (1981)
17. Hishida, M., et al.: A new slope limiter for fast unstructured CFD solver FaSTAR. In: Proceedings of 42nd Fluid Dynamics Conference/Aerospace Numerical Simulation Symposium. Japan Aerospace Exploration Agency, JAXA-SP-10-012, pp. 85–90 (2010). (in Japanese)
18. Takii, A., Yamakawa, M., Asao, S., Tajiri, K.: Six degrees of freedom flight simulation of tilt-rotor aircraft with nacelle conversion. J. Comput. Sci. **44**, 101164 (2020)
19. Ito, Y., et al.: Surface triangulation for polygonal models based on CAD data. Int. J. Numer. Methods Fluids 39(1), 75–96 (2002)
20. Ito, Y.: Challenges in unstructured mesh generation for practical and efficient computational fluid dynamics simulations. Comput. Fluids **85**(1), 47–52 (2013)
21. Paraview Homepage. https://www.paraview.org/. Accessed 25 Feb 2024

Toward Real-Time Solar Content-Based Image Retrieval

Rafał Grycuk[1]([✉])[iD], Giorgio De Magistris[2][iD], Christian Napoli[2,3][iD], and Rafał Scherer[1][iD]

[1] Czestochowa University of Technology, al. Armii Krajowej 36, Czestochowa, Poland
{rafal.grycuk,rafal.scherer}@pcz.pl
[2] Department of Computer, Control and Management Engineering, Sapienza University of Rome, Rome, Italy
{demagistris,cnapoli}@diag.uniroma1.it
[3] Institute for Systems Analysis and Computer Science, Italian National Research Council, Rome, Italy

Abstract. We present a new approach for real-time retrieval and classification of solar images using a proposed sector-based image hashing technique. To this end, we generate intermediate hand-crafted features from automatically detected active regions in the form of layer-sector-based descriptors. Additionally, we employ a small fully-connected autoencoder to encode and finally obtain the concise Layer-Sector Solar Hash. By reducing the amount of data required to describe the Sun images, we achieve almost real-time retrieval speed of similar images to the query image. Since solar AIA images are not labeled, for the purposes of the presented test experiments, we consider images produced within a short time frame (typically up to several hours) to be similar. This approach has several potential applications, including searching, classifying, and retrieving solar flares, which are of critical importance for many aspects of life on Earth.

1 Introduction

The Solar Dynamics Observatory (SDO) was launched by NASA in 2010 as a part of the Living with a Star program with the aim of providing data to study the interconnected Sun-Earth system and how the Sun impacts life on Earth. The Sun's activity, such as massive electromagnetic storms, can negatively affect various technologies including electronics, navigation systems, and electric power grids. Solar activity, which is influenced by the sunspot cycle and other transient aperiodic processes, plays a significant role in creating space weather that affects both space- and ground-based technologies, as well as the Earth's atmosphere. Furthermore, the Sun's behavior is partially responsible for climate fluctuations on a scale of centuries and longer. Comprehending and forecasting the sunspot cycle continues to be a significant scientific challenge, with far-reaching consequences for space science and our understanding of magnetohydrodynamic phenomena in the Solar System and on Earth.

The SDO is a 3-axis stabilized spacecraft equipped with three main sensoric instruments, one of which is the Atmospheric Imaging Assembly (AIA). AIA continuously captures full-disk observations of the solar chromosphere and corona in seven extreme ultraviolet (EUV) channels, producing high-resolution images with a 12-second cadence at 4096 × 4096 pixels. The commencement of the SDO program enabled the analysis of solar activity, despite the challenge of dealing with big data. SDO generates around 70 thousand images every day, which makes it impossible to manually search and annotate this vast collection of images. Additionally, the repetitiveness and monotony of these images make the annotation process even more difficult for humans. The images captured by the SDO are quite similar to one another, making it challenging to describe them using general-purpose visual features. Additionally, the images are only labeled by their timestamp, which further adds to the difficulty in analyzing and categorizing them.

For our research, we utilize a 4K resolution dataset that has been prepared by Kucuk et al. [19] specifically for image retrieval purposes. This dataset comprises hundreds of thousands of full-disk images of the Sun, with temporal and spatial event features included in the records. Such large datasets are difficult to search and detect changes anomalies [9,17].

Traditionally, hand-crafted features have been used to classify or predict the state of the Sun from its images. For instance, Banda et al. [2] identified ten distinct image parameters that are the best representation of the Solar state when extracted from Solar full-disk images. These parameters are also present in the dataset we use. In [4], the Lucene retrieval engine was modified to retrieve solar images based on descriptive solar features developed in [3]. These features have also been utilized by Boubrahimi et al. in [6] and Ma et al. in [20] to forecast solar event trajectories.

Our paper introduces a solar hash designed to locate similar solar images from a vast collection of solar images. We employ a fully-connected autoencoder that operates on preprocessed solar full-disk projections. Our emphasis is on optimizing the retrieval process, and our proposed method outperforms existing methods in terms of speed and accuracy. The proposed solar hashes are key to achieving faster retrieval times. It is worth noting that direct full-disk solar image hashing is computationally demanding (for one year period approx 10 d), which was the primary motivation for our work.

The rest of the paper is organized as follows. In Sect. 2 we describe shortly other content-based image retrieval methods. The proposed method is described in Sect. 3. The experiments are described in Sect. 4. Section 5 concludes the paper.

2   Related Works

The article [1] details a system for content-based image retrieval that covers the entire disk. The authors experimented with eighteen image similarity measures and a range of image features, resulting in one hundred and eighty unique combinations. Through these experiments, the authors identified suitable metrics for comparing solar images, which can aid in the retrieval and classification of various phenomena. The article referenced as [3] outlines a segmentation method for full-disk SDO images, where sub-images are created based on a 64 × 64 grid. Ten parameters are then calculated for each sub-image, including entropy, fractal dimension, mean intensity, third and fourth moments, relative smoothness, standard deviation of intensity, Tamura contrast, Tamura directionality, and uniformity.

In [4], the retrieval of solar images is performed using Lucene, a versatile retrieval engine. Each image is considered a distinct document, comprising 64 elements (rows of each image). To locate similar solar events, wild-card characters are used in the query strings. In [5], the effectiveness of the Lucene engine is compared to distance-based image retrieval approaches, but no clear winner is identified. The tested methods each exhibit advantages and drawbacks in terms of accuracy, speed, and suitability. The balance between accuracy and speed is considerable, with retrieval times of several minutes being necessary for precise outcomes. In [14], a sparse model representation was introduced for solar images, utilizing the approach from [21]. The proposed method surpassed previous solar image retrieval techniques in both accuracy and speed.

In [16], certain solar image parameters are selected to monitor various solar events across images with a 6-minute interval. In [15], sparse codes for AIA images are also utilized, with ten texture-based image parameters being employed to generate the code. The parameters are determined for regions identified by a 64 × 64 grid for nine wavelengths. A dictionary of k elements is learned for each wavelength, and a sparse representation is subsequently calculated. In order to address the issue of dimensionality that impacts solar data, the researchers employed the Minkowski norm and carefully selected an appropriate value for the parameter p. As a result of their efforts, they were able to utilize a 256-dimensional descriptor that demonstrated both efficiency and accuracy, surpassing previous methodologies. In recent years, significant progress has been made in image retrieval using learned semantic hashes [18]. The objective of semantic hashing [23] is to generate concise vectors that capture the semantic information of objects. By searching for similar hashes, we can quickly retrieve similar objects, a process that is considerably faster and requires less memory than direct manipulation of the objects themselves. Generating hashes from high-resolution full-disk solar images would not be feasible due to the sheer size of the image collections. As a result, we have devised a new rapid binary hash based on the engineered features we refer to as intermediate descriptors throughout this paper.

3   Proposed Method for Solar Image Hashing

In this section, we introduce an innovative approach to create a hash for solar images. This hash can be used later on to retrieve solar images from vast collections of solar image datasets. We obtained the solar images from the Solar Dynamics Observatory (SDO), which were refined and released through a Web

API by [19]. Although the API provides several resolutions, we opted for a resolution of 2048 × 2048. We performed evaluation study in order to determine the best resolution. The results, determine 2048 × 2048 as the best one. The algorithm we present comprises three primary stages: computing the descriptor of the solar image, hashing the descriptor, and retrieving the solar image.

3.1   Calculating Solar Image Descriptor

At this stage, we input a solar image and generate a corresponding solar image descriptor as output. After extensive research, we concluded that the most suitable image resolution to use is 2048 × 2048. Hence, we obtain a solar image descriptor as output by providing a solar image as input at a resolution of 2048 × 2048. The input image is obtained from the SDO's AIA (Atmospheric Imaging Assembly) instrument. The brighter areas visible in the input image correspond to Active Regions (ARs), which are of great significance in studying solar flares. These flares pose a significant risk to the safety of power grids, satellites, and other electronic devices situated in Earth's orbit or on its surface, making the study of ARs particularly important.

Active Regions (ARs) can exhibit a wide range of shapes and positions relative to the SDO telescope that can change as a result of the Sun's rotational movement. The first stage of the proposed method involves detecting and describing the shapes and positions of ARs. This stage comprises a sequence of steps, starting with the conversion of the image from RGB to grayscale to reduce the number of color channels from three to one. Subsequently, the pixel intensity values are in the range of $[0 : 255]$. A Gaussian blur filter, a widely used image filter, is then applied to remove insignificant and small regions. Next, we filter the pixel intensities using the threshold parameter th. After these preliminary steps, the resulting image undergoes thresholding. In this step, we compare the intensity of each pixel with the provided threshold value, th. If the intensity is greater than or equal to th, we classify the pixel as a part of the active region. The value of the th parameter was determined empirically based on experimental observations. For the given solar image dataset, we adjusted this parameter value to 180. After thresholding, the resulting image is subjected to common morphological operations, such as erosion and dilation.

The erosion operation eliminates small, isolated objects, also known as "islands", leaving only significant objects in the image. On the other hand, the dilation operation enhances the visibility of objects and fills small holes within them. By applying these two types of morphological operations, we are able to enhance the important features of active regions. Further information about morphological operations can be found in [10] or [24].

Figure 1 illustrates the results of the active region detection process. The applied operations successfully identify the active regions in the solar image. The precise detection of these regions, including their location and shape, is crucial for predicting Coronal Mass Ejections (CME) and solar flares.

After obtaining the active regions through the previous steps, the next step involves representing them mathematically in the form of a descriptor. Since the

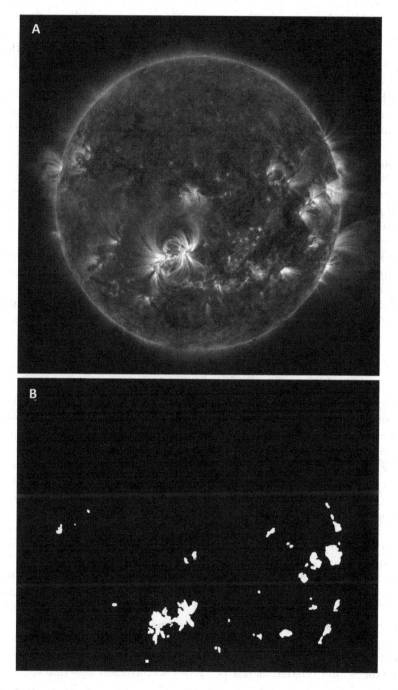

Fig. 1. Active region detection process. The top image (A) is the input image obtained from SDO Web API. The bottom image (B) is the output image obtained based on active region detection process.

dataset used in this study provides images with a 6-minute window cadence, it is assumed that the active regions change slightly between consecutive images due to the Sun's rotation movement. Therefore, a fast image hash is proposed that is resistant to small changes in perspective. To efficiently represent the shape and position of Active Regions on the Active Region Image (ARI), we propose a novel approach. We divide ARI into sectors (similarly to pizza slices) and calculate the sum of pixels for each sector. This method is efficient and allows for a compact representation of the shape and position of Active Regions, avoiding the need to compare high-resolution images. In order to describe the method in more detail, we begin by setting the coordinates of the image center, which we denote as cc. The radius r is fixed since the Sun's position on the image remains constant. Empirical experiments have led us to determine that the angle of $\theta = 30°$ yields optimal results. Then we need to divide the radius r to a number of layers of our descriptor. We used radius segment parameter rs for this purpose. The value of the rs can vary which has the significant impact on the descriptor. In Fig. 2 we divided the radius r only for four layers, in order to present the process transparently. After the extensive research we determined that the most optimal is to divide the radius on 10 segments. Then, we apply a cropping operation on the obtained sectors using the algorithm provided in Algorithm 1. To calculate the arc points of the slice (sector) aps and ape, we use the following formulas:

$$ape_x = cc_x - 1.5 * rs * \sin\theta, \tag{1}$$

$$ape_y = cc_y - 1.5 * rs * \cos\theta, \tag{2}$$

The arc points of the sector are calculated using the formulas that involve the trigonometric functions sin and cos. These formulas calculate the row and column coordinates of two points on the arc. To extend the arc slightly beyond the circle's radius, a factor of 1.5 is applied. The obtained arc points are then used to crop the slice from ARI. The cc is circle center position thus cc_x and cc_y are center position coordinates. The procedure for slicing the ARI is repeated for each layer circle sector, resulting in a list of ARI layer sectors (CARI) which contains a list of active region pixels for every CARI. In order to obtain the mathematical description of ARI we build an active region pixels histogram assigned to given circle sector histogram (LCSH). Next, the obtained histograms are concatenated into a single vector (**DV**). The entire process is illustrated in Fig. 2.

The initial step involves applying morphological operations of erosion and dilation, followed by thresholding of the input image, resulting in an image of detected active regions, as defined by Eq. 3. Subsequently, the image is sliced into layer circle sectors which allows us to obtain the CARI slices, using Eq. 4. In the next step the LCSH histograms are calculated, based on previously obtained CARI slices. Afterwards, we concatenate the histograms into the vector **DV**. This vector is later referred as LSBD – Layer-sector-based Descriptor.

$$t(ARI, i, j, th) = \begin{cases} 1, & ARI_{i,j} \geq th \\ 0, & \text{otherwise} \end{cases}, \tag{3}$$

INPUT: ARI - active region image
rs - radius segment
cc - center coordinates of ARI
θ - angle of the slice
Local Variables:
MC - mask circle matrix
$MARI$ - mask ARI matrix
ape - coordinates of starting point on the arc
OUTPUT: $CARI$ - cropped slice of ARI
$MC := CreateBooleanCircleMatrix(cc, rs)$
$MARI := CreatePolygonMatrix([cc_x, aps_x, ape_x, cc_x],$
$[cc_y, aps_y, ape_y, cc_y])$
$CARI := CombineMasks(MC, MARI)$
Algorithm 1: Algorithm for cropping the ARI slice.

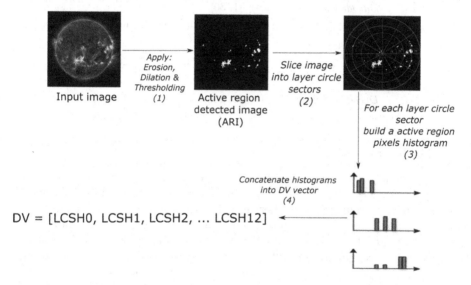

Fig. 2. Steps for calculating the layer circle sector vector (DV).

where th is the threshold value and ARI is the active region image.

$$\mathbf{CI}(CARI, th)_{k,l} =$$
$$= \sum_{i=k*ssx}^{(k+1)*ssx-1} \sum_{j=l*ssy}^{(l+1)*ssy-1} t(CARI, i, j, th) \,, \tag{4}$$

where ssx is the sector size in x-axis and ssy is the sector size in y-axis. The Layer-sector-based Descriptor (LSBD) calculation process serves to significantly reduce the data volume during the encoding stage. The primary objective of this process is to generate an intermediate, hand-crafted mathematical representation of AR images that can be utilized in the subsequent step. Thanks to this process we obtained the LSBD of 120-length (12 sectors and 10 layers), which is a significant reduction in comparison to the full-disc image. Despite that, we can reduce it even further, by using the fully-connected autoencoder described in Sec. 3.2.

Table 1. Tabular representation of the fully-connected autoencoder model.

Layer (type)	Output	Filters (in, out)	Params
$Input(InputLayer)$	[1, 120]		0
$Linear\_1(Linear)$	[1, 60]	120, 60	7,260
$ReLU\_1$	[1, 60]	0	
$Linear\_2(Linear)$	[1, 60]	60, 30	1,830
$ReLU\_2$	[1, 30]	0	
$Encoded(latent - space)$	[1, 30]		
$Linear\_4(Linear)$	[1, 30]	30, 60	1,860
$ReLU\_4$	[1, 60]	0	
$Linear\_5(Linear)$	[1, 120]	60, 120	7,320
$ReLU\_5$	[1, 120]	0	
$Decoded(Tanh)$	[1, 120]		

3.2   Hash Generation

This section details the process of generating a hash using a Layer-sector-based Descriptor (LSBD) as input. The objective of this step is to produce a representative hash that accurately describes the solar image, with a particular focus on its active regions at a specific timestamp. This step is critical as it enables the reduction of data during the retrieval stage, as discussed in Sect. 3.3. To execute this operation, we employed a fully-connected autoencoder (AE) to encode the acquired LSBD. Autoencoders are utilized in various machine learning tasks, including image compression, dimensionality reduction, feature extraction, and image reconstruction [7,11,22]. Autoencoders are ideal for generating semantic hashes as they utilize unsupervised learning. The architecture of the autoencoder model is presented in Table 1, and it should be analyzed from top to bottom. As illustrated, the model is relatively straightforward yet effective in reducing the hash length without significant information loss regarding magnetic regions of the magnetogram. It is worth noting that only the encoded portion of the trained AE's latent space is used for hash generation, while the decoding portion of the AE is solely utilized for training purposes. After conducting several experiments, we concluded that 40 epochs are adequate to achieve a satisfactory level of generalization without experiencing overfitting.

Table 1 demonstrates the utilization of a convolutional autoencoder for generating hashes, with the top layer serving as input. A one-dimensional autoencoder was utilized due to the fact that magnetic intensity descriptors are one-dimensional vectors, which helps to minimize computational complexity. By using this process, we are able to effectively reduce the hash length while retaining a substantial amount of information about the active regions of the solar image. The mean squared error function was employed as the loss function, and we determined that training the model for 40 epochs was sufficient for achieving the required level of generalization and avoiding network over-fitting. After the training process, each image descriptor was passed through the encoding layers of the autoencoder, resulting in a 30-element hash called the Layer-Sector Solar Hash (LSSH). This hash can be utilized in content-based retrieval applications that involve solar images. Furthermore, the selected autoencoder architecture was deliberately chosen to ensure optimal generalization.

3.3   Retrieval

In the final phase of the proposed method, we use the previously generated hashes for solar image retrieval. After completing the previous steps, we assume that each solar image in our database has been assigned a hash. The retrieval process involves executing an image query by comparing the distances between the hash of the query image and the hashes created for all images stored in the dataset. To perform this retrieval, we need to have a database of solar images that have undergone hash generation. In the subsequent step, we compute the distance (d) between the hash of the query image and every hash in the database. For this purpose, the cosine distance measure is employed. (see [13] for additional information).

$$\cos(QH_j, IH_j) = \sum_{j=0}^{n} \frac{(QH_j \bullet IH_j)}{\|QH_j\| \, \|IH_j\|}, \tag{5}$$

where $\bullet$ is a dot product, QH_j is the query image hash, and IH_j a consecutive image hash. After computing the cosine distance, the images in the database are sorted in ascending order based on their distance from the query hash. In the final step of the proposed method, the n images closest to the query are retrieved and returned to the user. The value of the parameter n needs to be provided by the user to execute the query. Algorithm 4 illustrates the complete process as pseudocode. An alternative method for image retrieval involves setting a threshold for the cosine distance. In this approach, the user provides a threshold parameter instead of n, and images are retrieved if their cosine distance to the query is below the threshold. The presented technique can also support the threshold-based image retrieval method, where images are retrieved if their cosine distance to the query is below the threshold value. However, the first method, which retrieves the top n images closest to the query, is more user-friendly and recommended. The retrieval process is presented in Algorithm 2.

INPUT: *ImageHashes, QueryImage, n*
OUTPUT: *RetrievedImages*
foreach *ImageHash ∈ ImageHashes* **do**
   | *QueryImageHash = CalculateHash(QueryImage)*
   | *D[i] = Cos(QueryImageHash, ImageHash)*
end
SortedDistances = SortAscending(D)
RetrievedImages = TakeFirst(n)

Algorithm 2: Image retrieval steps.

4 Experimental Results

In this section, we present the simulation results and a solution for evaluating unlabeled images using unsupervised learning for encoding descriptors. The lack of labeled data necessitated the use of this approach. As there was a lack of labeled data, evaluating the proposed method against state-of-the-art approaches was difficult. To overcome this issue, we utilized the rotation movement of the Sun to identify a set of similar images (SI). We speculated that consecutive images taken within a small-time window would display similar active regions, albeit with minor displacements. The solar images provided were captured at 6-minute intervals, which implied that they were similar due to the Sun's movement. The only necessary adjustment was to vary the time window. After experimentation, we found that images captured within a 48-hour time window could be considered similar. Let us analyze the following case. Suppose we have an image captured at 2011-02-15, 00:00:00. According to the assumptions mentioned earlier, we can consider every image captured 24 h before and after as similar. To identify images, we use their timestamps solely for evaluation purposes. Table 2 presents the process of determining similar images.

A set of experiments was conducted to assess image similarity using the proposed method. Each experiment comprised the following steps:

1. Execute an image query to retrieve images.
2. Comparing the retrieved images' timestamps with the query image timestamp.
3. Identifying images with timestamps that fell within a 48-h window as similar to the query.

After defining the set of similar images (SI), we can define the performance measures of precision and recall [8,25] based on the following sets:

- *SI* - set of similar images,
- *RI* - set of retrieved images for query,
- *PRI(TP)* - set of positive retrieved images (true positive),
- *FPRI(FP)* - false positive retrieved images (false positive),
- *PNRI(FN)* - positive, not retrieved images,
- *FNRI(TN)* - false, not retrieved images (TN).

Table 2. Defining image similarity. Based on experiments, we determined that images within a 48-hour window can be treated as similar. This allows to evaluate the method.

Timestamp	SI (similar image)/NSI (not similar image)
2011-02-13, 23:54:00	NSI
2011-02-14, 00:00:00	SI
2011-02-14, 00:06:00	SI
2011-02-14, 00:12:00	SI
2011-02-14, 00:18:00	SI
2011-02-14, 00:24:00	SI
2011-02-14, 00:30:00	SI
	SI
2011-02-15, 00:00:00	QI (query image)
	SI
2011-02-15, 23:24:00	SI
2011-02-15, 23:30:00	SI
2011-02-15, 23:36:00	SI
2011-02-15, 23:42:00	SI
2011-02-15, 23:48:00	SI
2011-02-15, 23:54:00	SI
2011-02-16, 00:00:00	NSI

We can then define the measures of precision, recall and F_1 for CBIR systems.

$$precision = \frac{|PRI|}{|PRI + FPRI|}, \tag{6}$$

$$recall = \frac{|PRI|}{|PRI + PNRI|}. \tag{7}$$

$$F_1 = 2\frac{precision \cdot recall}{precision + recall}. \tag{8}$$

The experimental results presented in Table 3 are promising, as demonstrated by the average value of F_1 and the high precision values. Our method demonstrated superior performance compared to previous works, with an average precision of 0.92186. In comparison, Banda et al. achieved a precision of 0.848, and Angryk et al. achieved a precision of 0.850 [3,5]. Moreover, our method also outperformed the results obtained in the study by Grycuk et al. [12]. Most of the solar images that had a small distance from the query image were retrieved successfully. However, for solar images with larger distances, they were classified as positive but not retrieved images ($PNRI$). Nevertheless, this value was considerably reduced compared to previous studies. The high values of $PNRI$ can

Table 3. Experiment results for the proposed algorithm. Due to lack of space, we present only a part of all queries.

Timestamp	RI	SI	PRI(TP)	FPRI(FP)	PNRI(FN)	Prec.	Recall	F_1
2011-01-01 00:00:00	199	241	187	12	54	0.94	0.78	0.85
2011-01-04 16:06:00	403	481	384	19	97	0.95	0.80	0.87
2011-01-06 19:12:00	412	481	366	46	115	0.89	0.76	0.82
...								
2011-01-15 18:18:00	386	481	361	25	120	0.94	0.75	0.83
2011-01-18 02:24:00	430	481	389	41	92	0.90	0.81	0.85
2011-01-20 12:24:00	404	481	379	25	102	0.94	0.79	0.86
...								
2011-02-03 07:36:00	404	481	373	31	108	0.92	0.78	0.84
2011-02-05 19:42:00	419	481	368	51	113	0.88	0.77	0.82
2011-02-13 17:48:00	420	481	387	33	94	0.92	0.80	0.86
Avg.						**0.922**	**0.788**	**0.849**

be attributed to the Sun's rotation movement, which may cause active regions to shift or disappear, even within the 48-hour time window.

5   Conclusions

We presented a new approach for very fast retrieving and classifying solar images using sector-based image hashing. Our initial attempt was to generate hashes directly from full-disc images, but we encountered computational complexity issues due to too large input data for the autoencoder. Consequently, we opted to create intermediate hand-crafted features to address this challenge. We utilized morphological operations to preprocess input images and detect active regions. Next, we compute the layer-sector-based descriptors. Once this step is completed, we employ a fully-connected autoencoder to encode the descriptors, resulting in the concise Layer-Sector Solar Hash. By undergoing a second encoding process, we are able to greatly reduce the length of the descriptors. Our experiments have shown a reduction of over four times compared to the layer-sector-based descriptor obtained in the initial stage. This reduction in hash length is crucial for improving the speed of calculating the distances between hashes, which in turn determines the similarity of solar images. Since solar AIA (Atmospheric Imaging Assembly) images are not labeled, we consider images produced within a short time frame of each other (typically up to several hours) to be similar. In reality, even at different times, the Sun's configuration may be similar. Hence, our precision and recall measures, which depend solely on the image content, are likely to have even higher values in practical use. The approach we have presented has several potential applications, including searching, classifying, and retrieving solar flares, which are of critical importance for many aspects of life on Earth.

References

1. Banda, J., Angryk, R., Martens, P.: Steps toward a large-scale solar image data analysis to differentiate solar phenomena. Sol. Phys. **288**(1), 435–462 (2013)
2. Banda, J.M., Angryk, R.A.: Selection of image parameters as the first step towards creating a CBIR system for the solar dynamics observatory. In: 2010 International Conference on Digital Image Computing: Techniques and Applications, pp. 528–534. IEEE (2010)
3. Banda, J.M., Angryk, R.A.: Large-scale region-based multimedia retrieval for solar images. In: Rutkowski, L., Korytkowski, M., Scherer, R., Tadeusiewicz, R., Zadeh, L.A., Zurada, J.M. (eds.) ICAISC 2014. LNCS (LNAI), vol. 8467, pp. 649–661. Springer, Cham (2014). https://doi.org/10.1007/978-3-319-07173-2_55
4. Banda, J.M., Angryk, R.A.: Scalable solar image retrieval with Lucene. In: 2014 IEEE International Conference on Big Data (Big Data), pp. 11–17. IEEE (2014)
5. Banda, J.M., Angryk, R.A.: Regional content-based image retrieval for solar images: traditional versus modern methods. Astron. Comput. **13**, 108–116 (2015)
6. Boubrahimi, S.F., Aydin, B., Schuh, M.A., Kempton, D., Angryk, R.A., Ma, R.: Spatiotemporal interpolation methods for solar event trajectories. Astrophys. J. Suppl. Ser. **236**(1), 23 (2018)
7. Brunner, C., Kö, A., Fodor, S.: An autoencoder-enhanced stacking neural network model for increasing the performance of intrusion detection. J. Artif. Intell. Soft Comput. Res. **12**(2), 149–163 (2022). https://doi.org/10.2478/jaiscr-2022-0010
8. Buckland, M., Gey, F.: The relationship between recall and precision. J. Am. Soc. Inf. Sci. **45**(1), 12 (1994)
9. Dolecki, M., et al.: On the detection of anomalies with the use of choquet integral and their interpretability in motion capture data. In: 2022 IEEE International Conference on Fuzzy Systems (FUZZ-IEEE), pp. 1–9. IEEE (2022)
10. Dougherty, E.R.: An introduction to morphological image processing. SPIE (1992)
11. Grycuk, R., Galkowski, T., Scherer, R., Rutkowski, L.: A novel method for solar image retrieval based on the Parzen Kernel estimate of the function derivative and convolutional autoencoder. In: 2022 International Joint Conference on Neural Networks (IJCNN), pp. 1–7. IEEE (2022)
12. Grycuk, R., Scherer, R.: Grid-based concise hash for solar images. In: Paszynski, M., Kranzlmüller, D., Krzhizhanovskaya, V.V., Dongarra, J.J., Sloot, P.M.A. (eds.) ICCS 2021. LNCS, vol. 12744, pp. 242–254. Springer, Cham (2021). https://doi.org/10.1007/978-3-030-77967-2_20
13. Kavitha, K., Rao, B.T.: Evaluation of distance measures for feature based image registration using alexnet. arXiv preprint arXiv:1907.12921 (2019)
14. Kempoton, D., Schuh, M., Angryk, R.: Towards using sparse coding in appearance models for solar event tracking. In: 2016 19th International Conference on Information Fusion (FUSION), pp. 1252–1259 (2016)
15. Kempton, D.J., Schuh, M.A., Angryk, R.A.: Describing solar images with sparse coding for similarity search. In: 2016 IEEE International Conference on Big Data (Big Data), pp. 3168–3176. IEEE (2016)
16. Kempton, D.J., Schuh, M.A., Angryk, R.A.: Tracking solar phenomena from the sdo. Astrophys. J. **869**(1), 54 (2018)
17. Kiersztyn, A., Karczmarek, P., Kiersztyn, K., Pedrycz, W.: The concept of detecting and classifying anomalies in large data sets on a basis of information granules. In: 2020 IEEE International Conference on Fuzzy Systems (FUZZ-IEEE), pp. 1–7. IEEE (2020)

18. Krizhevsky, A., Hinton, G.E.: Using very deep autoencoders for content-based image retrieval. In: ESANN, vol. 1, p. 2 (2011)
19. Kucuk, A., Banda, J.M., Angryk, R.A.: A large-scale solar dynamics observatory image dataset for computer vision applications. Scientific Data 4, 170096 (2017)
20. Ma, R., Boubrahimi, S.F., Hamdi, S.M., Angryk, R.A.: Solar flare prediction using multivariate time series decision trees. In: 2017 IEEE International Conference on Big Data (Big Data), pp. 2569–2578. IEEE (2017)
21. Mairal, J., Bach, F., Ponce, J., Sapiro, G.: Online learning for matrix factorization and sparse coding. J. Mach. Learn. Res. **11**, 19–60 (2010)
22. Najgebauer, P., Scherer, R., Rutkowski, L.: Fully convolutional network for removing DCT artefacts from images. In: 2020 International Joint Conference on Neural Networks (IJCNN), pp. 1–8. IEEE (2020)
23. Salakhutdinov, R., Hinton, G.: Semantic hashing. Int. J. Approx. Reason. **50**(7), 969–978 (2009). https://doi.org/10.1016/j.ijar.2008.11.006. Special Section on Graphical Models and Information Retrieval
24. Serra, J.: Image Analysis and Mathematical Morphology. Academic Press, Inc. (1983)
25. Ting, K.M.: Precision and recall. In: Encyclopedia of Machine Learning, pp. 781–781. Springer, Boston (2011). https://doi.org/10.1007/978-0-387-30164-8_652

Velocity Temporal Shape Affects Simulated Flow in Left Coronary Arteries

Justen R. Geddes(✉) ⓘ, Cyrus Tanade ⓘ, William Ladd, Nusrat Sadia Khan, and Amanda Randles(✉) ⓘ

Duke University, Durham, NC 27708, USA
{justen.geddes,amanda.rindles}@duke.edu

Abstract. Monitoring disease development in the coronary arteries, which supply blood to the heart, is crucial and can be assessed via hemodynamic metrics. While these metrics are known to depend on the inlet velocity, the effects of changes in the time-dependent inlet flow profile are not understood. In this study, we seek to quantify the effects of modulating temporal arterial waveforms to understand the effects of hemodynamic metrics. We expand on previous work that identified the minimum number of points of interest needed to characterize a left coronary artery inlet waveform. We vary these points of interest and quantify the effects on commonly used hemodynamic metrics such as wall shear stress, oscillatory shear index, and relative residence time. To simulate we use 1D Navier-Stokes and 3D lattice Boltzmann simulation approaches conducted on high performance compute clusters. The results allow us to observe which parts of the waveform are most susceptible to perturbations, and therefore also to measurement error. The impacts of this work include clinical insight as to which portions of velocity waveforms are most susceptible to measurement error, the construction of a method that can be applied to other fluid simulations with pulsatile inlet conditions, and the ability to distinguish the vital parts of a pulsatile inlet condition for computational fluid dynamic simulations.

Keywords: Computational Fluid Dynamics · Coronary Arteries · Temporal Velocity Profile

1 Introduction

Coronary arteries supply blood to the heart and are vital to healthy human function. When these arteries narrow, a condition known as coronary artery disease (CAD), they impede blood flow, posing a significant health risk. Therefore, the ability to monitor and predict the future of CAD is a crucial challenge. CAD and other diseases have been shown to correlate with hemodynamic metrics such as wall stress [13]. Therefore, gaining insights into the factors influencing hemodynamic metrics is vital for comprehending disease progression. Specifically, the temporal inlet velocity profile is a key determinant of these hemodynamic metrics in the vascular system.

© The Author(s), under exclusive license to Springer Nature Switzerland AG 2024
L. Franco et al. (Eds.): ICCS 2024, LNCS 14832, pp. 121–135, 2024.
https://doi.org/10.1007/978-3-031-63749-0_9

As with all measurements, the field should seek to quantify the effects of error in measurements such as the inlet velocity to understand their impact on both observed physiological states and simulations. In particular, state-of-the-art devices for measuring flow velocity, such as Doppler echocardiography, are prone to significant errors [7]. To understand these errors in the temporal domain, we study the effects of inlet velocity perturbations in computational fluid dynamics (CFD) simulations of the left coronary arteries as a proof of concept.

Previous works have successfully conducted pulsatile CFD in the left coronary arteries using a template waveform [2,19,23,24], while others use additional models, usually ordinary differential equations, to inform the coronary inlet from heart flow [5,8,11,16,22]. However, none of these studies considered perturbations or error in their inlet velocities. Rizzini et al. [18] quantified differences in inlet flow in the spatial development profile of the flow, but did not assess the temporal profile. Jiang et al. [9] studied the velocities of a temporal waveform by modulating a temporal waveform's minimum, average, and maximum velocities in the right coronary arteries. However, by only varying these general measurements of the profile, they were unable to quantify how differences in specific portions of the waveform can affect hemodynamics.

To our knowledge, no study has evaluated perturbations in the temporal shape of the inlet velocity of the left coronary artery or attempted to quantify the contributions of changes in specific parts of hemodynamic waveforms in general. Addressing this gap in the literature will allow for a better understanding of the impact of inlet conditions on hemodynamic metrics, measurement error of devices such as Doppler echocardiograms, and which portions of the inlet waveform are most important to hemodynamic metrics.

To accomplish this task, we use novel techniques from our previous study [6] to systematically stretch the inlet waveform to assess the effects of perturbations of portions of the inlet waveform. We simulate using 1D and 3D CFD approaches, thus allowing us to sample more plentifully with the low computational cost of 1D approaches, while also using 3D simulations to ensure the accuracy of simulations. We hypothesize that perturbation locations on the inlet waveform will not all produce the same effect on output metrics and that some points will be more important than others to capture accurate measurement and CFD results. Implications include the use of a non-uniform error tolerance in practice and identifying regions of increased interest when examining a patient's inlet velocity profile.

2   Methods

To assess the effect of waveform shape on hemodynamic metrics, we first choose the points that we wish to modulate to change the shape of the waveform. We then define our 1D and 3D models, as well as our geometries and numerical experiments. Lastly, we present the metrics that we use to characterize the hemodynamics.

2.1   Quantifying and Varying Temporal Waveform Shape

To assess how different waveform shapes modulate hemodynamic metrics, we must first specify how we will vary the shape of the waveform. To accomplish this task, we expand on our previous study [6] that used a novel "stretching" procedure to vary waveforms. This pipeline was applied to left coronary arteries and started with 20 possible points of interest (shown in Fig. 1) that could be stretched to change the shape of the waveform. The way these points are calculated is shown in Table 1. We denote the velocity value of point i as f_i and the time value of point i as t_i. Using this notation, our previous study found that the points t_5, t_6, f_4, f_5, f_6 and f_9 can well represent the variations in the waveforms. reducing points.

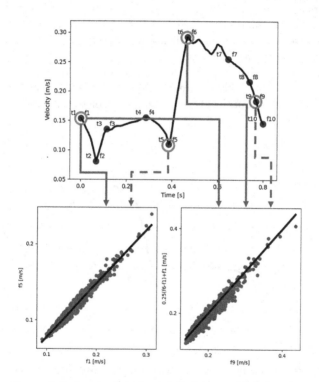

Fig. 1. Estimating points f_5 as a function of f_1 (bottom left) and f_9 as a function of f_1 and f_6 (bottom right). The top panel shows the points used to estimate, marked by solid arrows, and the points that are being estimated, marked by dashed arrows. The lower two panels show the linear relationship between the point and their estimators.

In the current study, our goal is to assess the effects of changing the first velocity point by vertically shifting the entire waveform, achieved by adding f_1 to the collection of points. Furthermore, we include the systolic valley for physiological significance (f_2).

Table 1. Definition of point locations on waveforms. Note that f_i refers to the velocity of point i. The location of all points can be seen in Fig. 1.

Point number	Description	Point number	Description
1	Beginning of systole	6	Diastolic peak
2	Systolic valley	7	$3/4\,(f_6 - f_{10}) + f_{10}$
3	Systolic shoulder	8	$1/2\,(f_6 - f_{10}) + f_{10}$
4	Systolic max	9	$1/4\,(f_6 - f_{10}) + f_{10}$
5	Beginning of diastole	10	End of cardiac cycle

Additionally, note that $f_9 = 0.25(f_6 - f_{10}) + f_{10}$. We observe that the first and last points are approximately equal ($f_1 \approx f_{10}$), and therefore $f_9 \approx 0.25(f_6 - f_1) + f_1$, and estimate f_9 as such. Lastly, we observe a strong ($R^2 = 0.97$) linear correlation between f_1 and f_5, and therefore estimate f_5 as $f_5 \approx 0.70797f_1 + 0.00665$. These relations are depicted in Fig. 1.

Finally, we have our set of points to modify: $s = \{t_5, t_6, f_1, f_2, f_4, f_6\}$, while using the modified values to estimate f_5 and f_9 to preserve physiological waveform shape while "stretching". These points are depicted in Fig. 2.

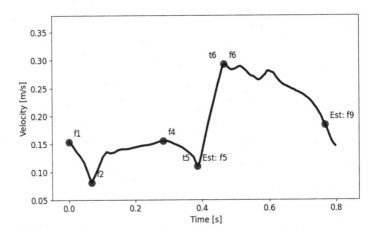

Fig. 2. The final selected points on a left coronary artery inlet velocity waveform. "Est" denotes points that are estimated using the values of other points, but are not independently modulated.

2.2   1D Simulation Approach

One-dimensional blood flow was simulated using the methods outlined in [3,4, 20,21] which are based on the equations

$$\frac{\partial A}{\partial t} + \frac{\partial Q}{\partial x} = 0 \tag{1}$$

$$\frac{\partial Q}{\partial t} + \frac{\partial}{\partial t}\left(\alpha \frac{Q^2}{A}\right) + \frac{A}{\rho}\frac{\partial P}{\partial x} = -C_f \frac{Q}{A} \tag{2}$$

where Q is the flow rate [m^3/s], t is time [s], A is cross-sectional area [m^2], α modulates the velocity profile, x is the spatial position [m], $\rho = 1{,}060\,\text{kg/m}^3$ is the density of blood, P is pressure [Pa], and $C_f = 22\pi\mu$ (where μ is the dynamic viscosity [Pa s]) introduces effects from friction. P relates to A via

$$P = P_{\text{ext}} + \beta(\sqrt{A} - \sqrt{A_0}), \beta = \frac{\sqrt{\pi}hE}{(1-\nu^2)A_0} \tag{3}$$

with P_{ext} denoting external pressure on the vessel and A_0 cross-sectional area under no pressure. β represents vessel stiffness and is computed using wall thickness, h, elastic modulus, E, and Poisson's ratio (ν). We use values from [20].

Boundary conditions were enforced using a 2-element Windkessel model to represent the effects of flow in the microvasculature as

$$Q = \frac{P}{R} + C\frac{dP}{dt} \tag{4}$$

with C [m^3/Pa] denoting compliance and R [(Pa s) / m^3] denoting resistance.

2.3   3D Simulation Approach

To simulate 3-dimensional blood flow, we use our in-house parallel fluid solver, HARVEY [17]. HARVEY uses the lattice Boltzmann method, which is a mesoscopic approach shown to recover the Navier-Stokes equations, to represent the fluid as a collection of particles that are quantified via particle distribution functions [12]. These distribution functions change in time and space per the lattice Boltzmann equation,

$$f_i(x + c_i\Delta t, t + \Delta t) = f_i(x, t) + \Omega_i(x, t), \tag{5}$$

with x denoting the spatial position, t representing time, c_i is a discrete velocity, and Ω_i is the Bhatnagar-Gross-Krook collision operator. HARVEY utilizes the D3Q19 velocity discretization with finite difference boundary conditions at inlets and outlets [14] as well as the halfway bounce-back condition, which enforces the no-slip condition at the rigid walls.

2.4   Image-Derived Coronary Geometries

It is critical to examine realistic geometries to assess waveform variation. We use image-derived 3D geometries reconstructed from 2D coronary angiograms by the methods described in [1]. This study was approved by the Duke University Medical Center Institutional Review Board (Pro00091022) and limited data can be provided on request from the authors. To obtain a 1D geometry, we calculated the centerlines using Mimics [15]. Once centerlines were obtained, we used techniques presented in [4] to create an interconnected tree of 1D domains. In total, we examined nine different patient-specific left coronary artery geometries.

2.5   Numerical Experimental Protocol

In order to assess how differences in the shape of the inlet waveform affect hemodynamic metrics, we use patient-specific waveforms created by scaling the template waveform using cardiac output and heart rate. To assess the effect of shape on these waveforms we first calculate the average value of each of the six points, $\overline{s_i}$ ($s = \{t_5, t_6, f_1, f_2, f_4, f_6\}$) across the nine different geometries.

For the 3D predictions of blood flow, we simulated the patient-specific waveform and also one trial for each selected point with the waveform "stretched" to increase the chosen point by 10% of the average, i.e. s_i becomes $s_i + 0.1\overline{s_i}$. 10% is chosen to align with the approximate Doppler echocardiogram error tolerance [7,10]. These modifications are illustrated in Fig. 3b. Thus, we conduct 7 simulations per geometry, 63 total.

The 1D model is much less computationally expensive than 3D, which allows for more simulations than 3D. On average, 3D simulations were run on 46 nodes and took between 4–12 hours, while 1D simulations take only a fraction of this time. For 1D simulations, for each geometry we simulate the patient-specific waveform, and then 40 simulations ranging from ±20% with one simulation for each percentage, e.g., point s_i becomes $s_i - 0.2\overline{s_i}$ for the -20% simulation. These trials are illustrated in Fig. 3a. We simulate ±20% for 1D simulations to cover the worst-case scenarios. Some waveforms were not physiological due to $t_5 > t_6$ or $t_5 \approx t_6$ and therefore were omitted. This protocol generates a maximum of 241 simulations per geometry (6 points of interest, up to 40 simulations per point of interest, and one baseline simulation), 2,169 simulations in total.

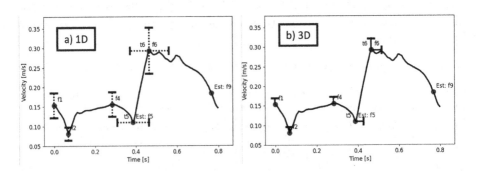

Fig. 3. How the waveforms were stretched in simulations. In panel a), the dashed lines denote 40 simulations were conducted between -20% of the average point value and +20%. panel b) shows the simulations for 3D simulations, with each solid bar denoting one simulation with the point increased by 10%.

2.6 Calculating Hemodynamic Metrics

For 1D simulations, we compute average and maximum wall shear stress (WSS) [Pa] over the geometry and the cardiac cycle, with WSS computed as

$$WSS = \frac{-4u\mu}{R} \tag{6}$$

with u [m/s] denoting the velocity at a particular spatial position, R [m] the radius of the vessel, and μ [Pa s] the dynamic viscosity of blood.

For 3D simulations, we calculate the maximum and average time-averaged wall shear stress (TAWSS) [Pa] over the geometry, with the TAWSS given by

$$TAWSS = \frac{1}{T} \int_0^T |\tau_w| dt \tag{7}$$

with T [s] denoting the time for a cardiac cycle and τ_w [Pa] the wall shear stress imposed by the moving fluid. To verify the results, we compare WSS (1D) and TAWSS (3D) for the same magnitude of perturbations.

We also compute the maximum and average oscillatory shear index (OSI) [N.D.] over the geometry, with OSI calculated as

$$OSI = 0.5 \left(1 - \frac{|\int_0^T \tau_w dt|}{\int_0^T |\tau_w| dt} \right) \tag{8}$$

which captures the effects of oscillatory flow on walls. In addition, we also calculate maximum and average relative residence time (RRT) [Pa^{-1}],

$$RRT = \frac{1}{(1 - 2\,OSI)\,TAWSS}. \tag{9}$$

We examine the relative difference of hemodynamic measurements as $r = (m_1 - m_0)/m_0$ where r is the relative difference [N.D.], m_1 is the metric computed from the perturbed waveform, and m_0 is the metric from the original patient-specific waveform.

2.7 Statistical Tests

We analyze the relative differences in two ways. First, we compare the output metric relative difference to the relative difference of the perturbation to observe whether the input perturbation is proportional to the output effects. These effects are denoted Fig. 4 and Fig. 5 as dashed lines. Secondly, we conduct an unpaired t-test with the null hypothesis of a mean equal to 0 for 3D simulations and that the slope of the line of best fit is equal to 0 for 1D simulations. We use a significance threshold of 0.005 for the unpaired t-tests.

3 Results

We first present results for 1D, followed by 3D, simulations. We then compared 1D and 3D results and, lastly, compared 3D results with inlet profile metrics.

3.1   1D Quantification of Wall Shear Stress

In our study, we conducted as many as 241 1D simulations for each geometry and show in Fig. 4 the 1D relative difference in maximum WSS and average WSS for each perturbed point compared to the original baseline simulation versus the percent change in the point of interest. The findings approximate a linear relationship, which allowed us to apply a linear regression analysis to the dataset. Subsequently, we compared the slopes of these lines to the null hypothesis that the slope is equal to 0.

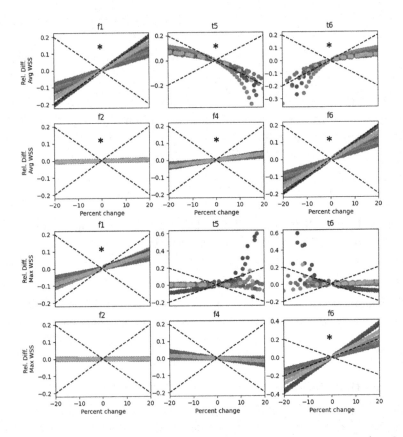

Fig. 4. Relative differences of average wall shear stress (WSS, top two rows) and maximum WSS (lower two rows) vs percent change of the point of interest for 1D simulations. Distinct geometries are noted by varying colors. Noting linear behavior, we compare the slope of the regression line of the line of best fit to a slope of 0, which implies no change due to varying the point of interest. A "*" denotes a p-value < 0.005 using an unpaired t-test. Dashed lines denote a magnitude of change in the output metric equal to the magnitude of change in the point of interest ($y = x$).

We observe that all average WSS comparisons are significant ($p < 0.005$), as well as f_1 and f_6 maximum WSS. We further note that the changes in f_1 and f_6

are approximately equal to the change in the average WSS. Additionally, there is nonlinear behavior for t_5 and t_6 as they become closer to each other. Lastly, while f_2 and f_4 have statistically significant slopes, the change in the output metrics is minor compared to the amount of change in the point of interest.

3.2 A Subset of Perturbed Points of Interest Resulted in Significant 3D Simulation Changes

Since 3D simulations are more computationally expensive than 1D, we conducted only one simulation for each point of interest per geometry, increasing the point of interest by 10% of the average value of points in all geometries. Figure 5 shows the results for TAWSS, OSI, and RRT. We observe that the average and maximum TAWSS are significant for each point of interest, with the largest effects observed for f_1 and t_5. Furthermore, the average OSI is significant for t_5, t_6, and f_2 while the average RRT is significant for f_1, f_4 and f_6.

3.3 Strong Agreement Between 1D and 3D Wall Shear Stress

We use 1D simulations to explore the parameter space more widely than computationally expensive 3D simulations. However, viewing the 1D results as an extension of 3D requires that the 1D and 3D produce similar relative differences where they overlap - in WSS/TAWSS at a +10% perturbation. To verify this agreement, Fig. 6 shows the Bland-Altman plot between the relative differences in the 1D and 3D simulations where the points of interest were increased by 10% of the average value between the geometries. We see that there is good agreement between 1D and 3D, with a bias of 0.0033 [N.D.] and a standard deviation of 0.02 [N.D.]. We also observe that as the magnitude of 1D predictions increases, the difference between predictions increases, thus causing the points on the left of the figure. Lastly, we see that 3D TAWSS is usually larger than 1D WSS, thus causing most of the points to be less than zero.

3.4 Correlation Between Metrics and Inlet Waveform Metrics

Given the above results, a natural question that follows is whether the calculated hemodynamic metrics can be predicted from the inlet waveform without simulation. Figure 7 shows the comparison of TAWSS with the area under the temporal inlet velocity curve, the systolic area under the curve (AUC), diastolic AUC, systolic duration, cardiac cycle duration, average systolic velocity, average diastolic velocity, and average velocity. Each point represents a 3D simulation - one point for each trial per geometry, thus resulting in 63 points.

We observe that TAWSS is correlated with the average velocity and other metrics that would increase the average velocity. This correlation aligns with expectations, given that velocity is a central component of the TAWSS calculation. We also detect a weak correlation with certain metrics like diastolic area under the curve (AUC), where we can identify distinct clusters of data points

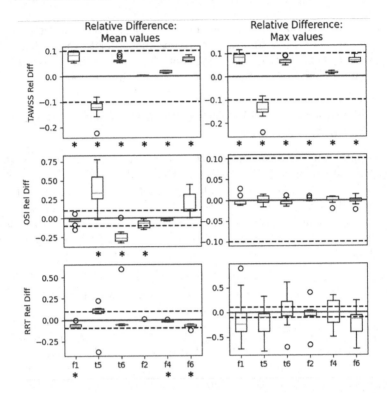

Fig. 5. 3D results for modulating points of interest by 10% of the average point value among geometries for time-averaged wall shear stress (TAWSS), oscillatory shear index (OSI), and relative residence time (RRT). Boxplots denote the distribution of 9 simulations, one for each geometry, for each point of interest. The solid blue line denotes the 0% change, while the dashed black lines mark ± 10% change. A "*" denotes a p-value < 0.005 using an unpaired t-test between the distribution and 0 (blue line), which would imply no output change for the perturbed point of interest. (Color figure online)

that are consistent across different perturbations and geometries. In contrast, we find no significant correlation between TAWSS and other metrics, such as systolic AUC or average systolic velocity. Additional results, which are not shown, indicate a lack of correlation between TAWSS and OSI or RRT.

4   Discussion

We explore the effects of alterations in the temporal profile of the coronary artery inlet velocity waveform on hemodynamic metrics, essential for tracking disease progression [13]. In particular, we elucidate which points of interest, which quantify the inlet waveform, are most sensitive to perturbations. We begin our discussion with 1D, and then 3D, simulation results and consider the relationship

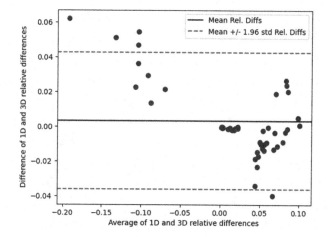

Fig. 6. Bland-Altman plot comparing 3D predicted time averaged wall shear stress (TAWSS) and 1D predicted wall shear stress (WSS). The solid black line denotes the mean of the difference between 1D and 3D, while the dashed red lines denote the mean $\pm$ 1.96 standard deviations. (Color figure online)

between the 1D and 3D results. Subsequently, we assert that relying solely on inlet waveform metrics is insufficient for precise predictions of hemodynamic metrics.

4.1 1D Linear Changes in Wall Shear Stress

Remarkably, perturbations of points of interest result in mostly linear changes in the 1D WSS. The exception to this linearity is when $0 < t_6 - t_5 \ll 1$. The linear change between point-of-interest perturbations and the relative change in WSS implies that changing point values does not greatly amplify input changes in point values, as would be suggested by an exponential relationship. Moreover, we see that the majority of the points result in relative differences of the output that are less than or equal to the relative differences of the input, as can be seen by the points in Fig. 4 being between the dashed black lines. This observation implies that the output effects are less than or equal to the corresponding input changes. Therefore, when considering how changing a waveform will affect the WSS of a new simulation, we can conclude that the WSS will change less than (or equal to) the change in the temporal waveform.

4.2 Three Points of Interest Cause 3D Statistically Significant OSI Differences

1D simulations are adequate to predict WSS, however, only 3D simulations can predict OSI and RRT. We see that in Fig. 5 that, as predicted by the 1D simulations, the change in input is not amplified to the output (marked by dashed

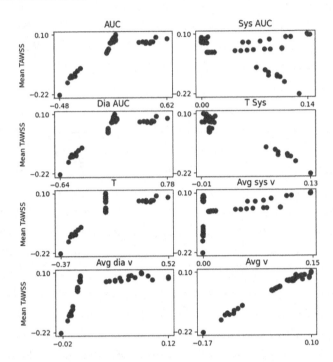

Fig. 7. Plots of the mean relative difference of perturbed simulation to original for TAWSS against relative differences for inlet waveform metrics: area under curve (AUC), systolic area under curve (sys AUC), diastolic area under curve (dia AUC), time for systole (T Sys), Cardiac cycle length (T), average systolic velocity (Avg sys v), average diastolic velocity (Avg dia v), and average velocity (v). Each color represents a distinct geometry.

lines) in 3D simulations except for t_5. This amplification of the t_5 input arises because, by increasing t_5, we make $t_6 - 5_5 \approx 0$ as mentioned above.

It is expected and true that all TAWSS comparisons result in statistically significant differences since velocity is closely related to TAWSS. It is perhaps more interesting that t_5, t_6, and f_2 all affect the mean OSI differences. Furthermore, these differences represent sudden changes in velocity. This relation suggests that inlet velocity waveforms can have an effect on OSI and that OSI is not only geometry dependent. Additionally, we see that RRT is only statistically significant when OSI is not, likely because TAWSS is significant for all points and RRT is a function of TAWSS and OSI.

4.3   Inlet Waveforms Are Not Enough to Predict All Metrics

Given the above analysis, we wish to evaluate whether metrics that quantify the inlet velocity waveform can predict hemodynamic metrics without the need to simulate. For OSI and RRT (results not shown), no inlet waveform correlated whatsoever with OSI or RRT. Additionally, the 1D WSS results displayed weak correlations. However, the average velocity in the 3D simulations was able to correlate well with the mean TAWSS (Fig. 7). Thus, in general, we require patient-specific modeling to quantify hemodynamic metrics.

4.4 Limitations

The presented work has limited 3D simulations due to the computational expense of the simulations. As a result, only seven 3D simulations were conducted for each of the nine geometries. We attempted to remedy this limited number of simulations by conducting up to 241 1D simulations. Furthermore, we only considered one template waveform for each geometry. As such, it is possible that these results could differ for other waveforms that, for example, are recorded in a hyperemic state.

4.5 Clinical and Research Implications

This work influences how waveforms are quantified in the clinic by measuring the effects of varying points of interest. Specifically, we see that the acceleration of flow may be most important to OSI, and that OSI depends on both geometry and inlet conditions. Additionally, we have shown that some points have a greater impact on hemodynamic metrics than others - such as peak diastolic velocity (f_6). Furthermore, while TAWSS can be estimated from inlet waveform measurements (such as average velocity), our findings advocate for the necessity of 1D and/or 3D simulations to accurately determine precise WSS, OSI, and RRT.

Lastly, the approximate linear behavior suggests that one could potentially predict the mean and maximum WSS for varying waves by simply simulating one wave of interest and quantifying how far points of interest on additional waves are from the original waves. Further work can apply this method to other portions of the vasculature.

5 Conclusion

The findings of our research underscore the critical influence of the waveform shape, beyond mean velocity, on key hemodynamic parameters such as WSS, TAWSS, OSI, and RRT. This nuanced understanding advances our capabilities in flow simulations of coronary hemodynamics, emphasizing the importance of time-domain inlet conditions. Furthermore, our work delineates the waveform segments most susceptible to inaccuracies and identifies the specific aspects of the temporal waveform that significantly impact hemodynamic metrics, offering vital insights for clinical assessments related to disease progression. This analysis was carried out in the left coronary arteries, but, as with the pipeline that precedes this work [6], can be generalized to other geometries. The impacts of this work include an increased understanding of inlet conditions in CFD simulations, the identification of sensitive portions of the temporal waveform to error, and clinical insight as to which areas of the temporal waveform have the most effect on hemodynamic metrics and, therefore, also hold importance to disease states.

Acknowledgments. Research reported in this publication was supported by the National Institute On Aging of the National Institutes of Health under Award Number DP1AG082343. The content is solely the responsibility of the authors and does not necessarily represent the official views of the National Institutes of Health. Funding was also given under NSF GRFP under Grant No. DGE 164486. An award of computer time was provided by the INCITE program. This research also used resources of the Oak Ridge Leadership Computing Facility, which is a DOE Office of Science User Facility supported under Contract DE-AC05-00OR22725.

References

1. Chen, S.J., Carroll, J.D.: 3-d reconstruction of coronary arterial tree to optimize angiographic visualization. IEEE Trans. Med. Imaging **19**(4), 318–336 (2000)
2. Chidyagwai, S.G., Vardhan, M., Kaplan, M., Chamberlain, R., Barker, P., Randles, A.: Characterization of hemodynamics in anomalous aortic origin of coronary arteries using patient-specific modeling. J. Biomech. **132**, 110919 (2022)
3. Feiger, B., Adebiyi, A., Randles, A.: Multiscale modeling of blood flow to assess neurological complications in patients supported by venoarterial extracorporeal membrane oxygenation. Comput. Biol. Med. **129**, 104155 (2021)
4. Feiger, B., Kochar, A., Gounley, J., Bonadonna, D., Daneshmand, M., Randles, A.: Determining the impacts of venoarterial extracorporeal membrane oxygenation on cerebral oxygenation using a one-dimensional blood flow simulator. J. Biomech. **104**, 109707 (2020)
5. Fleeter, C.M., Geraci, G., Schiavazzi, D.E., Kahn, A.M., Marsden, A.L.: Multilevel and multifidelity uncertainty quantification for cardiovascular hemodynamics. Comput. Methods Appl. Mech. Eng. **365**, 113030 (2020)
6. Geddes, J.R., Randles, A.: Optimizing temporal waveform analysis: a novel pipeline for efficient characterization of left coronary artery velocity profiles. In: 46th Annual International Conference of the IEEE Engineering in Medicine & Biology Society (EMBC). IEEE, (2024)
7. Ghorbannia, A., et al.: Clinical, experimental, and computational validation of a new doppler-based index for coarctation severity assessment. J. Am. Soc. Echocardiogr. **35**(12), 1311–1321 (2022)
8. Gutierrez, N.G., et al.: Hemodynamic variables in aneurysms are associated with thrombotic risk in children with Kawasaki disease. Int. J. Cardiol. **281**, 15–21 (2019)
9. Jiang, Y., Zhang, J., Zhao, W.: Effects of the inlet conditions and blood models on accurate prediction of hemodynamics in the stented coronary arteries. AIP Adv. **5**(5), 057109 (2015)
10. Kebed, K., Sun, D., Addetia, K., Mor-Avi, V., Markuzon, N., Lang, R.M.: Measurement errors in serial echocardiographic assessments of aortic valve stenosis severity. Int. J. Cardiovasc. Imaging **36**, 471–479 (2020)
11. Kim, H.J., Vignon-Clementel, I., Coogan, J., Figueroa, C., Jansen, K., Taylor, C.: Patient-specific modeling of blood flow and pressure in human coronary arteries. Ann. Biomed. Eng. **38**, 3195–3209 (2010)

12. Krüger, T., Kusumaatmaja, H., Kuzmin, A., Shardt, O., Silva, G., Viggen, E.M.: The Lattice Boltzmann Method. Springer (2017)
13. Kumar, A., et al.: Low coronary wall shear stress is associated with severe endothelial dysfunction in patients with nonobstructive coronary artery disease. JACC: Cardiovasc. Intervent. 11(20), 2072–2080 (2018)
14. Latt, J., Chopard, B., Malaspinas, O., Deville, M., Michler, A.: Straight velocity boundaries in the lattice boltzmann method. Phys. Rev. E 77(5), 056703 (2008)
15. Materialise: Mimics. https://www.materialise.com/en/healthcare/mimics-innovation-suite/mimics
16. Ramachandra, A.B., Kahn, A.M., Marsden, A.L.: Patient-specific simulations reveal significant differences in mechanical stimuli in venous and arterial coronary grafts. J. Cardiovasc. Transl. Res. 9(4), 279–290 (2016)
17. Randles, A.P., Kale, V., Hammond, J., Gropp, W., Kaxiras, E.: Performance analysis of the lattice Boltzmann model beyond Navier-Stokes. In: 2013 IEEE 27th International Symposium on Parallel and Distributed Processing, pp. 1063–1074. IEEE (2013)
18. Rizzini, M.L., et al.: Does the inflow velocity profile influence physiologically relevant flow patterns in computational hemodynamic models of left anterior descending coronary artery? Med. Eng. Phys. 82, 58–69 (2020)
19. Sharma, P., et al.: A framework for personalization of coronary flow computations during rest and hyperemia. In: 2012 Annual International Conference of the IEEE Engineering in Medicine and Biology Society, pp. 6665–6668. IEEE (2012)
20. Tanade, C., Chen, S.J., Leopold, J.A., Randles, A.: Analysis identifying minimal governing parameters for clinically accurate in silico fractional flow reserve. Front. Med. Technol. 4, 1034801 (2022)
21. Tanade, C., Feiger, B., Vardhan, M., Chen, S.J., Leopold, J.A., Randles, A.: Global sensitivity analysis for clinically validated 1d models of fractional flow reserve. In: 2021 43rd Annual International Conference of the IEEE Engineering in Medicine and Biology Society (EMBC), pp. 4395–4398. IEEE (2021)
22. Taylor, C.A., Fonte, T.A., Min, J.K.: Computational fluid dynamics applied to cardiac computed tomography for noninvasive quantification of fractional flow reserve: scientific basis. J. Am. Coll. Cardiol. 61(22), 2233–2241 (2013)
23. Vardhan, M., et al.: Non-invasive characterization of complex coronary lesions. Sci. Rep. 11(1), 1–15 (2021)
24. Vardhan, M., Gounley, J., Chen, S.J., Kahn, A.M., Leopold, J.A., Randles, A.: The importance of side branches in modeling 3d hemodynamics from angiograms for patients with coronary artery disease. Sci. Rep. 9(1), 1–10 (2019)

Simulating, Visualizing and Playing with de Sitter and Anti de Sitter Spacetime

Eryk Kopczyński[(✉)] [iD]

Institute of Informatics, Institute of Warsaw, Warsaw, Poland
erykk@mimuw.edu.pl

Abstract. In this paper we discuss computer simulations of de Sitter and anti de Sitter spacetimes, which are maximally symmetric, relativistic analogs of non-Euclidean geometries. We present prototype games played in these spacetimes; such games and visualizations can help the players gain intuition about these spacetimes. We discuss the technical challenges in creating such simulations, and discuss the geometric and relativistic effects that can be witnessed by the players.

Keywords: de Sitter spacetime · relativity · science game

1 Introduction

Science-based games are games based on a real scientific phenomenon. For example, there are games based on special relativity [8,14], quantum mechanics [20], orbital physics [22], non-Euclidean geometry [4,11,24,25]. Compared to typical educational games, where the concept explained and the gameplay are not related, science games take the scientific phenomenon and use it to create interesting gameplay [20]. Other than providing entertainment, science games provide intuitive understanding of difficult science concepts; many of them let the player perform their own experiments, by including sandbox elements, level editors, or being open source. Such a better understanding helps young players consider a scientific career, and also gives ideas for new scientific developments to mature researchers [9]. Science games require a specialized engine for efficient modelling and visualization of the given scientific concept; such engines may be later used for other applications than just games [2].

In this paper, we describe a game (or rather, a collection of two games) taking place in de Sitter and anti de Sitter spacetimes. These spacetimes could be seen as relativistic analogs of spherical and hyperbolic geometry. De Sitter spacetime is of interest as an asymptotic approximation of our universe [16], and anti-de Sitter spacetime is of interest for its correspondence with conformal field theory [15]. While many games exist based on special relativity and non-Euclidean geometry, the unintuitive properties of spacetimes combining these two concepts seems to be still unexplored.

© The Author(s), under exclusive license to Springer Nature Switzerland AG 2024
L. Franco et al. (Eds.): ICCS 2024, LNCS 14832, pp. 136–150, 2024.
https://doi.org/10.1007/978-3-031-63749-0_10

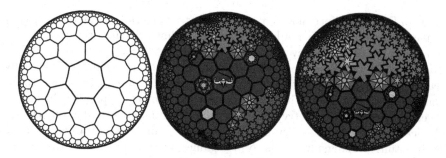

Fig. 1. On the left, the {7,3} hyperbolic tessellation in Poincaré disk model. The Poincaré disk model is conformal: it does not distort small shapes, so all heptagons look close to regular; however, it distorts scale: all the heptagons shown are of the same size. On the right, the same scene in *HyperRogue* viewed from two points. In the Poincaré model, straight lines are projected as circular arcs orthogonal to the disk boundary; moving the center shows the player that the walls (orange) are indeed straight lines. (Color figure online)

Players can try our prototype game, named *Relative Hell*, by downloading the Microsoft Windows binary[1]. The source code, based on the non-Euclidean engine RogueViz [12], is also available.

2   Background

Our prototype helps non-experts understand scientific concepts intuitively. In this section, we provide an intuitive introduction to non-Euclidean geometry and relativity theory, and explain how games can provide such understanding.

The history of non-Euclidean geometry starts with Euclid's *Elements*. This book has changed teaching geometry by giving it structure: starting with very basic *postulates* and *axioms*, such as the space never ending and being the same everywhere and in every direction, and continuing with all the more complex geometric facts known to Euclid, which followed from the postulates and axioms [5]. However, some geometric facts related to *parallel lines* did not actually seem to follow from such basic postulates. For example, if we have two points $A_1 B_1$ in distance d on a straight line l_1 and two points $A_2 B_2$ in distance d on a parallel straight line l_2, the distance between A_1 and A_2 should be always the same as between B_1 and B_2. Euclid has solved this by declaring (an equivalent formulation of) this statement as his *fifth postulate* (or *parallel postulate*).

Mathematicians have been trying to prove the *parallel postulate* from the other postulates, until Lobachevsky and Bolyai have discovered *hyperbolic geometry* $\mathbb{H}^d$, which satisfied all Euclid's postulates except his *fifth postulate* [5]. (In $\mathbb{H}^d$, d is the number of dimensions.) To explain how the parallel postulate could not be true, imagine drawing great circles on a sphere; such great circles are an

---

[1] https://zenorogue.itch.io/relative-hell, source code: https://github.com/zenorogue/hyperrogue/blob/master/rogueviz/ads/ads-game.cpp. Last accessed April 9, 2024.

analog of Euclidean lines, however, if we look at two meridians, we discover that the parallel postulate does not hold. In our three-dimensional world, great circles are obviously curved; however, we can imagine that our Universe is actually a hypersphere in four dimensions, and what we perceive as straight lines is actually great circles on this hypersphere. This is the *spherical geometry* $\mathbb{S}^d$; hyperbolic geometry is the opposite of it (while, in spherical geometry, "parallel" lines converge, in hyperbolic geometry they diverge). While many everyday objects are spherical, hyperbolic geometry is more difficult to understand; a good way to obtain intuitive understanding is to play games, aiming to simulate the experience of a inhabitant of a non-Euclidean world. Games in $\mathbb{H}^2$ [11,17,24] typically display the view in the *Poincaré disk model* (Fig. 1, which is a projection of hyperbolic plane $\mathbb{H}^2$ to a Euclidean disk, centered at the player character's position; this lets the player see that straight lines are indeed straight and acting strangely. There are also immersive visualizations of $\mathbb{H}^3$ [4,10,11,18,23].

While the world was originally assumed to be Euclidean, further experiments have shown this to not to be the case. The Morley-Michelson experiment has shown that the speed of the light measured by a moving observer is always the same (independent by the observer's velocity), which was in conflict with theories at that time. This was resolved in the theory of special relativity: Euclidean space and Newton's notion of time were replaced by *Minkowski spacetime*; there was no absolute time – for two observers O_1 and O_2 meeting at position 0 in time 0, an event at position x and time t according to O_1 would be at position x' and t' according to O_2, with both space and time being changed by Lorentz transformations ($x \neq x'$ and $t \neq t'$), a bit similar to how spatial rotations change both x and y coordinates, and causing effects such as contraction of lengths and dilation of time. Special relativity effects normally require large speed to observe in the real world, but again, they can be experimented with in games [8,14]. Special relativity could not explain gravity; for this, we need general relativity: the notion of curved spacetime, similar to the curved space of non-Euclidean geometry. Furthermore, as explained in Sect. 3 below, Minkowski spacetime geometry can be used to provide an elegant model of $\mathbb{H}^d$.

Most existing games and simulations, including non-Euclidean ones, do not take relativity into account. At every point, they represent the current state of the simulation (at time t) in the computer memory for all objects using the chosen internal model of the geometry, and to compute the further states of the simulation (at times $t' > t$), a chosen model of physics is used, usually some simplification and adaptation of Newtonian physics. Interestingly, while such a non-relativistic model of Euclidean spacetime obeys the Galilean principle of relativity (a moving object having no access to external reference point cannot determine that it is moving), this is no longer the case for non-Euclidean geometries. For example, in $\mathbb{H}^d$, lines diverge, so a large object moving experiences apparent centrifugal force. This is the reason why, for example, the flock of boids in the flocking simulations in RogueViz [12] cannot keep its shape, contrary to a Euclidean flocking simulation [19]. For a similar reason, HyperRogue [11] could not feature large, freely moving objects – the boundary of such an

Fig. 2. Relative Hell. $\widetilde{ad\mathbb{S}^2}$ is displayed in the Poincaré disk model on the left, and the Beltrami-Klein disk model in the center. $d\mathbb{S}^2$ on the right, in stereographic projection.

object would have to move significantly faster than the center, in a curved line, which would be unintuitive; so while the game does feature some somewhat large creatures (such as snakes and krakens), they are narrow enough to avoid this problem.

This can be solved by using *de Sitter* ($d\mathbb{S}^d$) and *anti-de Sitter* ($\widetilde{ad\mathbb{S}^d}$) spacetimes, which are relativistic analogs of $\mathbb{S}^d$ and $\mathbb{H}^d$, respectively. They are symmetric under the Lorentz transformations used to simulate the change of velocity, and therefore, they obey the Galilean principle of relativity. Intuitively, while $\mathbb{H}^d$ and $\mathbb{S}^d$ stretch the distances in space, $d\mathbb{S}^d$ and $\widetilde{ad\mathbb{S}^d}$ also stretch the time. The game *Relative Hell* (Fig. 2) lets the player fly a two-dimensional spaceship in these spacetimes. Due to the nature of these spacetimes, the gameplay is different: in $d\mathbb{S}^2$, objects are naturally pulled apart, so the goal is to keep close to the *main star* while avoiding bullets, as in the *bullet hell* genre; while in $\widetilde{ad\mathbb{S}^2}$, objects are naturally pulled together, so the whole space is rotating in order to generate centrifugal force to balance that, and the goal is to shoot, similar to classic omnidirectional shooters such as *Asteroids*. The player can also gain insight into the usual non-Euclidean geometric and relativistic (time dilation, space contraction) phenomena, such as the exponential expansion of $\mathbb{H}^2$, and time dilation and Lorentz contraction.

3   Preliminaries

We briefly recall the definitions of basic spaces and spacetimes. For more details, see, e.g., [1] for hyperbolic geometry, and [7] for $d\mathbb{S}^d$ and $\widetilde{ad\mathbb{S}^d}$.

The *Euclidean space* of dimension d, $\mathbb{E}^d$, is $\mathbb{R}^d$ equipped with the Euclidean inner product $g_E(x,y) = \sum_i x_i y_i$. For a vector $x \in \mathbb{R}^d$, we use the notation $x[a/i]$ for the vector x with i-th coordinate replaced by a. In particular, $0[1/i]$ is the point whose i-th coordinate is 1 and the other coordinates are 0. Our Euclidean inner product defines the distance between points: $d_E(x,y) = \sqrt{g_E(x-y, x-y)}$. The length of a differentiable curve $\gamma : [a,b] \to \mathbb{E}^d$ can be computed as $\int_a^b \sqrt{g_E(\dot{\gamma}(t), \dot{\gamma}(t))}dt$; note that $d_E(x,y)$ is also the length of the shortest curve from x to y. Orientation-preserving isometries of the Euclidean space are generated by translations $T_i^a(x) = x[x_i + a/i]$ and rotations $R_{i,j}^\alpha(x) = x[x_i \cos \alpha + x_j \sin \alpha/i][x_j \cos \alpha - x_i \sin \alpha/j]$ for $i \neq j$. Isometries preserve g_E, and thus distances and curve lengths. The basic translations and rotations can be composed to obtain other isometries. Isometries of $\mathbb{E}^d$ correspond to looking at $\mathbb{E}^d$ from another frame of reference.

In video games and computer graphics, it is convenient to use the *homogeneous coordinates*, that is, $\mathbb{R}^{d+1}$, where the extra $(d+1)$-th coordinate is always equal to 1. This method lets us represent translations and rotations as matrices.

The *spherical space (sphere)* of dimension d, $\mathbb{S}^d$, is $\mathbb{E}^{d+1}$ restricted to $\{x : g_E(x, x) = 1\}$. (For convenience we consider only spheres of radius 1 here.) The distance $d_S(x, y)$ between two points $x, y \in \mathbb{S}^d$ is the length of the shortest curve (geodesic) connecting them on the sphere, $c : [a, b] \to \mathbb{S}^d$; we have $d_S(x, y) = \arccos(g_E(x, y))$. Orientation-preserving isometries of the sphere are generated by $R_{i,j}^\alpha$ (while in the Euclidean space we needed the extra coordinate for translations, in $\mathbb{S}^d$ we can instead just use the one extra coordinate we already have). A sphere is maximally symmetric: any isometry of the underlying $\mathbb{E}^{d+1}$ that maps 0 to 0 maps the sphere to itself.

A *signature* is a $\sigma \in \{-1, +1\}^d$; we will drop the 1 and just write the signs, for example $(+, +, +, -)$; if a sign repeats, we use an exponent, for example $(+^3, -)$. The *pseudo-Euclidean spacetime* of signature σ, $\mathbb{E}^\sigma$, is $\mathbb{R}^d$ equipped with the inner product $g_\sigma(x, y) = \sum_i \sigma_i x_i y_i$. In this paper, coordinates x_i with $\sigma_i = 1$ correspond to space dimensions, and coordinates with $\sigma_i = -1$ correspond to time dimensions. Positive values of $g_\sigma(v, v)$ correspond to space-like intervals, and negative values correspond to time-like intervals. A curve is *space-like* if the integral in equation (Sect. 3) is well-defined (that is, the value under the square root is always non-negative), and *time-like* if this value is always non-positive. The proper time of a time-like curve γ is defined as $\int_a^b \sqrt{-g_E(\dot{\gamma}(t), \dot{\gamma}(t))} dt$.

In special relativity, our spacetime is modelled as a pseudo-Euclidean space of signature $(+, +, +, -)$; the fourth coordinate corresponds to time, and the units of time and distance are chosen so that Einstein's constant c equals 1. Isometries of pseudo-Euclidean spacetime are similar, but for $\sigma(i) \neq \sigma(j)$ we replace rotations $R_{i,j}^\alpha$ by Lorentz boosts $L_{i,j}^\alpha(x) = x[x_i \cosh \alpha + x_j \sinh \alpha/i][x_j \cosh \alpha + x_i \sinh \alpha/j]$. The parameter α is called *rapidity*; in spacetimes with 1 time coordinate, Lorentz boosts correspond to changing the velocity of our frame of reference. Objects generally move along time-like curves; Proper time is interpreted as the time measured by a clock, or intuitively felt by a sentient creature, moving along such a curve. Time-like geodesics have the longest possible proper time. While the relative speed of moving objects $\tanh \alpha$ is bounded by $c = 1$, the rapidity is not bounded. The interaction of space and time coordinates causes well-known relativistic effects such as Lorentz contraction and dilation of time. The set of points connected to v with time-like geodesics γ such that the time coordinate of $\gamma(t)$ is increasing on is called the *future light cone* of v, and the set of points connected with time-descreasing time-like geodesics is the *past light cone*. These light cones are preserved when we apply the isometries of $\mathbb{E}^\sigma$ that keep v. The other points are neither in the future or the past, but rather elsewhere – their time coordinate may be greater or smaller than the time coordinate of v, depending on the chosen isometry. What happens at v causally depends only on the spacetime events in the past light cone of v, and may affect only the spacetime events in the future light cone of v.

The *hyperbolic space* of dimension d, $\mathbb{H}^d$, is $\mathbb{E}^\sigma$ for $\sigma = (+^d, -)$ restricted to $\{x : g_\sigma(x, x) = -1, x_{d+1} > 0\}$. This is the Minkowski hyperboloid model of hyperbolic geometry. This is a space, not a spacetime, that is, all the curves on $\mathbb{H}^d$ are space-like. While hyperbolic geometry is usually taught in the Poincaré disk model, the Minkowski hyperboloid model is generally easier to understand (assuming familiarity with Minkowski geometry) and work with computationally because of its similarity to the natural model of the sphere. In particular, orientation-preserving isometries are generated by rotations $R_{i,j}^\alpha$ and Lorentz boosts $L_{i,d+1}^\alpha$ (corresponding to translations of $\mathbb{H}^d$), and the distance between x and y is $\operatorname{arcosh}(g_\sigma(x, y))$. While in special relativity the time coordinate is usually indexed as x_0, in this paper we prefer to make it the last coordinate, for consistency with the usual indexing of homogeneous coordinates of the Euclidean space in computer graphics.

Taking $z \in \mathbb{R}$, we can project $x \in \mathbb{H}^d$ to $x' \in \mathbb{R}^d$ by $x_i' = x_i / (x_{d+1} + z)$. This is called the *general perspective projection*. For $z = 1$ we get the *Poincaré ball model* (also called the *Poincaré disk model* in 2 dimensions). It is the hyperbolic analog of the stereographic projection of the sphere, which uses the same formula. It is the most popular method of visualizing hyperbolic geometry; just like the stereographic projection, it is conformal, meaning that the angles and small shapes are mapped faithfully. Another one is the *Beltrami-Klein ball (disk) model*, obtained for $z = 0$. Beltrami-Klein disk and Poincaré disk are called models because they are alternative mathematical representations of hyperbolic geometry; in this paper, we use only the Minkowski hyperboloid model for mathematical representation, and the disk models are used as projections for visual representation. More possible projections exist which are not special cases of the general perspective projection, for example the azimuthal equidistant projection, which renders the distances and angles from the chosen central point correctly [17]. Dozens of spherical projections are used in cartography [21]; many of them have hyperbolic analogs, available in the HyperRogue engine [11].

The *de Sitter spacetime* with d space dimensions and 1 time dimension, dS^d, is $\mathbb{E}^\sigma$ for $\sigma = (+^{d+1}, -)$ restricted to $\{x : g_\sigma(x, x) = 1\}$.

The *wrapped anti-de Sitter spacetime* with d space dimensions and 1 time dimension, adS^d, is $\mathbb{E}^\sigma$ for $\sigma = (+^d, -^2)$ restricted to $\{x : g_\sigma(x, x) = -1\}$. Note that adS^d has closed time-like loops, for example, $\gamma(t) = 0[\sin t/d + 1][\cos t/d + 2]$. Such closed-time loops are obtained by going around the axis $A = \{x \in \mathbb{E}^\sigma : \forall i \leq d\ x_i = 0\}$. Due to the closed time-like loops, adS^d has no causal structure (everything is simultaneously in the past and future of everything else and itself). This problem is resolved by taking the universal cover of adS^d, i.e., the point reached by going $n \neq 0$ times around the axis A is considered a different point in the spacetime. We will call this universal cover the *unwrapped anti-de Sitter spacetime* $\widetilde{ad\mathbb{S}^d}$, or just *anti-de Sitter spacetime* for short.

Spacetimes $d\mathbb{S}^d$ and $\widetilde{ad\mathbb{S}^d}$ will be discussed in detail in the following sections. Here, we will only remark that time-like and space-like curves and their lengths and proper times are defined as above, and that both $d\mathbb{S}^d$ and $\widetilde{ad\mathbb{S}^d}$ are maximally symmetric spacetimes, with their isometries generated from $R_{i,j}^\alpha$ and $L_{i,j}^\alpha$.

4   Simulation of the Anti-de Sitter Spacetime

For simplicity, we will start with $ad\mathbb{S}^d$. When necessary, we fix $d = 2$.

The point $O = 0[1/d + 2]$ is considered the origin of the $ad\mathbb{S}^d$. Every object b in the simulation, at a specific point of its proper time t, is represented by an isometry $T_{b,t}$ of $ad\mathbb{S}^d$, which maps the coordinates relative to (b, t) to the world coordinates. Usually, (b, t)-relative coordinates of b itself at time t is O, so the world coordinates of b at time t are $T_b(O)$. The player controls a ship s, which is one of the objects in the game. To display an object at world coordinates x on the screen at time t, we need to map x into ship-relative coordinates, $x' = T_{s,t}^{-1} x$. The state of the game universe at the current time is a slice of the spacetime. This slice is $S = \{x \in ad\mathbb{S}^d : x_{d+1} = 0, x_{d+2} > 0\}$, which is isometric to $\mathbb{H}^d$, and can be rendered e.g. in the Poincaré disk/ball model.

The ship s is controlled in real time by a player, so for every time moment t displayed in an animation frame, the game has to compute the next frame transform $T_{s,t+\epsilon}$ depending on both $T_{s,t}$ and player's decisions. Objects in spacetime move along timelike geodesics if no force is acting on them. Timelike geodesics in $ad\mathbb{S}^d$ are generally of form $T R_{d+1,d+2}^t O$. Thus, if the player does not accelerate from time t_1 to t_2, we have $T_{s,t_2} = T_{s,t_1} R_{d+1,d+2}^{t_2 - t_1}$. Changing the camera speed in dimension $i \in \{1, \ldots, d\}$ corresponds to changing the frame of reference by multiplying it by $L_{i,d+2}^\alpha$. Thus, if the player accelerates, we also need to multiply the formula for T_{s,t_2} by $L_{i,d+2}^\alpha$ on the right.

One possible objection to displaying the slice S is that the player should not see the state of the universe at the current time – the ship at time t only knows the past light cone of $T_{s,t} O$. In the current game prototype, we assume that all the objects other than s behave deterministically, so this is not an issue— the ship could compute the current state of other objects depending on what it knows. While displaying S is less immersive, it is useful for understanding how the spacetime works. (One can change the options to get the actual view.)

The simplest deterministic movement is geodesic movement. Let $b \neq s$. The object b at any given time is not a point, but it is rather a subset $X(b) \subseteq S$. The formula $T_{b,t} = T_{s,0} R_{d+1,d+2}^t$ turns out not to be satisfactory – while the object b's origin O moves geodesically, other points in $X(b)$ do not. In $d = 2$, this issue can be fixed by using the formula $T_{b,t} = T_{b,0} R_{3,4}^t R_{1,2}^t$, which enforces geodesic movement of every point of b. For short, we denote the isometry $R_{3,4}^t R_{1,2}^t$ with M^t. (It is possible to change the options to also use M^t for the ship, but with such a setting, the game becomes more confusing to the player.) Every object $b \neq s$ starts its lifetime at some time t_1 and ends at time t_2; the time t_1 may be $-\infty$ if the object did exist forever, and t_2 may depend on player's actions, and be ∞ if it is not yet known to be ever destroyed. We can compute $T_{b,t}$ at every time t knowing $T_{b,0}$. So, to display the point $x \in X(b)$ of object b at the ship's proper time t, we need to find t_b such that $x_b := T_{s,t}^{-1} T_{b,0} M^{t_b} x \in S$. If $t_b \in [t_1, t_2]$, we apply the chosen projection to x_b. We display the point x there.

So far, we have been assuming adS^d for simplicity, which does not really work due to the time-like loops – technically, all of spacetime is in the past, so the player would be able to perceive the results of the actions they have not performed yet (this is a problem with all games featuring a powerful enough form of player control and time travel). Another issue is numerical precision: hyperbolic space (S in our case) is characterized by its exponential growth, which causes numerical errors to accumulate quickly, and using world coordinates relative to some fixed origin does not really work when we travel far enough from that origin. Our solution of these issues is based on the existing implementation of $\widetilde{G}$, the universal cover of Lie group G of orientation-preserving isometries of $\mathbb{H}^2$, also called $\widetilde{SL(2,\mathbb{R})}$ or the twisted product of $\mathbb{H}^2$ and $\mathbb{R}$. This implementation is described in paper [13]. The space of isometries of $\mathbb{H}^2$ is a three-dimensional space; the three dimensions correspond to the two dimensions of $\mathbb{H}^2$ itself (translations), plus one extra dimension which corresponds to rotation by some angle α. In [13], this space of rotations is represented using unit split quaternions (this is the hyperbolic analog of the fact that isometries of $\mathbb{S}^2$ are represented using quaternions, which is well known in computer graphics). The set of unit split quaternions corresponds exactly to adS^d – the only difference is that in [13] the dimension corresponding to rotation is considered spacelike, while in adS^d it is timelike; specifically, rotation by angle α corresponds to M^α.

In the universal cover, we consider the isometries whose rotation components are described by different angles $\alpha, \beta \in \mathbb{R}$ to be different, even if they are actually the same rotation (that is, $\alpha - \beta$ is a multiple of 2π). We have a natural projection of the universal cover to the underlying space $\pi : \widetilde{adS^d} \to adS^d$. The space $\widetilde{adS^d}$ has its origin O', $\pi(O') = O$. We pick a lift $j : adS^d \to \widetilde{adS^d}$ such that $\pi \circ j$ is identity; the point $j(x)$ is chosen among the points x' such that $\pi(x') = x$ in a natural way: the path from x' to O' does not cross $\pi^{-1}\{x \in adS^(d) : x_1 < 0, x_2 = 0\}$. We reuse the same notation M^α, $R_{i,j}^\alpha$ and $L_{i,j}^\alpha$ for the isometries of $\widetilde{adS^d}$. Note that the pass-of-time isometry M^α (as well as the underlying $R_{2,3}^\alpha$) will now be different for every $\alpha \in \mathbb{R}$, while in adS^d we had $M^{2\pi} = M^0 = Id$. If T is an isometry of $\widetilde{adS^d}$, we likewise have uniquely defined $\pi(T)$, which is an isometry of adS^d such that $\pi(Tx) = \pi(T)(\pi(x))$. If T is an isometry of adS^d, let $j(T)$ be the isometry of $\widetilde{adS^d}$ which takes O to $j(T(O))$ and such that $\pi(j(T)) = T$.

In our simulation engine, the points of $\widetilde{G}$, and equivalently $\widetilde{adS^d}$ are represented as *shift points*. A shift point (x, h) consists of a point $x \in adS^d$ and a shift $h \in \mathbb{R}$, and represents $x' = M^h j(x)$. Note that this representation is not unique; the *canonical* representation of x' is the one in which $x_3 = 0$ and $x_4 > 0$. Similarly, a isometry T' of $\widetilde{adS^d}$ is represented as *shift matrices*: (T, h), where

T is an isometry matrix and h is a shift, represents $M^h j(T)$. The canonical representation of T' is the one in which (TO, h) is canonical.

To avoid the numerical precision issues, we tessellate $\widetilde{adS^d}$. That is, $\widetilde{adS^d}$ is subdivided into a number of tiles τ, and every object is described not by a shift matrix relative to some origin of the whole space, but by a tile τ it lives in, and a shift matrix (x, d) relative to the center of the tile τ. As previously mentioned, $\widetilde{adS^2}$ corresponds to the Lie group of isometries of $\mathbb{H}^2$; the centers of our tiles correspond to the isometries which map the order-3 heptagonal tessellation (Fig. 1) to itself. Knowing the shift matrices transforming the coordinates relative to tile τ_1 into the coordinates relative to adjacent tile τ_2, it is straightforward to compute the coordinates of all the object nearby to the player, relative to the player. The tessellations themselves can be computed using discrete methods such as automata theory [3,6], thus avoiding numerical precision issues arising otherwise when the player makes their ship travel a large distance from the start.

Due to the nature of time-like geodesics in $\widetilde{adS^d}$, objects appear to move in circles. This suggests a game design somewhat reminiscent of the classic arcade multidirectional shooter game *Asteroids* (which took place in a space with torus topology, so the objects also did not escape the playing area). The player has to shoot rocks for resources, such as gold (increasing score), health (used up when hit by an asteroid), ammo (used up when shot), fuel (used up when accelerating), and oxygen (used up proportionally to proper time elapsed). These are standard resources well-known to gamers. One interesting consequence of relativity is that the player may use up the fuel resource to save a bit of the oxygen resource, since acceleration can be used to reduce the proper time necessary to reach another point in the spacetime (similar to the twin paradox). Shooting the missile creates a new object m (missile); we compute if the worldline of the missile m intersects the world line of some rock r, and if so, the life of both m and r end at this time, and we create a collectible resource at the same spacetime event.

The implementation of such a simulation requires us to implement the necessary operations for shift points and shift matrices: compose isometries (multiply shift matrices), apply isometry to a point (multiply shift matrix by a shift point), find t_b and x_b for rendering objects. While in adS^d we would just use the well-known matrix and vector multiplications, in $\widetilde{adS^d}$ these operations are somewhat more involved due to the necessity of computing shifts correctly (we need to be careful to obtain the correct shift value h instead of, e.g., $h \pm 2\pi$). To keep the paper short, we do not include the full formulas in this paper; they can be found in the source code of our simulation (file `math.cpp`).

5 Simulation of the de Sitter Spacetime

The general ideas of our de Sitter simulation are similar to the anti-de Sitter case described in the previous section, so we only list the differences.

The origin is now $O = 0[1/d + 1]$. The slice corresponding to current time is $S = \{x \in dS^d : x_{d+2} = 0\}$. It is isometric to $\mathbb{S}^d$, so it can be rendered e.g. in the stereographic model. The pass of time is now represented using isometry $M^\alpha = L^\alpha_{d+1,d+2}$ (we use the same isometry for the pass of time for s and the other objects). If the player accelerates, we multiply T_{s,t_2} by $L^\alpha_{i,d+2}$.

The equivalence of S to $\mathbb{S}^d$ might suggest *Asteroids*-like design again: rocks flying around the sphere. However, the spacetime $d\mathbb{S}^d$ works very differently: the space appears to be expanding as time passes, and objects which fly too far away from s are impossible to reach anymore. In particular, the other side of the sphere is unreachable. If the game started with a number of randomly pre-placed objects at time 0, after some time all of them would depressingly fly away from each other, with at most one of them remaining in the part of universe reachable to the player. So the world of our de Sitter game is constructed differently. There is a main star (black-and-yellow in Fig. 2), and the goal is to remain close to the main star as long as possible. Other objects in the universe make this task harder. Every object (bullet) b gets close to the main star at some time t_b; avoiding being hit is the main challenge of the game, thus making the game an example of a game in the *bullet hell* genre. As usual in bullet hell games, the bullets arrive in waves (sharing similar value of t_b) arranged in various patterns.

Numerical precision issues caused by the space expanding exponentially also arise in $d\mathbb{S}^d$. This time, the solution we use in our simulation is not generally applicable, but rather tailored to the design of our game, that is, based on the assumption that the player will be always forced to remain close to the main star. We again use the concept of shift matrices: the shift matrix describing an object b will be of form (T, t_b), which is equivalent to matrix $M^{t_b}T$ relative to the main star at time 0, or equivalently, T relative to the main star at time t_b. Objects need not be rendered if their t_b is not close enough to the proper time of the main star currently visible to the player (if we tried to render them, we would run into precision issues due to the exponential growth of sinh and cosh).

In the anti-de Sitter case we assumed that every point of the object b, $x \in X(b) \subseteq S$, moves geodesically. This does not work in $d\mathbb{S}^d$ – if every point moved geodesically, the objects would expand. Instead, we only assume that the center O of the object moves geodesically. Instead, after computing t_b^O and x_b^O for the center O ($x_b = T_{s,t}^{-1}T_{b,0}M^{t_b}O \in S$), for every other point $x \in X(b)$ we find a geodesic γ'_x which is correct at time t_b using the formula $\gamma'_x(t_b + t) = T_{b,0}M_{t_b}L^{x_1}_{1,4}L^{x_2}_{2,4}$ where (x_1, x_2) are the coordinates of point x, and find x'_b and t'_b such that $x'_b = T^{-1}s, t\gamma'_x(t'_b)$. The point is now rendered at x'_b. We compute x'_b for every vertex of the polygonal model describing the object b, and render b as a polygon with vertices obtained by mapping x'_b using the chosen projection.

6   Visualizations and Insights

Science-based games should allow the player to not only experience the challenge provided by the gameplay, but also play with the simulation. One aspect of this is that the player can change the parameters of the game (amount of resources, rocks, scale, speed, etc.) to explore more diverse scenarios without being bothered with the challenge, making the game accessible to players who are not skilled at

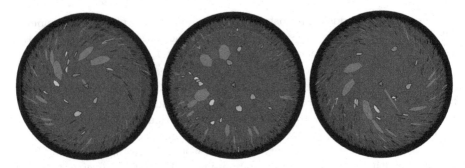

Fig. 3. Anti-de Sitter spacetime: past light cone view (left), present (middle), and future light cone view (right). Note the stretched missile in the future light cone view. Taken from a replay; the red circle is the boundary of the light cone relative to a future position of the ship. Poincaré disk model. (Color figure online)

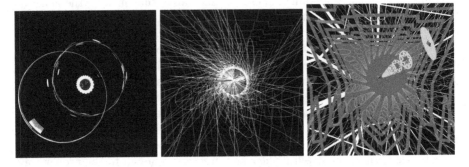

Fig. 4. De Sitter spacetime: present in stereographic projection (left), spacetime view (center), and $\mathbb{H}^3$ view (right). Unfortunately, the spacetime and $\mathbb{H}^3$ views are difficult to capture in a still image (and also in video due to the compression).

the given game genre. Another aspect is that we should keep the history of every object in the game. Such history keeping is necessary by the basic construction of a relativistic game – for example, even if the game knows that some object b disappears at proper time t (for example, a rock is hit by player's missile), and the player has seen that, the player might accelerate, causing them to see the object b still existing, due to how Lorentz transformations work. While it would be safe to remove b from memory if the life of b ended in the past light cone, for the discovery purposes it is better to just remember everything. Since the whole history is kept, the player can replay their game, possibly from other frame of reference. While viewing such a replay (or possibly even during the actual game), the player can change aspects of how the in-game universe is visualized, such as:

- In $\widetilde{ad\mathbb{S}^d}$, let the time α pass according to $R_{d+1,d+2}^\alpha$ or M^α. The first option is generally more easier to understand (the camera is not rotating), but both views are interesting. In the second view, the world appears to stop spinning around the player, but the movement starts looking like wrapping the spacetime, which is interesting but unintuitive.

- The player can choose to use the current S as explained so far, or they can take S to be the boundary of the past light cone (corresponding to the events the ship would be actually seeing at that precise moment), or the boundary of the future light cone (Fig. 3).
- Proper time of every object could be displayed, to let the player see whether the relativistic effects (time dilation) work as they expect.
- There is also an option to view the 2+1-dimensional spacetime using the perspective projection. In this visualization, it is assumed that light travels along geodesics of all kinds. We see a circle of radius 1, the interior of that circle corresponds to time-like geodesics (space-time events the ship could reach), and the exterior corresponds to events happening elsewhere. The circle is essentially a hyperbolic plane in the Beltrami-Klein model; changing the ship's velocity transforms the interior of the circle according to the isometries of this hyperbolic plane. See Fig. 4 for an example.
- Interestingly, the isometries of $d\mathbb{S}^d$ are also the isometries of $\mathbb{H}^3$. Both of them live in $\mathbb{E}^\sigma$ for $\sigma = (+^3, -)$, but $\mathbb{H}^3$ is the set of points x where $g_\sigma(x, x) = -1$, while $d\mathbb{S}^2$ is the set of points x where $g_\sigma(x, x) = 1$. This observation leads to another visualization: take the isometry describing the ship's view, and use the same isometry to view $\mathbb{H}^3$ using Klein-Beltrami ball model, or equivalently, perspective. Such a dual view has interesting properties (pass of time corresponds to moving the dual camera forward; accelerating corresponds to moving the dual camera sidewise; points of $d\mathbb{S}^d$ are represented in this perspective as points outside of the Klein-Beltrami ball, in particular points of S being represented as points in infinity; etc.). See Fig. 4 for an example.

The following insights into de Sitter and anti-de Sitter spaces are gained.

- The space in the anti-de Sitter game appears to be rotating. This is because the fixed parts of the map (the walls based on the tessellation) move along the geodesics $M^{t_1}v$, while the progress of time is modelled by multiplying the spacetime coordinates $T_{3,4}^t$. So, if $v \in S$ and the player moves by t units, the object is displayed at (taking $t = t_1$) $T_{3,4}^{-t}M^t v = R^{1,2}(t)$. An object cannot simply stay in a fixed place v on screen, because that would not be geodesic movement: the time interval between $v \in S$ and $T_{3,4}^t v$ is greater and greater when we increase the distance from v to O, and thus timelike geodesics are pulled towards the center. So in general objects appear to be orbiting around the ship in circular or elliptical orbits.
- In the de Sitter game, if we imagine S as a sphere with the main star at the north pole N, objects moving along geodesics will get close to N at some point, then escape towards the equator. Similarly, if we reversed the time, we discover they started towards the equator too. If we view the spacetime relative to the main star, the equator is well visible due to all the objects close to it. Normally we view the spacetime relative to the ship, which is a

different frame of reference, so the two equators are distinct. Figure 2 shows one of them, while Fig. 4 shows the equator circle made by bullets we have avoided in the past, the equator circle made by bullets we will have to avoid in the future, and also a small circle of bullets we are avoiding right at the moment when the screenshot was taken.

- Another observation is the Lorentz contraction. All objects appear compressed in the direction of the movement. For example, while all the tessellation tiles are of regular heptagonal shape and the Poincaré model is conformal, but they appear more and more narrow as we move further away from the center (Fig. 2). Due to the relationship between split quaternions and the underlying hyperbolic plane, the distance between tiles in $\widetilde{adS^2}$ are half the distances between the respective tiles in $\mathbb{H}^2$; after switching the view to Beltrami-Klein model, the heptagons look regular again.
- Dilation of time is best observed in the de Sitter game, by comparing the proper time of the ship with the score, which is the proper time of the main star we are close to. Generally the score will be smaller, due to the geodesic movement of the main star, and non-geodesic movement of the ship. Interestingly, when the player loses the de Sitter game by getting too far away from the main star, their score no longer increases – the main star is escaping so fast that its proper time seen by us does not change.

7   Further Work

The engine described in this paper can be extended to other games and experiments in $\widetilde{adS^2}$ and dS^2. To conclude the paper, we describe some extensions.

- An active enemy that attempts to predict where the player's ship is going to go (for example, assuming that the ship is simply going along a time-like geodesic), and shoot there. Such enemies are popular in video games. This concept would be naturally more interesting in a relativistic game, since such an enemy would have to predict the player's ship position based on the past (that is, if the enemy shoots at time t, it only sees what the player did at some time in the past, due to the limited speed of light). This is still deterministic. Non-deterministic enemies are of course also a possibility. However, interestingly, making the game multi-player introduces an unexpected challenge, at least if we want the proper time of both players proportional to the real time elapsed. One interesting way to avoid this issue would be to make the game end in the case when causality is lost, i.e., one of the players falls inside the past light cone of another player.
- A map editor to let the players construct their own scenarios and puzzles, for example, to explore the twin paradox (using fuel to conserve oxygen).
- We have chosen the game to have two spatial dimensions. Two-dimensional games generally showcase the experimental gameplay better (multi-directional shooters and bullet hell games work better in two dimensions),

and also allow our three-dimensional visualizations of the space-time. However, immersive three-dimensional variants of our games would also be worthwhile. Our implementation of $\widetilde{ad\mathbb{S}^2}$ crucially depends on two-dimensionality, in particular, the trick of using M^α to pass time while keeping all the world stable, due to all dimensions being rotated, works only if the number of space dimensions is even. In three dimensions, one of the dimensions would not be rotated, and thus not stable. A simple way to add another dimension, keeping our tessellations of $\mathbb{H}^2$ and interpretation of M^α, would be to make our universe still roughly two-dimensional (a bit like real-world galaxies), just adding one extra dimension of our space that extrudes the tessellation and is not rotated.

Acknowledgments. This work has been supported by the National Science Centre, Poland, grant UMO-2019/35/B/ST6/04456.

Disclosure of Interests. Authors have no conflict of interest to declare.

References

1. Cannon, J.W., Floyd, W.J., Kenyon, R., Walter, Parry, R.: Hyperbolic geometry. In: In Flavors of geometry, pp. 59–115. University Press (1997). http://www.msri.org/communications/books/Book31/files/cannon.pdf
2. Celińska, D., Kopczyński, E.: Programming languages in Github: a visualization in hyperbolic plane. In: Proceedings of the Eleventh International Conference on Web and Social Media, ICWSM, Montréal, Québec, Canada, 15–18 May 2017, pp. 727–728. The AAAI Press, Palo Alto, California (2017). https://aaai.org/ocs/index.php/ICWSM/ICWSM17/paper/view/15583
3. Celińska-Kopczyńska, D., Kopczyński, E.: Generating tree structures for hyperbolic tessellations (2021)
4. CodeParade: Hyperbolica (2022)
5. COXETER, H.S.M.: Non-Euclidean Geometry, 1 edn. Mathematical Association of America (1998). http://www.jstor.org/stable/10.4169/j.ctt13x0n7c
6. Epstein, D.B.A., Paterson, M.S., Cannon, J.W., Holt, D.F., Levy, S.V., Thurston, W.P.: Word Processing in Groups. A. K. Peters Ltd., Natick (1992)
7. Griffiths, J.B., Podolský, J.: Exact Space-Times in Einstein's General Relativity. Cambridge Monographs on Mathematical Physics, Cambridge University Press, Cambridge (2009)
8. Hall, A.: Velocity raptor (2011). https://testtubegames.com/velocityraptor.html. Accessed 9 Apr 2024
9. Hamilton, L., Moitra, A.: A no-go theorem for robust acceleration in the hyperbolic plane. In: Annual Conference on Neural Information Processing Systems 2021. NeurIPS 2021, 6–14 December 2021, Virtual, pp. 3914–3924 (2021). https://proceedings.neurips.cc/paper/2021/hash/201d546992726352471cfea6b0df0a48-Abstract.html
10. Hart, V., Hawksley, A., Matsumoto, E.A., Segerman, H.: Non-euclidean virtual reality I: explorations of $\mathbb{H}^3$. In: Proceedings of Bridges: Mathematics, Music, Art, Architecture, Culture, pp. 33–40. Tessellations Publishing, Phoenix, Arizona (2017)

11. Kopczyński, E., Celińska, D., Čtrnáct, M.: HyperRogue: playing with hyperbolic geometry. In: Proceedings of Bridges: Mathematics, Art, Music, Architecture, Education, Culture, pp. 9–16. Tessellations Publishing, Phoenix, Arizona (2017)
12. Kopczyński, E., Celińska-Kopczyńska, D.: RogueViz: non-Euclidean geometry engine for visualizations, games, math art, and research, October 2023. https://github.com/zenorogue/hyperrogue/
13. Kopczyński, E., Celińska-Kopczyńska, D.: Real-time visualization in anisotropic geometries. Exp. Math. 1–20 (2022). https://doi.org/10.1080/10586458.2022.2050324
14. Kortemeyer, G., Tan, P., Schirra, S.: A slower speed of light: developing intuition about special relativity with games. In: International Conference on Foundations of Digital Games (2013)
15. Maldacena, J.: The large-n limit of superconformal field theories and supergravity. Int. J. Theor. Phys. **38**(4), 1113–1133 (1999). https://doi.org/10.1023/A:1026654312961
16. Medved, A.J.M.: How not to construct an asymptotically de sitter universe. Classical Quant. Gravity **19**(17), 4511 (2002). https://doi.org/10.1088/0264-9381/19/17/303
17. Osudin, D., Child, C., He, Y.-H.: Rendering non-Euclidean space in real-time using spherical and hyperbolic trigonometry. In: Rodrigues, J.M.F., et al. (eds.) ICCS 2019. LNCS, vol. 11540, pp. 543–550. Springer, Cham (2019). https://doi.org/10.1007/978-3-030-22750-0_49
18. Phillips, M., Gunn, C.: Visualizing hyperbolic space: unusual uses of 4x4 matrices. In: Proceedings of I3D, pp. 209–214. Association for Computing Machinery, New York, NY, USA (1992. https://doi.org/10.1145/147156.147206
19. Reynolds, C.W.: Flocks, herds and schools: a distributed behavioral model. In: Proceedings of the 14th Annual Conference on Computer Graphics and Interactive Techniques. SIGGRAPH '87, pp. 25–34. Association for Computing Machinery, New York, NY, USA (1987). https://doi.org/10.1145/37401.37406
20. Seskir, Z.C., et al.: Quantum games and interactive tools for quantum technologies outreach and education. Opt. Eng. **61**(8), 081809 (2022). https://doi.org/10.1117/1.OE.61.8.081809
21. Snyder, J.: Flattening the Earth: Two Thousand Years of Map Projections. University of Chicago Press (1997). https://books.google.pl/books?id=0UzjTJ4w9yEC
22. Squad: Kerbal space program - create and manage your own space program (2022). https://www.kerbalspaceprogram.com/. Accessed 9 Apr 2024
23. Weeks, J.: Real-time rendering in curved spaces. IEEE Comput. Graph. Appl. **22**(6), 90–99 (2002). https://doi.org/10.1109/MCG.2002.1046633
24. Weeks, J.: Geometry games (2009–2021). https://www.geometrygames.org/HyperbolicGames/. Accessed 9 Apr 2024
25. Weeks, J.: Non-Euclidean billiards in VR. In: Yackel, C., Bosch, R., Torrence, E., Fenyvesi, K. (eds.) Proceedings of Bridges 2020: Mathematics, Art, Music, Architecture, Education, Culture, pp. 1–8. Tessellations Publishing, Phoenix, Arizona (2020)

Enhancing the Realism of Wildfire Simulation Using Composite Bézier Curves

I. González[1]([⊠]) [iD], C. Carrillo[2] [iD], A. Cortés[1] [iD], and T. Margalef[1] [iD]

[1] Computer Architecture and Operating Systems Department,
Universitat Autónoma de Barcelona, Carrer de les Sitges s/n, Cerdanyola del Vallès,
Spain
{irene.gonzalez.fernandez,ana.cortes,tomas.margalef}@uab.cat
[2] MITIGA Solutions S.L., Passeig del Mare Nostrum, 15, Barcelona, Spain
carles.carrillo@mitigasolutions.com

Abstract. One of the consequences of climate change is the increase in forest fires around the world. In order to act quickly when this type of natural disaster occurs, it is important to have simulation tools that allow a better approximation of the evolution of the fire, especially in Wildland Urban Interface (WUI) areas. Most forest fire propagation simulators tend to represent the perimeter of the fire in a polygonal way, which often does not allow us to capture the real evolution of the fire in complex environments, both at the terrain and vegetation levels. In this work, we focus on Elliptical Wave Propagation (*EWP*) based simulators, which represent the perimeter of the fire with a set of points connected to each other by straight lines. When the perimeter grows and new points must be added, the interpolation method used is linear interpolation. This system generates unrealistic shapes of fires. In this work, an interpolation method leveraging *Composite Bézier Curves* (*CBC*) is proposed to generate fire evolution shapes in a more realistic way. The proposed method has been incorporated into FARSITE, a well-known *EWP*-based forest fire spread simulator. Both interpolation methods have been applied to ideal scenarios and a real case. The results show that the proposed interpolation method (*CBC*) is capable of generating more realistic fire shapes and, in addition, enables the simulator the ability to better simulate the spread of fire in WUI zones.

Keywords: Interpolation · Forest Fire perimeter · Bezier curve

1 Introduction

The role of computational science in addressing environmental challenges such as wildfires, is increasingly recognised as part of the broader pursuit of sustainability. Wildfires are increasingly recognised as a major environmental and societal challenge. The frequency and intensity of these events have risen notably in recent years, leading to significant ecological, economic, and social impacts.

Climate change, marked by rising temperatures and changing precipitation patterns, has exacerbated the conditions that lead to wildfires [8,10,13]. Effective wildfire management, facilitated by advanced simulation tools, contributes significantly to the safety and well-being of societies, as well as to the protection and conservation of natural ecosystems. The accuracy and realism of fire behaviour simulations are crucial for this effective wildfire management. Traditional simulation methods, while useful, often lack the fine scale detail necessary to capture the complex nature of wildfire spread. This gap in simulation fidelity can lead to challenges in predicting fire behaviour, especially in heterogeneous landscapes with variable fuel and weather conditions.

In response to this challenge, numerous mathematical models and simulators have been developed in the last decades [6,7], which are broadly categorized based on their spread strategy into three types: *Cellular Automata (CA)*, *Elliptical Wave Propagation (EWP)* and *Level Set Method (LSM)*. Fire spread in the *CA* models is performed based on a grid of cells, where the state of each cell could be either burned or unburned [1]. *EWP* models treat the fire perimeter as a discretized curve (set of points) offering detailed and dynamic representations of fire spread [2]. Finally, *LSM* employs the Hamilton-Jacobi equation to define the fire front implicitly through a level-set function [4,11].

Focusing on the realism of the simulations provided by the simulators that use *CA* and *EWP* as fire front propagation strategies, a relevant issue can be found, the polygonal shape of the results. Both, *CA* and *EWP* methods tend to generate fire perimeters with polygonal shapes instead of curved shapes. This issue is not so notable in *LSM* since this approach does not discretize the fire front. The reason for this behavior in the case of *CA*-based simulators, is the use of cells as a propagation unit, while in *EWP*-based models, the main problem lies in the interpolation method used. As the forest fire evolves, new points must be added to the representation of the fire perimeter to keep the resolution of the simulation limited by a predetermined value. This point addition is done through linear interpolation, which generates straight shapes instead of smooth curves.

The main objective of this work is to emulate the dynamic and curved characteristics inherent in the spread of a forest fire by applying *Composite Bézier Curves*. This concept has been widely applied in the area of computing graphics but, in this paper, we are transferring its applicability to a completely different research field. The proposed methodology uses an interpolation technique that strategically introduces points on the front obtained from the composition of the generated curves. The resulting addition of points gracefully articulates a more realistic depiction of the fire's boundary. What distinguishes this approach from traditional polygonal methods is that it allows to capture the authentic, non-polygonal behavior exhibited by a wildfire. Furthermore, the proposed methodology does not imply extra computing time since it has been designed to have the same complexity as current linear interpolations ($O(n)$). In order to analyze the behaviour of this proposal when simulating the behaviour of real forest fires in complex scenarios, we have used FARSITE as a simulation framework. FARSITE has been chosen for being the most widely used forest fire simulator that incorporates the *EWP* spread method. The proposed interpolation method has been codified in FARSITE changing its linear interpolation method to the proposed one but keeping the rest of the code intact. FARSITE has been used to simulate the evolution of a real wildfire using both interpolation methods separately. The results show that the proposed method not only provides more realistic forest fire perimeters, but also enables the simulator the ability to spread the fire through areas that would not otherwise be reached.

The rest of the paper is organized as follows. In Sect. 2 a basic description of FARSITE is introduced. Section 3 includes the description of the proposed *CBC* interpolation method. The experimental study is reported in Sect. 4 and, finally, Sect. 5 summarises the main conclusions of this work.

2   Forest Fire Spread Modelling

Modelling and understanding the behaviour of wildfire is a complex process that involves a lot of different fields (physics, forestry, chemistry, etc.). However, with in special conditions (flat terrain, no wind and homogeneous fuel), the propagation of the fire front can be simplified by a circle [12]. Other well known special (ideal) cases are those with either constant slope or, flat terrain with constant wind speed and direction. Assuming these conditions, the propagation of the fire front describes an ellipse [3]. This behaviour is shown in Fig. 1 where the fire evolution in controlled laboratory experiments are depicted [12].

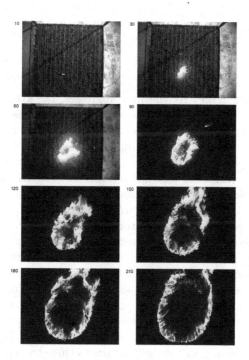

Fig. 1. Elliptical behaviour observed in laboratory experiments [12].

FARSITE is a widely employed simulator that adopts *Elliptical Wave Propagation* as its primary method for modelling fire spread across landscapes. Combining this propagation scheme with the Huygens' principle, the fundamental framework of FARSITE's methodology is obtained. FARSITE works iteratively by modifying the location of the fire front in time steps of preset duration. In each iteration, FARSITE dynamically updates the fire front by strategically placing points along elliptical waves. This adaptability allows the simulator to navigate diverse landscapes, capturing the nuanced path and intensity of the spreading fire. The Huygens' principle, a cornerstone of FARSITE's approach, involves points along the elliptical wavefront acting as sources of secondary waves. This collective influence shapes the evolving wavefront, enhancing the precision of fire spread predictions. Figure 2 shows a basic scheme of how the *EWP* method is used to spread an initial fire front to an evolved fire perimeter.

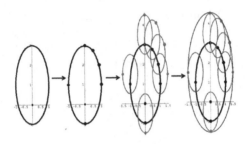

Fig. 2. Basic scheme of the *Elliptical Wave Propagation* (*EWP*) approach.

However, despite its robust iterative dynamics, FARSITE encounters challenges related to the *EWP*. As the fire progresses, the gradual drift of points necessitate careful consideration. Maintaining simulation accuracy and resolution becomes crucial, additionally, the autonomy in the spread of each point contributes to a lack of knowledge regarding previously burned areas. The decentralized nature of point propagation creates a potential gap in understanding the fire's history, necessitating innovative solutions to overcome this information limitation and enhance the overall accuracy of the simulation.

In facing the lack of knowledge regarding previously burned area, FARSITE introduces the usage of a normal vector to the perimeter, which is crucial in guiding the fire spread. The normal vector of a given point is computed based on its surrounding points and aligned with the existing momentum. Figure 3 illustrates schematically how the normal vectors for two different points (grey and red coloured) are obtained from its neighbouring points. To evaluate the direction of the normal vector for both points, the perpendicular direction of the segment that joins the corresponding two neighbouring points is used. Therefore, the location of the two neighbours of a given point have a direct impact in the normal vector direction. Later we will return to this characteristic since, as we will see, it is a relevant point in the proposal of this work.

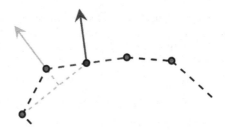

Fig. 3. Normal vector for two points (grey and red) when interpolation has been done linearly. (Color figure online)

To address spatial accuracy, FARSITE introduces a rediscretization process to counteract the gradual point density drift. FARSITE incorporates the `perimeter_resolution` (PR) parameter that defines the maximum distance between two consecutive points within the fire front, determining when a new point should be added to maintain simulation resolution limited by the PR value. The method employed by FARSITE to introduce new points along the elliptical wavefront during the rediscretization stage is linear interpolation. This technique adds a new perimeter point at the midpoint along the segment connecting two existing points on the wavefront when required. While this rediscretization strategy effectively addresses spatial accuracy concerns, it introduces a challenge related to the Huygens' principle. As new points are inserted within the area enclosed by the elliptical wavefront, the smoothness or curvature of the wavefront is disrupted. This disruption not only impacts the smoothness of the wavefront but also has implications for the computation of the normal vector for the neighbour points, so this disruption propagates throughout the iterations. Later on, in the section devoted to explain the experimental study carried out in this work, a deep analysis of this issue is done.

3   Methodology

In this section, a detailed exploration of the mathematical intricacies behind *Composite Bézier Curves* (*CBC* for short) is introduced. The main emphasis lies not only in understanding their application but also in presenting an efficient computational approach. The proposed *CBC* method exhibits a complexity of $O(n)$, ensuring that the computational time aligns with the basic linear interpolation akin to what FARSITE uses. This strategic approach allows us to maintain computational efficiency while enhancing the simulation's fidelity in capturing the nuanced behavior of a forest fire through *Bézier* curves.

3.1   Composite Bézier Curves

The foundation of the *CBC* method lies in the composition of Bézier curves. The Bézier curves under consideration are cubic Bézier curves, defined based on four points, as it is illustrated in Fig. 4. Two of these points (P_0 and P_1) act as the starting and ending points of the curve, while the other two (σ_0 and ρ_0) serve as anchor points that influence the curve's shape and directionality.

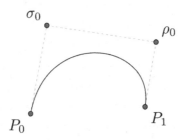

Fig. 4. Representation of a Cubic Bézier Curve.

The Cubic Bézier curve (B) depicted in Fig. 4 has the following generic formula:

$$B(t) = (1 - t)^3 \cdot P_0 + 3t(1 - t)^2 \cdot \sigma_0 + 3t^2(1 - t) \cdot \rho_0 + t^3 \cdot P_1$$
$$0 \leq t \leq 1 \in \mathbf{R} \tag{1}$$

In the context of *Composite Bézier Curves*, each curve is defined by a pair of points from the fire perimeter - marking the start and end - coupled with two anchor points yet to be determined. The task at hand involves determining the anchor points, a process we will elucidate later in this section. With as many curves as there are points on the fire perimeter, the general formulation for each Bézier curve (Γ_i) describing the perimeter is expressed through Eq. (2).

$$\Gamma_i(t) = (1 - t)^3 \cdot P_i + 3t(1 - t)^2 \cdot \sigma_i + 3t^2(1 - t) \cdot \rho_i + t^3 \cdot P_{i+1}$$
$$i \in \{0, \ldots, n - 1\}, \ 0 \leq t \leq 1 \in \mathbf{R} \tag{2}$$

In Eq. (2), σ_i and ρ_i represent the anchor points to be determined, P_i denotes the i-th point on the perimeter, P_{i+1} is the following point clockwise for all curves, and n is the number of points in the perimeter. As the perimeter forms a closed shape, the point following the last perimeter point seamlessly connects to the first point of the perimeter, so P_n is P_0.

The challenge is to find the σ_i and ρ_i for each curve in a way that leaves us with integrated curves and smooth composition. As we currently have $2n$ unknowns factors, we need to produce $2n$ equations to obtain a compatible determined system. Furthermore, since we want the composition to be differentiable for smoothness, we propose it to be twice differentiable to obtain the $2n$ equations. This differentiability condition is expressed in Eqs. (3) and (4).

$$\Gamma_i'(1) = \Gamma_{i+1}'(0), \ i \in \{0, \ldots, n - 1\} \tag{3}$$

$$\Gamma_i''(1) = \Gamma_{i+1}''(0), \quad i \in \{0, \dots, n-1\} \tag{4}$$

In Eqs. (3) and (4), it should be noted that since we are working with a circular system, Γ_n is Γ_0, as it will happen elsewhere in our methodology.

Using the definition of $\Gamma_i(t)$ from Eq. (2), we can calculate its first and second derivatives. By using these definitions and Eqs. (3) and (4), we can derive Eqs. (5) and (6).

$$\Gamma_i'(1) = \Gamma_{i+1}'(0) \iff \sigma_{i+1} + \rho_i = 2P_{i+1} \tag{5}$$

$$\Gamma_i''(1) = \Gamma_{i+1}''(0) \iff \sigma_i + 2\sigma_{i+1} = 2\rho_i + \rho_{i+1} \tag{6}$$

Therefore, we obtain the system of equations in Eq. (7) with $2n$ unknowns and $2n$ equations.

$$\begin{cases} \sigma_{i+1} + \rho_i = 2P_{i+1}, & i \in \{0, \dots, n-1\} \\ \sigma_i + 2\sigma_{i+1} = 2\rho_i + \rho_{i+1}, & i \in \{0, \dots, n-1\} \end{cases} \tag{7}$$

Applying the substitution method, we obtain a system containing only n unknowns, which are the σ_i with $i \in \{0, \dots, n-1\}$, and with n equations.

$$\{ \sigma_{i-1} + 4\sigma i + \sigma_{i+1} = 2(2P_i + P_{i+1}), \quad i \in \{0, \dots, n-1\} \tag{8}$$

It is essential to note that this system is a circular tridiagonal system, and for this reason, it can be expressed in the following generic form of Eq. (9). We express it in this generic form because the resolution methodology presented below is expressed using this form, as it can be used for all circular tridiagonal systems.

$$\{ b_i \cdot \sigma_{i-1} + a_i \cdot \sigma i + c_i \cdot \sigma_{i+1} = r_i, \quad i \in \{0, \dots, n-1\} \tag{9}$$

Based on the methodology of [9], which uses Gaussian elimination and a iterative reparameterization, we obtain an equivalent system. The reparametrization used starts with the initial parameters in Eq. (10),

$$\begin{cases} \alpha_0 = \dfrac{-1}{a_0} \\ \beta_0 = b_1 \alpha_0 \\ \delta_0 = b_0 \\ \epsilon_0 = -c_{n-1}\alpha_0 \\ y_0 = r_0 \end{cases} \tag{10}$$

Then, we compute the iterative parameters, based on the previous ones, in Eq. (11).

$$\begin{cases} \alpha_i = \dfrac{-1}{a_i + \beta_{i-1}c_{i-1}} \\ \beta_i = b_{i+1}\alpha_i \\ \delta_i = \beta_{i-1}\delta_{i-1} \qquad\qquad i \in \{1,\ldots,n-2\} \\ \epsilon_i = \epsilon_{i-1}c_{i-1}\alpha_i \\ y_i = r_i + y_{i-1}\beta_{i-1} \end{cases} \tag{11}$$

Finally, we calculate the final parameters of Eq. (12).

$$\begin{cases} \alpha_{n-1} = a_{n-1} + \beta_{n-2}(\gamma_{n-2} + c_{n-2}) - \epsilon_{n-2}c_{n-2} - \displaystyle\sum_{j=0}^{n-2} \epsilon_j \gamma_j \\ y_{n-1} = r_{n-1} + \beta_{n-2}y_{n-2} - \displaystyle\sum_{j=0}^{n-2} \epsilon_j y_j \end{cases} \tag{12}$$

Therefore, we obtain the following system of equations equivalent to Eq. (9) using the described reparametrization:

$$\begin{cases} \sigma i + -c_i\alpha_i \cdot \sigma_{i+1} - \gamma_i\alpha_i\sigma_{n-1} = -y_i\alpha_i, \qquad i \in \{0,\ldots,n-3\} \\ \sigma_{n-2} - \alpha_{n-2}(c_{n-2} + \gamma_{n-2})\sigma_{n-1} = -y_{n-2}\alpha_{n-2} \\ \alpha_{n-1}\sigma_{n-1} = y_{n-1} \end{cases} \tag{13}$$

Iterative starting from σ_{n-1}, we can solve the system, and the solution is given by Eq. (14).

$$\begin{cases} \sigma_{n-1} = \dfrac{y_{n-1}}{\alpha_{n-1}} \\ \sigma_i = (c_i\sigma_{i+1} - y_i + \gamma_i\sigma_{n-1})\alpha_i \qquad i \in \{n-2,\ldots,0\} \end{cases} \tag{14}$$

Finally, using Eq. (5), we also obtain the variables ρ_i. After solving the system and obtaining the variables σ_i and ρ_i, we have determined the curves that constitute the *Composite Bézier Curves*. These curves smoothly and cohesively describe and follow the fire perimeter, accurately capturing the dynamics and directionality of the fire spread.

To exemplify how this methodology works, a toy example is used. Figure 5a shows a set of points (grey points) that represent a certain fire perimeter just before the rediscretization process starts. The points have been represented by joining them with a line just to clarify the shape they form. Figure 5b depicts the two set of points that represent the fire front once the two interpolation methods have been applied. More precisely, the red points (red perimeter) correspond to the points obtained when using FARSITE's linear interpolation, whereas the green points (green perimeter) are the points obtained with the proposed *CBC* interpolation method. As it can be observed, the red perimeter exhibits a polygonal shape, meanwhile the green perimeter draws a smoother curve shape. Furthermore, if we observe the framed area in Fig. 5b in more detail, we can see that

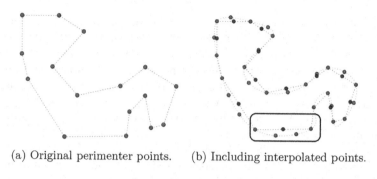

(a) Original perimenter points.   (b) Including interpolated points.

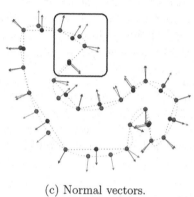

(c) Normal vectors.

Fig. 5. Original perimeter points before any interpolation (grey) (a). Perimenter including interpolated points, the green ones corresponds to the interpolation done with the proposed method *CBC* and the red ones correspond to the points obtained applying the linear interpolation (b). Normal vectors of all perimeter points (c). (Color figure online)

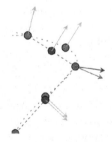

Fig. 6. Normal vectors for interpolated points (linear interpolation in red and *CBC* method in green) and original perimeter points before interpolation in grey (Color figure online)

the interpolation process, in the case of the standard method, only introduces one point (red) between the two points of the original perimeter (grey points), while the *CBC* method adds two points (green). This example highlights that the interpolation used by FARSITE only adds one point in the center of the segment that joins the perimeter points between which it is detected that points need to be added. This process is carried out in this way even if the distance between points exceeds PR. However, the proposed CBC interpolation adds to the perimeter all required extra points to keep the perimeter resolution bounded by PR. Figure 5c shows the normal vectors for all points. Although at first glance it seems that the normal vectors for the two interpolation methods coincide, if we zoom in the framed area of Fig. 5c (Fig. 6), we can see that there is a slight difference between them. This difference is more pronounced in the grey points, that is, in the perimeter points before the rediscretization stage (interpolation process) starts. As it has been previously mentioned, since the two interpolation methods add points to the perimeter at different locations, the resulting neighbouring points for a given grey point could be very different. Figure 6 depicts this situation. In this figure, the two normal vector obtained when the interpolation method used is the linear one (red points) and the CBC method is applied (green point) are shown. As it can be observed, the obtained normal vectors, especially for the grey point, are quite different since the two neighbouring points in each case are in locations significantly different. This difference will propagate at each simulation step leading to relevant differences at the end of the complete simulations process. This issue will be later on analysed in this work, in the discussion of the real case.

4   Experimental Study and Results

In this section, the experimental study carried out is reported. In order to analyze the behaviour of the proposed method, we have done two kind of experiments. On the one hand, ideal scenarios where most of the environmental conditions are quite controlled have been simulated to determine the effectiveness of the proposed interpolation method against the current basic approach. Being able to control variables such as wind and terrain allows the study to focus on the behavior of fire spread. On the other hand, a real wildfire scenario has been simulated to test the CBC methodology under complex scenarios. The application of these methods to a real wildfire event is crucial for assessing their practical utility and accuracy in replicating complex fire dynamics. This comparison aims to reveal not only how each method performs in theory but also their effectiveness in actual wildfire situations. In addition, the real wildfire event introduces complexities absent in ideal scenarios, presenting an opportunity to assess the practical utility of the CBC method under dynamic and heterogeneous conditions.

4.1   Ideal Cases

To test the effectiveness of the proposed rediscretization method within the context of the FARSITE forest fire spread simulator, a set of experiments based on ideal cases have been designed. Ideal cases are those that the relevant environment conditions (slope, wind and vegetation) are controlled and constant.

That is, for example, the basic ideal case consists of simulating the spread of a forest fire in a completely flat terrain, with homogeneous vegetation and no wind. Under this conditions, it is well known that the evolution of the fire develops in concentric circles [5, 12]. In particular, the results shown in this section corresponds to this ideal basic case, however, other ideal cases have been tested such as flat terrain and constant wind, terrain with a constant slope with and without constant wind and so on. The results of all these experiments exhibit similar behaviour to the basic one, therefore, we have chosen this case to compare the linear interpolation method used by FARSITE with running FARSITE including the proposed *CBC* interpolation method.

FARSITE exhibits the capability to maintain a circle-like fire shape when simulating the basic ideal case. However, as it has been previously mentioned, the current rediscretization method included in FARSITE adds new points at the midpoint between two neighbouring points when required. This method could locally alter the elliptical or circular shape of the fire perimeter turning it into a polygonal shape. To show such a behaviour, the ideal case has been simulated using, on the one hand, FARSITE with its standard linear interpolation method and, on the other hand, FARSITE has been executed by changing the linear interpolation to the proposed *CBC* interpolation method. Figure 7a shows the simulations results when using linear interpolations, meanwhile Fig. 7b depicts the evolution of the fire when using *CBC* interpolation approach. The circular fire spread pattern, inherent to FARSITE's underlying principles, are faithfully reproduced by both methods. Although at first glance, the results obtained when simulating the basic ideal case seem identical, if we analyze both figures in more detail a subtle difference can be seen as it can be observed in Fig. 7c. The *CBC* method introduces refinements, optimizing the representation of circular fire spread generating smoother perimeters. These results corroborate that the new interpolation methodology is capable of reproducing the circular evolution of the spread of a fire in an ideal scenario. Furthermore, it is capable of eliminating the polygonal appearance of the perimeters generated by FARSITE with its basic interpolation implementation.

4.2   Real Case

This section is devoted to study the results obtained in terms of realism of the fire perimeter, when applying both interpolation methods, the original scheme included in FARSITE and the proposal method (*CBC*) in a real wildfire. The study case corresponds to a forest fire that took place in *Pont de Vilomara* in Catalonia (north-east of Spain). This forest fire started on July 17, 2022, at 13:04 and it was completely controlled on July 18 at 00:20. It rapidly burnt around 14 hectares during the first 15 min, reaching 100 hectares in less than one hour. The final burned area was 1,697 ha. This final area is shown in Fig. 8 with the orange shape. The yellow shape corresponds to one intermediate perimeter, which has been used as the initial perimeter for the experiments reported in this section. This intermediate perimeter corresponds to the area burned by the fire until July 17, 2022 at 16:30. This scenario has been chosen for being characterized by

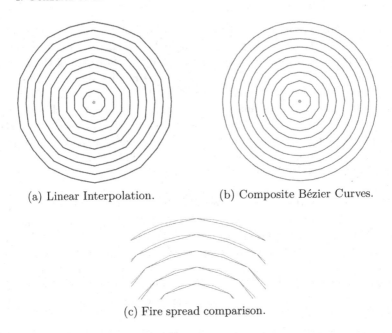

(a) Linear Interpolation.   (b) Composite Bézier Curves.

(c) Fire spread comparison.

Fig. 7. Basic ideal case propagation results when linear interpolation (a) and *CBC* interpolation (b) are used in FARSITE. Comparison of the results obtained by both methods (c).

intricate terrain and a complex Wildland-Urban Interface (WUI). This scenario allows to introduce complexities that were absent in the ideal settings, aiming to replicate scenarios frequently encountered in wildfire management, where urban and wildland areas intersect, contributing to heterogeneous landscapes. Figure 9 shows the simulations results in terms of area burned when executing FARSITE including its basic linear interpolation scheme (Fig. 9a) and including the *CBC* method (Fig. 9b). As it can be observed, the *CBC* method exhibits a better capacity to represent the behavior of wildfires in a more accurate manner, particularly in areas featuring a Wildland-Urban Interface (WUI) and intricate terrain, like the framed area of the fire. FARSITE method tends to underestimate fire spread in those contexts, emphasizing the limitations of conventional linear interpolation in capturing the nuances of complex fire dynamics. The framed area within Fig. 9 is characterised for its terrain complexity. The curve perimeter description allows the *CBC* method to recognize accelerated fire propagation, providing a more accurate representation of fire behavior in this area. On the contrary, linear interpolation method fails to capture this nuanced behavior, resulting in a discrepancy in the representation of fire spread.

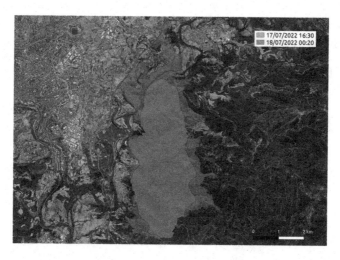

Fig. 8. Initial perimeter (yellow shape) an final burned area (orange shape) of the *Pont de Vilomara* fire. (Color figure online)

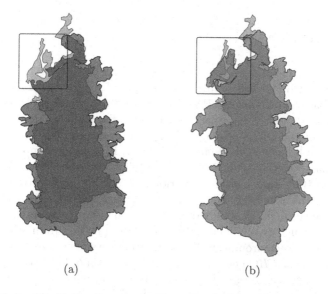

(a)   (b)

Fig. 9. Final simulated burned area using linear interpolation (red shape) (a) and final simulated burned area using the proposed *CBS* method (green shape) (b). The grey shape corresponds to the final real burned area (Color figure online)

The detailed visual analysis of this framed area (Fig. 10) pinpoints specific areas where the *CBC* method excels during the rediscretization stage. The impact of the *CBC* method on the normal vector that describes the inertial direction of the fire becomes evident. The curve perimeter description afforded by *CBC* enables the simulator to discern faster fire propagation in strategic areas, a subtlety not effectively captured by the original method. This knowledge, introduced during rediscretization, propagates between iterations, illustrating the significant influence of the *CBC* method on normal vectors and subsequent fire spread.

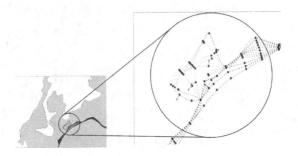

Fig. 10. Detailed aspects of the perimeter evolution when applying linear interpolation (red) and the *CBC* method (green) (Color figure online)

5   Conclusions

In this work, an alternative interpolation method for rediscretizing the front of a wildfire in *EWP*-based simulators is proposed. In particular, the proposed approach is based on *Composite Bézier Curves* (*CBC*), which captures more precisely the smooth and circle aspect of the forest fire shape than classical linear interpolation method. The proposed interpolation method has been included in FARSITE, a well-known *EWP*-based simulator, in order to compare its performance compared to the original linear interpolation scheme. The *CBC* method generates fire front that exhibits superior realism compared to the traditional FARSITE interpolation method. This improvement arises for the *CBC* method capacity in capturing intricate fire dynamics, with a particular emphasis on challenging terrains such as Wildland-Urban Interfaces (*WUIs*). In these areas, where urban and wildland environments intersect, accurate simulations are paramount. The adaptability of *CBC* ensures a nuanced depiction of fire behavior, allowing for more effective decision-making in wildfire management and mitigation. The importance of this precision is underscored in *WUIs*, where complex landscapes demand accurate simulations to develop strategies that safeguard both human settlements and natural ecosystems. The successful integration of *CBC* marks a significant step forward in addressing the unique challenges posed by intricate terrains, providing a valuable tool for enhancing the realism and efficacy of wildfire simulations in complex environments.

Acknowledgments. This work has been granted by the Spanish Ministry of Science and Innovation MCIN AEI/10.13039/501100011033 under contracts PID2020-113614RB-C21 and CPP2021-008762 and by the European Union-*NextGenerationEU*/PRTR. It also has been partially granted by the Catalan Government under grant 2021-SGR-574. We would like to thank *Bombers de la Generalitat de Catalunya* and *Agencia Estatal de Meteorología* (AEMET) of Spain for providing valuable data on fire evolution and meteorological data of the *Pont de Vilomara* fire.

References

1. Avolio, M.V., Di Gregorio, S., Lupiano, V., Trunfio, G.A.: Simulation of wildfire spread using cellular automata with randomized local sources. In: Sirakoulis, G.C., Bandini, S. (eds.) ACRI 2012. LNCS, vol. 7495, pp. 279–288. Springer, Heidelberg (2012). https://doi.org/10.1007/978-3-642-33350-7_29

2. Finney, M.A.: FARSITE: fire area simulator-model development and evaluation. U.S. Department of Agriculture, Forest Service, Rocky Mountain Research Station (1998). https://doi.org/10.2737/rmrs-rp-4

3. Liu, N.: Wildland surface fire spread: mechanism transformation and behavior transition. Fire Saf. J. **141**, 103974 (2023). https://doi.org/10.1016/j.firesaf.2023.103974

4. Mandel, J., et al.: Recent advances and applications of WRF-SFIRE. Nat. Hazards Earth Syst. Sci. **14**(10), 2829–2845 (2014). https://doi.org/10.5194/nhess-14-2829-2014

5. Meng, Q., Lu, H., Huai, Y., Xu, H., Yang, S.: Forest fire spread simulation and fire extinguishing visualization research. Forests **14**(7) (2023). https://doi.org/10.3390/f14071371

6. Or, D., et al.: Review of wildfire modeling considering effects on land surfaces. Earth-Sci. Rev. **245**, 104569 (2023). https://doi.org/10.1016/j.earscirev.2023.104569

7. Pastor, E., Zárate, L., Planas, E., Arnaldos, J.: Mathematical models and calculation systems for the study of wildland fire behaviour. Prog. Energy Combust. Sci. **29**(2), 139–153 (2003). https://doi.org/10.1016/S0360-1285(03)00017-0

8. Pyne, S.: Fire: A Brief History. Cycle of fire, University of Washington Press (2019). https://books.google.es/books?id=FPu8xAEACAAJ

9. Reuter, R.: Solving (cyclic) tridiagonal systems. SIGAPL APL Quote Quad **18**(3), 6-12 (1988). https://doi.org/10.1145/44164.44165

10. Senande-Rivera, M., Insua-Costa, D., Miguez-Macho, G.: Spatial and temporal expansion of global wildland fire activity in response to climate change. Nat. Commun. **13**(1), 1208 (2022). https://doi.org/10.1038/s41467-022-28835-2

11. Sullivan, A.L.: Wildland surface fire spread modelling, 1990–2007. 3: simulation and mathematical analogue models. Int. J. Wildland Fire **18**(4), 387 (2009). https://doi.org/10.1071/wf06144

12. Viegas, D.X.: Slope and wind effects on fire propagation. Int. J. Wildland Fire **13**(2), 143–156 (2004). https://doi.org/10.1071/WF03046

13. Westerling, A.L., Hidalgo, H.G., Cayan, D.R., Swetnam, T.W.: Warming and earlier spring increase Western U.S. forest wildfire activity. Science **313**(5789), 940–943 (2006). https://doi.org/10.1126/science.1128834

Learning Mesh Geometry Prediction

Filip Hácha[(⊠)] [ID] and Libor Váša [ID]

Department of Computer Science and Engineering, Faculty of Applied Sciences,
University of West Bohemia, Univerzitní 8, 301 00 Plzeň, Czech Republic
{hachaf,lvasa}@kiv.zcu.cz

Abstract. We propose a single-rate method for geometry compression of triangle meshes based on using a neural predictor to predict the encoded vertex positions using connectivity and an already known part of the geometry. The method is based on standard traversal-based methods but uses a neural predictor for prediction instead of a hand-crafted prediction scheme. The parameters of the neural predictor are learned on a dataset of existing triangle meshes. The method additionally includes an estimate of the prediction uncertainty, which is used to guide the encoding traversal of the mesh. The results of the proposed method are compared with a benchmark method on the ABC dataset using both mechanistic and perceptual metrics.

Keywords: Mesh compression · Computer graphics · Deep learning

1 Introduction

The compression of triangle meshes is a very often solved problem in the field of computer graphics. Triangle meshes are a widely used representation of shapes, mainly due to their efficient rendering, simple structure, and good representation capabilities. In order to accurately represent even complex shapes using triangle meshes, triangle meshes reach a large number of vertices that carry information about the geometry of the shape. As the number of vertices increases, it is advisable to compress the triangle meshes to achieve compactness of representation.

A triangle mesh is the set of vertices, edges, and triangle faces $\mathcal{M} = (V, E, F)$, where the set of edges can be derived from the set of faces, and the mesh can thus only be described by the set of vertices and faces $\mathcal{M} = (V, F)$. The vertices of the mesh carry information about the *geometry* of the surface since they have a position in space. The triangle faces form the *connectivity* of the mesh and tell which triplets of vertices are to be connected into the resulting triangles.

Today, there is a large number of algorithms for compressing the geometry of triangle meshes that achieve a good ratio of bitrate to distortion. A significant group of these algorithms is connectivity-driven approaches that use the information contained in the connectivity to reduce the bitrate since the connectivity itself partially indicates the mesh geometry. These approaches also exploit the fact that, as the mesh is incrementally decoded, the decoder can use the already

L. Franco et al. (Eds.): ICCS 2024, LNCS 14832, pp. 166–180, 2024.
https://doi.org/10.1007/978-3-031-63749-0_12

decoded part of the geometry to predict the rest of it. How the decoder performs such a prediction is determined by the prediction scheme of the method. One of the simplest ways to predict a decoded vertex is the *parallelogram* prediction [21]. However, there are also more sophisticated methods that use, in addition to the vertex positions of adjacent triangles, e.g., vertex valence or already decoded interior angles of a triangle, such as the *weighted parallelogram* [22].

In this paper, we propose a prediction scheme based on encoded vertex prediction using a neural network that, similar to conventional prediction schemes, performs prediction based on an already decoded part of the geometry. We exploit the neural network's ability to better predict the position of the encoded vertex across different meshes. We compare our results in terms of rate-distortion ratio with the *weighted parallelogram* as the reference method using mechanistic and perceptual mesh quality assessment metrics.

2   Related Work

In the domain of compression single triangle meshes, we typically consider lossy compression, where the vertex coordinates are encoded with distortion while the connectivity is encoded losslessly. There is also a number of mesh-simplification approaches that allow for a lossy compression of connectivity. However, we do not consider these methods in this paper. When compressing a mesh, we can choose whether to encode the connectivity or the geometry first. Although there are methods that encode the mesh geometry first and then use its knowledge to encode the connectivity more efficiently [7,15], most methods encode the connectivity first and then the geometry.

One of the simplest approaches for geometry compression is based on predicting the position of the encoded vertex using parallelogram prediction, proposed by Touma and Gotsman [21] in 1998. When encoding a vertex $\mathbf{X}$ that incidents with an already encoded base triangle $\triangle\mathbf{BLR}$ over an edge $(\mathbf{L}, \mathbf{R})$ (gate), we can predict its position using Eq. 1 as shown in Fig. 1. Then, instead of encoding the coordinates of the vertex $\mathbf{X}$, only the correction vector $\mathbf{X} - \mathbf{X}_{pred}$ is encoded. This correction is typically further quantized, where the resolution of the quantization affects the distortion of the decoded mesh, and then it is compressed using an entropy encoder such as an arithmetic encoder.

$$\mathbf{X}_{pred} = \mathbf{L} + \mathbf{R} - \mathbf{B} \tag{1}$$

Also, in 1998, Taubin and Rossignac [20] proposed a compression method *topological surgery*. This method works with a spanning tree of the connectivity of triangle mesh. Based on the spanning tree, the encoded vertices of the mesh are then predicted from 2, 3, or 4 of its ancestors in the tree using linear extrapolation.

The connectivity of a triangle mesh can be efficiently encoded based on the encoding of the valences of vertices [1,21]. Due to the non-uniform probability for different vertex valences, it is possible to use shorter *words* to compress more

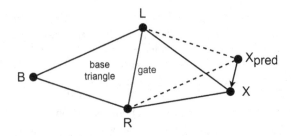

Fig. 1. Diagram showing parallelogram prediction of the vertex.

probable valences and longer *words* to compress less probable valences. Rossignac proposed a method for connectivity compression called *Edgebreaker* [17], which is based on compressing a sequence of *CLERS* symbols that are formed by sequentially traversing the mesh and encoding the different ways in which an encoded vertex can be attached to an already decoded part of the mesh.

Geometry prediction was improved by Lee et al. with the Angle-Analyzer method [13], which encodes the vertex position using the interior angles of the triangle and the dihedral angle. This work also considers a priority-based mesh traversal, in which the priority of the decoded triangle is controlled based on the already decoded angles. Sim et al. [18] proposed an extension of the parallelogram prediction to dual parallelogram prediction, which predicts the position of an encoded vertex by making a prediction based on each pair of successive triangles in a triangle fan. Gumhold and Amjoun [8] proposed a prediction scheme based on separating the tangential and normal components of the correction. Their method first encodes the tangential coordinates of the vertex, which are predicted by the parallelogram. Subsequently, a higher-order surface is fitted to predict the normal component.

Kälberer et al. proposed *FreeLence*, a connectivity coding method based on free valences [12]. Free valence codes benefit from a geometry-guided mesh traversal. Mesh traversal is guided by the so-called *opening angle*, which is given by the difference of 2π and the sum of the known angles around a given vertex. This information is also used to separate the encoded symbols into different contexts, which exploits the fact that each free valence has its own distribution of correction values. Courbet and Hudelot [6] derived a modified set of weights for various linear geometry predictors using Taylor expansion of the mesh geometry function. Váša and Brunnett [22] proposed *weighted parallelogram*, which uses a parallelogram predictor with weights, which are derived based on known angles that are already decoded and the valences of vertices.

In contrast to the above-mentioned single-rate compression methods, Pajarola and Rossignac [16] proposed a progressive coding method. This method first transmits a simplified coarse version of the original mesh and then encodes the vertex splits into batches to refine the coarse mesh progressively. The displacement of a new vertex that is created during vertex splitting is predicted based on the Butterfly subdivision scheme, which further leads to a reduction in the data rate. Cohen and Irony proposed another progressive method [5]. They also use the idea of averaging multiple parallelogram predictions to obtain a more accurate estimate of the position of the encoded vertex.

In addition to methods based on a prediction scheme that uses mesh connectivity and the decoded part of the geometry, there are also methods based on spectral analysis, such as Karni and Gotsman [9], which use a projection of the mesh geometry onto an orthonormal basis derived from the mesh topology. This approach offers the possibility of separating low and high-frequency signals and encoding them with different precision, given that different frequencies of distortion are perceived with different intensities.

Another popular geometry compression approach is the encoding of delta-coordinates [4, 19, 23], which are obtained as a result of discrete Laplace operators of the geometry function. For a more detailed look into the compression of triangle meshes, we refer the reader to the surveys of Alliez and Gotsman [2] or Maglo et al. [14].

When comparing algorithms for triangle mesh compression, it is essential how the distortion caused by lossy geometry compression is measured. For this purpose, quality assessment metrics are used. Mechanistic metrics, such as *Mean Squared Error* (MSE) of the vertices, Hausdorff distance, or *Average Absolute Surface Distance* (AvgD), are typically easy to compute. However, mechanistic metrics depend on properties such as the distance between points of a surface and often do not correlate well with the human perception of the distortion. In contrast, perceptual metrics such as *Dihedral Angle Mesh Error* (DAME) [24] or *Fast Mesh Perceptual Distance* [25] (FMPD) describe the distortion better in the context of human perception. These metrics typically depend on features such as curvature, roughness, or dihedral angles.

3   Compression Method

Our compression method, as well as many other connectivity-first compression methods, uses the *Edgebreaker* method [17] to compress the connectivity of a triangle mesh as a sequence of *CLERS* symbols. The connectivity of the triangle mesh is encoded first and can thus be used to remove redundancy in geometry encoding.

An improvement in compression ratio over other single-rate state-of-the-art methods, primarily the *Weighted parallelogram*, is achieved by our method at the geometry compression level. We use an approach similar to *FreeLence* or *Weighted parallelogram*. The method is based on predicting the position of the encoded vertex. This then makes it unnecessary to encode the positions themselves; only the corrections, i.e., the difference between the actual and predicted vertex positions, are encoded. Similarly to the above-mentioned methods, our method uses the base triangle together with information about vertex valences and already decoded angles to generate the vertex position prediction. While the prediction of the weighted parallelogram is given by an explicit equation, we learn this prediction through a neural network (further denoted as a neural predictor). As a result, among other things, we are able to incorporate additional features that would be difficult to include in the *weighted parallelogram*. The acquired correction vectors are then quantized and further compressed using an arithmetic encoder.

3.1 Input Features

While training the neural network, it is necessary to properly describe the current configuration, which is mainly determined by the positions of the vertices of the base triangle but also by other available information about the already decoded geometry around the gate (edge over which a new vertex is encoded). For the purposes of the neural predictor, it is convenient that this data is suitably normalized in terms of its range. It is also desirable that the resulting feature vectors are invariant to transformations that should not, in principle, affect the shape of the predicted triangle. These are primarily rigid transformations (translation and rotation), but likewise, the description of the gate should be invariant to uniform scaling, which we would expect to affect the size of the predicted triangle in the same way, but not its shape.

A natural way of constructing feature vectors representing the shape of the base triangle, which is also offered, given the other information used, is the inner angles of the triangle. We can describe the shape of a base triangle using any two inner angles. We choose the pair of angles α and β that lie next to the gate. The shape of the predicted triangle can also be described by the two inner angles γ and δ and the dihedral angle of the base and predicted triangle ω (see Fig. 2). Given these five features, it is possible to reconstruct the shape of both triangles.

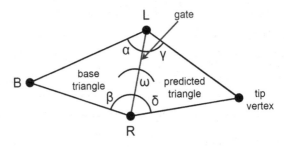

Fig. 2. Diagram showing features describing the base and the predicted triangles.

In addition to the shape of the base of the triangle described by the inner angles, the feature vectors contain information about the valence of the vertices **B, L, R** and **X**. Since the mesh connectivity is encoded before the geometry, this information is also available to the decoder side at the moment of computing the prediction. In order to normalize the vertex valences n_i of the i-th vertex, we convert them to the relative angle ξ_i from the total angle 2π of a given vertex, as shown in Eq. 2.

$$\xi_i = \frac{2\pi}{n_i} \tag{2}$$

Another information used by the neural predictor is the angle estimation based on the already decoded angles around the vertices **B**, **L** and **R**. The feature vector contains the proportion ϵ_i of the difference between the angle 2π and the sum of the k already decoded angles φ_j to the remaining number of angles that have not yet been decoded, as shown in Eq. 3.

$$\epsilon_i = \frac{2\pi - \sum_{j=1}^{k} \varphi_k}{n_i - k} \tag{3}$$

To allow the neural predictor to work with the normal component of the prediction by predicting the dihedral angle ω, the feature vector also contains information about the curvature of the surface around the base triangle. This information is represented by three angles. During mesh traversal, the vertex normal is estimated at the mesh vertices as the average of the normals of the adjacent triangles. These vertex normals $\mathbf{n}_B$, $\mathbf{n}_L$, $\mathbf{n}_R$ of the base, left and right vertices are then compared with the normal of the base triangle. At each prediction, the angle between vertex normal and triangle normal is determined. We denote these angles as *normal angles* $(\kappa_B, \kappa_L, \kappa_R)$.

As a result, the neural predictor uses feature vectors $\mathbf{x} \in \mathbb{R}^{2+4+3+2}$ as input and predicts a triplet of angles based on which the predicted vertex $\mathbf{y} \in \mathbb{R}^3$ is reconstructed. The predictor neural network is then described by the function $\mathbf{X}_{\mathrm{pred}} : \mathbb{R}^{2+4+3+2} \to \mathbb{R}^3$. Figure 3 shows a diagram with an outline of the normalization of the input features and the computation of the coded correction.

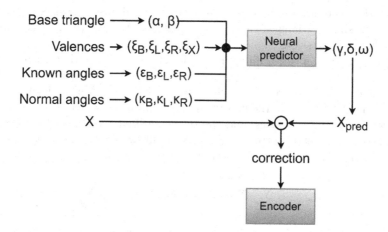

Fig. 3. Diagram showing an outline of the normalization of the input features and the computation of the coded correction.

3.2 Optimization

The neural network that represents our neural predictor is a fully connected feedforward network with 4 layers and 256 units in hidden layers. For the hidden layers, we use the *ReLU* activation function, and the output layer contains a modified hyperbolic tangent activation function such that the range of values matches the interval $(0, 2\pi)$ (see Eq. 4) given by the extreme values of dihedral angle.

$$\sigma(x) = \pi \cdot (\tanh(x) + 1) \tag{4}$$

The optimization algorithm *Adam* [10] with learning rate $lr = 1 \cdot 10^{-4}$ is used to optimize the weights of the neural network. As a loss function for training the neural network, we use the *Mean Absolute Error* between the predicted triplet of angles $(\gamma_{\text{pred}}, \delta_{\text{pred}}, \omega_{\text{pred}})$ and the true triplet of angles (γ, δ, ω) from the encoded triangle (see Eq. 5).

$$\mathcal{L} = \frac{1}{n} \sum_{i=1}^{n} (|\gamma_{\text{pred}} - \gamma| + |\delta_{\text{pred}} - \delta| + |\omega_{\text{pred}} - \omega|) \tag{5}$$

The neural predictor is trained on a dataset acquired from a wide range of triangle meshes. The dataset is divided into the training and test sets. The training set is used to optimize the weights of the neural network. Part of the training dataset is used as a validation set for the early stopping of the optimization process. The test set is used to evaluate the rate-distortion measures. The test set data are separated from the rest at the full-mesh level, so that data from the meshes that were used by the neural network during learning are not used to measure the accuracy of the neural predictor. The dataset consists of pairs of base and predicted triangles that are found during the traversal of the meshes during their compression.

For the validation set, a different loss function is used during learning than for the training set. The validation set is not evaluated using the Mean Absolute Error between the predicted and true angles but the Mean Absolute Error between the coordinates of the predicted vertex position, which is reconstructed based on the predicted angles, and the true coordinates of the encoded vertex. This loss function better corresponds to the resulting corrections that are finally encoded during the actual mesh compression (see Eq. 6).

$$\mathcal{L}_{\text{val}} = \frac{1}{n} \sum_{i=1}^{n} \|\mathbf{X}_{\text{pred}}(\gamma_{\text{pred}}, \delta_{\text{pred}}, \omega_{\text{pred}}) - \mathbf{X}\|_1 \tag{6}$$

3.3 Uncertainty Prediction

In many connectivity-driven geometry compression methods, the order of encoded vertices is arbitrarily determined by the connectivity encoding method used, such as *Edgebreaker*. Since the connectivity and geometry decoding can be done separately, there is no obstacle to choosing a different order of vertex encoding that will lead to an improved bitrate.

We propose an uncertainty-driven traversal in which the order of the encoded vertices is given by the estimated uncertainty of the prediction of the position of the encoded vertex. The goal of this approach is first to encode the vertices that are more likely to achieve a more accurate prediction and, therefore, necessitate the encoding of shorter correction vectors. In addition, a single vertex can generally be encoded over different edges (gates) with varying degrees of uncertainty. In this way, we are partially able to arrange shorter correction vectors before longer ones in the output stream and also encode vertices over more convenient gates.

Since the magnitude of the correction vectors is not normalized, it is necessary to normalize them before training the neural network. The magnitude of the correction vector can be expected to increase with the size of the triangle. Therefore, we divide the magnitudes of the correction vectors by the square root of the base triangle surface area. In turn, when the uncertainty is evaluated during the encoding process, the output of the neural network is then multiplied by that value:

$$e = \frac{\|\mathbf{X} - \mathbf{X}_{\text{pred}}\|}{\frac{1}{2}\|(\mathbf{L} - \mathbf{B}) \times (\mathbf{R} - \mathbf{B})\|} \tag{7}$$

In order to be able to estimate the uncertainty of the coded vertex prediction, we train a separate neural network $u : \mathbb{R}^7 \to \mathbb{R}$ that uses the same feature vector that is used for the vertex position prediction itself to predict the uncertainty, which is correlated with the size of the correction vector. To optimize this neural network, we use Concordance Correlation Loss [3], which is used to optimize the correlation between ground-truth values and predicted values. It is defined as $1 -$ Concordance Correlation Coefficient as shown in the following Equation:

$$\mathcal{L}_{\text{unc}} = 1 - \frac{2\rho_{eu}\sigma_e\sigma_u}{\sigma_e^2 + \sigma_u^2 + (\mu_e - \mu_u)^2}, \tag{8}$$

where e is the relative error in the prediction, u is estimated uncertainty, ρ_{eu} denotes Pearson correlation coefficient, σ is the standard deviation, and μ is a mean value. The structure of the uncertainty estimation neural network is the same as the structure of the vertex position prediction network, except that it contains a different activation function at the output, specifically the square function.

For training the neural network estimating uncertainty, we use the same data as for training the neural predictor. However, outliers are excluded from this data. The filtering of outliers is done based on the Z-score of relative error e, which is defined as the distance from the mean divided by the standard deviation. For our purposes, we only keep samples with Z-scores less than 3. This helps to provide a more stable learning process, as degenerated triangles (having zero area) and nearly degenerated ones that are close to singular are removed in this step.

$$e_{\text{train}} = \left\{ e : \frac{e - \mu_e}{\sigma_e} < 3 \right\}_{i=1}^{N} \qquad (9)$$

When using uncertainty estimation, the encoding traversal is controlled by a priority queue that contains the gates and their corresponding uncertainty estimations. In each iteration, the gate with the minimum uncertainty is dequeued, a new vertex is encoded, and then the uncertainty of the newly created gates is estimated and enqueued.

4   Experimental Results

To evaluate the proposed neural predictor, we compare our method with the implementation of the *weighted parallelogram* [22], a state-of-the-art single-rate method for connectivity-driven compression of triangle meshes. We use the *Mean Squared Error* (MSE) of mesh vertices, a commonly used metric for evaluating mesh compression algorithms that do not modify mesh connectivity, to measure mesh geometry distortion. In addition, we also use *Dihedral Angle Mesh Error* (DAME) [24], which is a perceptual metric that better correlates with how humans perceive the distortion of the compressed mesh.

The proposed compression method was tested in a scenario of a general geometry prediction, where the neural network of the predictor is trained for a general triangle mesh. The weights of the neural network are then part of the encoder implementation itself, and therefore, the encoder and decoder have these weights available without having to be transferred through the stream with the encoded mesh. For geometry prediction experiments, meshes from the ABC dataset [11] were used. Since both *weighted parallelogram*, as a reference method, and our proposed method require the input mesh to be orientable 2-manifold, meshes that the reference implementation could not handle were filtered out from the dataset.

We evaluated the rate-distortion function on the ABC dataset using the two metrics mentioned above on 100 meshes that were not used in learning the neural predictor, which was trained using 385 meshes. Figure 4 contains a representative example of RD curves comparing the neural predictor and *weighted parallelogram*. This graph shows the reduction of the data rate of the geometry using the proposed method over a large portion of the bitrate interval. To compare the distortion of different compression methods at the same bitrate, an RD-curve with sufficiently dense sampling was computed for each mesh, and the desired bitrate was found by interpolating the RD-curve. These statistics were measured for all mesh test sets containing 100 meshes, and the data rate was compared with that of the *weighted parallelogram*. Figure 5 contains the average relative improvement in bitrate of the neural predictor over the *weighted parallelogram* over the entire test dataset. These experiments show that the proposed method provides a better rate-distortion ratio for most of the test data.

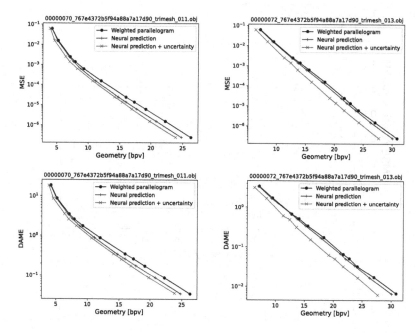

Fig. 4. Representative comparison of RD curve *weighted parallelogram* and neural predictor on ABC dataset. The x-axis shows the bitrate of the compressed geometry (excluding connectivity), and the y-axis contains the bias of the compressed mesh as measured by the chosen metric. Top: MSE. Bottom: DAME.

To verify the contribution of the normal component prediction of the encoded vertex, which is represented by the dihedral angle, we compared the performance of the same neural predictor with and without the normal component. This experiment was evaluated across the entire test set of the ABC dataset. Figure 6 contains a plot of the relative distortion of the coded mesh versus the *weighted parallelogram* measured using both MSE and DAME. These results show that it is advantageous to include the normal prediction since without the prediction of the normal component, no better results were obtained with respect to using *weighted parallelogram*.

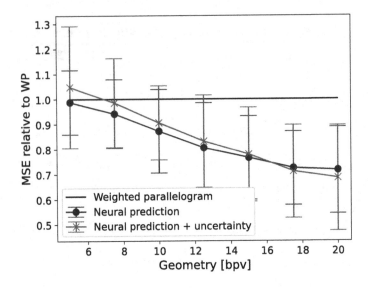

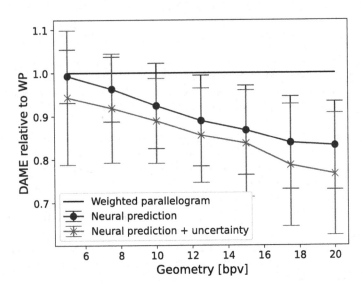

Fig. 5. Comparison of the ratio of bitrate and distortion of compressed meshes of the proposed method relative to *weighted parallelogram*. The x-axis shows the bitrate of the compressed geometry (excluding connectivity), and the y-axis contains relative distortion to the *weighted parallelogram* measured by the chosen metric. Relative distortion smaller than one means lower distortion than the reference method. The chart contains the average relative distortion values across the test dataset, and the length of the bar corresponds to the variance of the relative distortion.

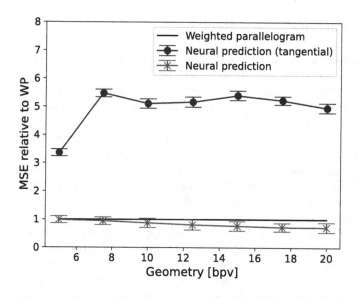

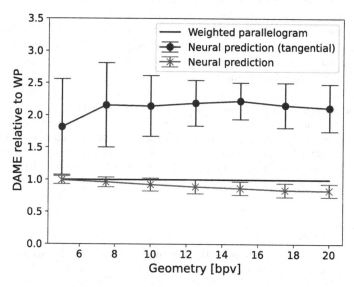

Fig. 6. Comparison of the relative bias of the neural predictor to WP using normal component prediction (green) and tangential component only prediction (blue). Top: MSE. Bottom: DAME. (Color figure online)

5 Conclusions

In this paper, a new single-rate method for geometry compression of triangle meshes was proposed. The method uses an artificial neural network to predict the mesh geometry based on its connectivity and the already decoded part of the geometry around the vertex. The method uses principles similar to some existing state-of-the-art methods, such as parallelogram prediction, prediction based on the valence of mesh vertices, and estimation of interior angles of triangles.

The performed experiments show that the proposed method is able to achieve a better ratio between bitrate and mesh distortion on a large dataset of static meshes than the tested state-of-the-art method, using both mechanistic and perceptual metrics to assess mesh quality.

Although static mesh compression is already a well-researched area, it can be seen that using modern machine learning approaches, such as neural networks, can improve the existing state-of-the-art methods. In future work, it is possible to try to improve the neural predictor further. It might be interesting to try to incorporate other known local geometric properties of the compressed mesh into the feature vectors of the neural predictor. At the same time, it is worth exploring whether some of the global features could be used as latent code that would be constant for all predictions within a single mesh, further helping to improve the accuracy of the predictor. Global features describing meshes could be particularly useful in compressing larger sets of meshes with similar properties, such as sequences of triangle meshes with varying connectivity.

Acknowledgement. The authors have no competing interests to declare that are relevant to the content of this article. This work was supported by the project 23-04622L, Data compression paradigm based on omitting self-evident information - COMPROMISE, of the Czech Science Foundation. Filip Hácha was partially supported by the University specific research project SGS-2022-015, New Methods for Medical, Spatial and Communication Data.

References

1. Alliez, P., Desbrun, M.: Valence-driven connectivity encoding for 3D meshes. Comput. Graph. Forum **20**(3), 480–489 (2001). https://doi.org/10.1111/1467-8659.00541
2. Alliez, P., Gotsman, C.: Recent advances in compression of 3d meshes. In: Dodgson, N.A., Floater, M.S., Sabin, M.A. (eds.) Advances in Multiresolution for Geometric Modelling. Mathematics and Visualization, pp. 3–26. Springer, Heidelberg (2005). https://doi.org/10.1007/3-540-26808-1_1
3. Atmaja, B.T., Akagi, M.: Evaluation of error- and correlation-based loss functions for multitask learning dimensional speech emotion recognition. J. Phys. Conf. Ser. **1896**(1), 012004 (2021). https://doi.org/10.1088/1742-6596/1896/1/012004
4. Chen, D., Cohen-Or, D., Sorkine, O., Toledo, S.: Algebraic analysis of high-pass quantization. ACM Trans. Graph. **24**(4), 1259–1282 (2005). https://doi.org/10.1145/1095878.1095880
5. Cohen, R., Irony, R.: Multi-way geometry encoding. Trans. Comput. Sci. (2002)

6. Courbet, C., Hudelot, C.: Taylor prediction for mesh geometry compression. Comput. Graph. Forum **30**, 139–151 (2011). https://doi.org/10.1111/j.1467-8659.2010.01838.x
7. Dvořák, J., Káčereková, Z., Vaněček, P., Váša, L.: Priority-based encoding of triangle mesh connectivity for a known geometry. Comput. Graph. Forum **42**(1), 60–71 (2023). https://doi.org/10.1111/cgf.14719
8. Gumhold, S., Amjoun, R.: Higher order prediction for geometry compression. In: 2003 Shape Modeling International, pp. 59–66 (2003). https://doi.org/10.1109/SMI.2003.1199602
9. Karni, Z., Gotsman, C.: Spectral compression of mesh geometry. In: Proceedings of the 27th Annual Conference on Computer Graphics and Interactive Techniques. SIGGRAPH '00, pp. 279–286. ACM Press/Addison-Wesley Publishing Co., USA (2000). https://doi.org/10.1145/344779.344924
10. Kingma, D., Ba, J.: Adam: a method for stochastic optimization. In: International Conference on Learning Representations (2014)
11. Koch, S., et al.: ABC: a big cad model dataset for geometric deep learning. In: The IEEE Conference on Computer Vision and Pattern Recognition (CVPR) (2019)
12. Kälberer, F., Polthier, K., Reitebuch, U., Wardetzky, M.: Freelence - coding with free valences. Comput. Graph. Forum **24**(3), 469–478 (2005). https://doi.org/10.1111/j.1467-8659.2005.00872.x
13. Lee, H., Alliez, P., Desbrun, M.: Angle-analyzer: a triangle-quad mesh codec. Comput. Graph. Forum **21**(3), 383–392 (2002). https://doi.org/10.1111/1467-8659.t01-1-00598
14. Maglo, A., Lavoué, G., Dupont, F., Hudelot, C.: 3D mesh compression: survey, comparisons, and emerging trends. ACM Comput. Surv. **47**(3) (2015). https://doi.org/10.1145/2693443
15. Marais, P., Gain, J., Shreiner, D.: Distance-ranked connectivity compression of triangle meshes. Comput. Graph. Forum **26**(4), 813–823 (2007). https://doi.org/10.1111/j.1467-8659.2007.01026.x
16. Pajarola, R., Rossignac, J.: Compressed progressive meshes. IEEE Trans. Vis. Comput. Graph. **6**(1), 79–93 (2000). https://doi.org/10.1109/2945.841122
17. Rossignac, J.: Edgebreaker: connectivity compression for triangle meshes. IEEE Trans. Vis. Comput. Graph. **5**(1), 47–61 (1999). https://doi.org/10.1109/2945.764870
18. Sim, J.Y., Kim, C.S., Lee, S.U.: An efficient 3D mesh compression technique based on triangle fan structure. Sig. Process. Image Commun. **18**(1), 17–32 (2003). https://doi.org/10.1016/S0923-5965(02)00090-5
19. Sorkine, O., Cohen-Or, D., Toledo, S.: High-pass quantization for mesh encoding. In: Kobbelt, L., Schroeder, P., Hoppe, H. (eds.) Eurographics Symposium on Geometry Processing. The Eurographics Association (2003). https://doi.org/10.2312/SGP/SGP03/042-051
20. Taubin, G., Rossignac, J.: Geometric compression through topological surgery. ACM Trans. Graph. **17**(2), 84–115 (1998). https://doi.org/10.1145/274363.274365
21. Touma, C., Gotsman, C.: Triangle mesh compression. In: Proceedings of the Graphics Interface 1998 Conference, 18–20 June 1998, Vancouver, BC, Canada, pp. 26–34 (1998). http://graphicsinterface.org/wp-content/uploads/gi1998-4.pdf
22. Vasa, L., Brunnett, G.: Exploiting connectivity to improve the tangential part of geometry prediction. IEEE Trans. Visual. Comput. Graph. **19**(09), 1467–1475 (2013). https://doi.org/10.1109/TVCG.2013.22
23. Váša, L., Dvořák, J.: Error propagation control in Laplacian mesh compression. Comput. Graph. Forum **37**(5), 61–70 (2018). https://doi.org/10.1111/cgf.13491

24. Váša, L., Rus, J.: Dihedral angle mesh error: a fast perception correlated distortion measure for fixed connectivity triangle meshes. Comput. Graph. Forum **31**(5), 1715–1724 (2012). https://doi.org/10.1111/j.1467-8659.2012.03176.x

25. Wang, K., Torkhani, F., Montanvert, A.: A fast roughness-based approach to the assessment of 3d mesh visual quality. Comput. Graph. **36**(7), 808–818 (2012). https://doi.org/10.1016/j.cag.2012.06.004. Augmented Reality Computer Graphics in China

Data-Efficient Knowledge Distillation with Teacher Assistant-Based Dynamic Objective Alignment

Yangyan Xu[1,2], Cong Cao[1], Fangfang Yuan[1(✉)], Rongxin Mi[3(✉)],
Dakui Wang[1], Yanbing Liu[1,2], and Majing Su[4]

[1] Institute of Information Engineering, Chinese Academy of Sciences, Beijing, China
{xuyangyan,caocong,yuanfangfang,wangdakui,liuyanbing}@iie.ac.cn
[2] School of Cyber Security, University of Chinese Academy of Sciences,
Beijing, China
[3] National Computer Network Emergency Response Technical Team/Coordination
Center of China, Beijing, China
mirongxin@cert.org.cn
[4] The 6th Research Institute of China Electronic Corporations, Beijing, China
sumj@ncse.com.cn

Abstract. Pre-trained language models encounter a bottleneck in production due to their high computational cost. Model compression methods have emerged as critical technologies for overcoming this bottleneck. As a popular compression method, knowledge distillation transfers knowledge from a large (teacher) model to a small (student) one. However, existing methods perform distillation on the entire data, which easily leads to repetitive learning for the student. Furthermore, the capacity gap between the teacher and student hinders knowledge transfer. To address these issues, we propose the Data-efficient Knowledge Distillation (DeKD) with teacher assistant-based dynamic objective alignment, which empowers the student to dynamically adjust the learning process. Specifically, we first design an entropy-based strategy to select informative instances at the data level, which can reduce the learning from the mastered instances for the student. Next, we introduce the teacher assistant as an auxiliary model for the student at the model level to mitigate the degradation of distillation performance. Finally, we further develop the mechanism of dynamically aligning intermediate representations of the teacher to ensure effective knowledge transfer at the objective level. Extensive experiments on the benchmark datasets show that our method outperforms the state-of-the-art methods.

Keywords: Pre-trained language model · Model compression · Knowledge distillation

1 Introduction

Large-scale pre-trained language models (PLMs), such as BERT [1], XLNet [24], RoBERTa [5], T5 [9], and GPT-4 [8], have reached very competitive performance

ⓒ The Author(s), under exclusive license to Springer Nature Switzerland AG 2024
L. Franco et al. (Eds.): ICCS 2024, LNCS 14832, pp. 181–195, 2024.
https://doi.org/10.1007/978-3-031-63749-0_13

and simply require fine-tuning of downstream natural language processing (NLP) tasks [4,23]. However, PLMs require large computational resources for huge amounts of model parameters, which leads to overloaded GPU usage and slow inference speeds in real-world production. To reduce computation and carbon footprint, knowledge distillation (KD) [2] has emerged as an effective method to compress large models into small ones and has gradually become the most popular choice among various compression methods.

Table 1. The performance comparison between BERT-base and BERT-large under different numbers of layers and varying data conditions.

Teacher	Layer	SST-2	QNLI
		Acc	Acc
BERT-base	3-layer (50%)	85.44	82.24
	3-layer	85.55	82.46
	12-layer	92.55	91.32
BERT-large	3-layer (50%)	83.25	78.94
	3-layer	83.71	79.22
	24-layer	93.00	92.66

The core concept of KD is based on the teacher-student learning framework, in which the teacher transfers knowledge to the student via soft targets. Existing KD methods [3,6,12,13,15,21] mainly focus on transferring knowledge from the teacher model to the student model in the form of single or multiple teacher models. However, these methods have two major drawbacks: (1) They do not take into account the student's mastery of knowledge during the distillation process, so the student continues to learn instances that contain repeated information. (2) They ignore the capability gap between the small student and the large teacher, which degrades the distillation performance. For example, as shown in Table 1, part of the data can produce the similar distillation performance as all the data. This is because the student model can gain important knowledge from a portion of the informative data. Furthermore, the 3-layer student model distilled from the stronger teacher model is weaker than the same student model distilled from the weaker teacher model on the same tasks. Generally speaking, BERT-large performs better than BERT-base on the SST-2 and QNLI tasks, but a stronger teacher model does not always lead to a better student model. The reason is that the competency of the small student model cannot match that of the large teacher model, which weakens distillation performance.

Based on the above insights, in this paper, we propose the Data-efficient Knowledge Distillation (DeKD) with teacher assistant-based dynamic objective alignment, which promotes knowledge transfer from the teacher model to the student model and improves the distillation performance as the competency of the student evolves. On the one hand, as distillation progresses, the student's learning on a downstream task is gradually deepened, and examples that the student has already learned should be eliminated, which inspires us to investigate which data is more important to distillation. On the other hand, the capacity gap between the weak student and the strong teacher motivates us to overcome the limitations of the student. We strive to answer the following research questions: (RQ1) Which data is actually useful for the student model during the distillation process? (RQ2) How to consider the evolution of the student model to realize efficient distillation? Specifically, we first choose representative instances to learn based on entropy to maximize data efficiency and prevent the student from repeating learning at the data level. Then, we introduce a teacher assistant model at the model level, which allows the student to decide whether to query the teacher or the teacher assistant for enhancing the performance of KD. Moreover, at the objective level, we further design a dynamic objective alignment strategy that aligns the informative layers to alleviate the objective supervision problem between the large teacher and the small student. We conduct extensive experiments on several benchmark datasets to validate the effectiveness of our method. Experimental results clearly show that our DeKD significantly boosts the performance of the student model.

As a summary, the contributions of this paper are threefold:

- We are the first to consider efficient KD from the data, model, and objective levels, which is critical but overlooked by existing KD methods.
- We choose informative instances based on the prediction entropy of the student to achieve a competitive performance on part of the data. Meanwhile, we introduce the teacher assistant model and dynamic supervision alignment to improve the performance of the student as it evolves.
- We conduct extensive experiments on the benchmark datasets, and the results demonstrate that our method outperforms the state-of-the-art distillation methods.

2   Related Work

Knowledge distillation [2, 22] aims to compress the knowledge of a large and computationally complex model into a simple and computationally efficient model. The KD approach has been widely used in the compression of pre-trained language models. Existing KD methods for compressing large-scale language models are divided into general distillation and task-specific distillation.

General distillation refers to conducting KD on the universal text corpus. For example, DistilBERT [10] presents a method for pre-training a smaller general-purpose language representation model, which can subsequently be fine-tuned to perform well on a variety of tasks. PD [14] demonstrates that pre-training is still crucial in the setting of smaller architectures, and that fine-tuning pre-trained compact models may compete with more complicated strategies suggested in concurrent work. In addition, the large Transformer-based pre-trained models can be compressed using a straightforward method called deep self-attention distillation, which is presented in MiniLM [19]. The self-attention module of the large model (teacher), which is crucial to Transformer networks, is deeply imitated to train the small model (student). In the subsequent work, MiniLMv2 [18] uses the self-attention relation distillation to generalize and streamline the deep self-attention distillation of MiniLM [19]. The above studies of general distillation require extra training time and computational resources, and they are not applicable in resource-limited scenarios.

Instead, task-specific distillation trains the student model on specific downstream tasks. In particular, BERT-PKD [13] encourages the student model to extract knowledge from the intermediate layers of the teacher model, rather than just learning parameters from the last layer of the teacher model. Recently, the idea of combining the knowledge from models with different capacities has been explored [6,12,15]. Besides, MUKI [3] broadens the concept of KD from mimicking teachers to integrating teacher knowledge, and proposes Knowledge Integration (KI) for PLMs. KI attempts to train a flexible student capable of making predictions over the union of teacher label sets given multiple fine-tuned teacher-PLMs, each of which is capable of conducting classification over a unique label set. Nevertheless, the effectiveness of distilling the knowledge from a large language model into a small one has not yet been well studied. The phenomenon of the student repeating learning and the gap in capacity between the large-scale teacher and the compact student still exist.

In this study, we focus on the task-specific distillation, which is widely used in practice. Compared to previous KD approaches, we further investigate efficient KD and consider the competency evolution of the student model to comprehensively improve the distillation effect.

3   Methodology

We propose the DeKD framework, and Fig. 1 depicts its general design. Firstly, we choose informative data using an entropy-based approach to prevent repeated learning of the student model. Then, we introduce the teacher assistant to make the student model match the competency of the teacher model, thus alleviating the ability gap to boost the distillation performance. In addition, objective alignment can also bring additional performance improvement. The specific implementation strategy is to select the informative layers from the teacher and then let our student align with the teacher's hidden representations of these layers.

3.1   Preliminary

The goal of knowledge distillation is to train the student model S not just using the information supplied by true labels but also by studying how the teacher model T represents and interacts with the data.

KD [2] uses the teacher's model outputs, for instance, as a soft learning target for the student. We represent $S(x)$ and $T(x)$ as the output logit vectors of the student and the teacher for input x, respectively. Then, the Kullback-Leibler

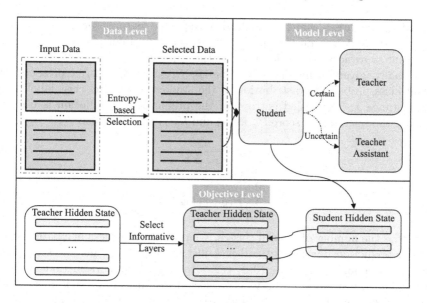

Fig. 1. Our proposed DeKD framework.

(KL) divergence loss between the student and teacher output is calculated as follows:

$$\mathcal{L}_{KL} = \mathrm{KL}(\phi(S(x)/\tau) \| \phi(T(x)/\tau)), \qquad (1)$$

where $\phi(\cdot)$ refers to the softmax function, and τ is included as a temperature hyperparameter to provide additional control over signal softening from the output of the teacher model.

The distillation loss and the original classification loss (i.e., the cross-entropy loss) over the ground-truth label y are used to update the parameters of the student:

$$\mathcal{L}_{CE} = -y \log \phi(S(x)), \qquad (2)$$

$$\mathcal{L} = (1 - \lambda)\mathcal{L}_{CE} + \lambda\mathcal{L}_{KL}, \qquad (3)$$

where λ is the hyperparameter that regulates the trade-off between the two losses. It should be noted that training of the original distillation is performed indiscriminately on all instances based on the given objectives and the weights corresponding to different objectives. However, it is unreasonable to ignore the evolution of the student model during the training process. This motivates us to explore an efficient distillation framework from three aspects: data, model, and objective, which will improve the learning efficiency of the student model.

3.2 Data Selection (DS)

In response to the first research question, we explore which data is more beneficial to the performance of the student model. The student becomes stronger as the distillation progresses, which easily leads to repeated learning for those instances that the student has mastered. Therefore, selecting informative instances is important to avoid repeated learning by the student.

Formally, given N instances in a batch, $P(y|x) = \phi(S(x))$ represents the output class probability distribution of the student across the class label y for each instance x. The scaled entropy of the probability distribution is used to compute the uncertainty score U_x for x, and U_x is calculated as:

$$U_x = \frac{u_x}{\log |Y|}, \tag{4}$$

$$u_x = -\sum_{y=1}^{|Y|} P(y \mid x) \log P(y \mid x), \tag{5}$$

where Y is the number of labeled classes. We rank the instances in a batch based on their prediction uncertainty and pick just the top $N \times r$ instances to query the teacher model. Here, $r \in (0, 1]$ refers to the selection ratio that controls the number of instances to query. The selected instances have high uncertainty scores, indicating that they are informative instances that the student should learn from.

3.3   Teacher Assistant (TA)

In this part, we respond to the second research question from the model level. The capacity gap between the teacher and the student is an inherent issue. Our solution is to introduce the teacher assistant model and dynamically query the teacher or the teacher assistant according to the evolution of the student's competency during the training process. The core idea behind this is to empower the student to adjust the learning process based on its current state.

We assume that the student can rely on the teacher assistant during the initial training stage and turn to the teacher for more accurate supervision signals as the student becomes stronger. More specifically, we sort the instances in a batch in accordance with the prediction confidence of the student model. The confidence C_x is measured by entropy, as:

$$C_x = Entropy(\phi(S(x))). \tag{6}$$

Here, the higher the confidence C_x is, the more uncertain the student is. Therefore, we can evenly divide instances into the certain and uncertain ones for the student. For the certain part, the student learns the supervision signals from the teacher, while the teacher assistant provides soft labels for the instances about which the student is uncertain. The loss function is determined as:

$$\mathcal{L}_{TA} = \mathcal{L}_{KL}^T + \mathcal{L}_{KL}^A, \tag{7}$$

where $\mathcal{L}_{KL}^T$ refers to the KL divergence distance between the student and the teacher, and $\mathcal{L}_{KL}^A$ denotes the KL divergence distance between the student and the teacher assistant.

Algorithm 1. Training of the Student Model

Input: Training data $\mathcal{D}$, number of epochs E, set of parameters Ω needed to train the student model

Output: An optimized student

1: **for** *epoch* $e = 1$ to E **do**
2: **for** *each batch* $\mathcal{D}_b \in \mathcal{D}$ **do**
3:   Select informative instances via the uncertainty score: $U_x = \frac{u_x}{\log |Y|}$.
4:   Divide the instances into two parts: the certainty part and the uncertainty part by confidence: $C_x = Entropy(\phi(S(x)))$.
5:   Loss $\mathcal{L}_{TA}$ becomes: $\mathcal{L}_{TA} = \mathcal{L}_{KL}^T + \mathcal{L}_{KL}^A$.
6:   Compute the entropy of the hidden representations H: $R_x = \frac{Entropy(\phi(H(x)))}{\log |Y|}$.
7:   Select the M layers via the R_x value to align the teacher.
8:   Update parameters Ω by: $\mathcal{L}_{total} = \lambda_1 \mathcal{L}_{CE} + \lambda_2 \mathcal{L}_{TA} + \lambda_3 \mathcal{L}_{DA}$.
9: **end for**
10: **end for**

3.4 Dynamic Objective Alignment (DOA)

We finally deal with the second research question from the objective level. Inspired by the previous studies [13] on the alignment of the intermediate layers between the teacher and the student, we further investigate dynamic objective alignment to boost the distillation performance. According to the previous studies, the corresponding aligned objective weights are determined by hyperparametric search and remain constant during the training. To address the aforementioned issue, we choose the informative layers based on the entropy calculation of the corresponding layer representations to dynamically align the teacher, thus preventing unnecessary alignment.

We first compute the entropy R_x of the hidden representations H of the teacher:

$$R_x = \frac{Entropy(\phi(H(x)))}{\log |Y|}, \tag{8}$$

the greater the value of R_x is, the more informative the hidden representations of this layer are. We sort the R_x of each layer from large to small and select M layers with higher R_x values. Then, the loss of dynamic objective alignment becomes

$$\mathcal{L}_{DA} = \sum_{i=1}^{M} \left\| \frac{\mathbf{h}_i^s}{\|\mathbf{h}_i^s\|_2} - \frac{\mathbf{h}_{I(j)}^t}{\left\|\mathbf{h}_{I(j)}^t\right\|_2} \right\|_2^2, \tag{9}$$

where M represents the number of layers in the student model, $I(j)$ denotes that the i-th layer of the student is aligned with the j-th layer of the teacher; $\mathbf{h}^s$ and $\mathbf{h}^t$ are hidden representations of the student and the teacher, respectively.

3.5 Total Loss

Finally, the total loss is determined as follows:

$$\mathcal{L}_{total} = \lambda_1 \mathcal{L}_{CE} + \lambda_2 \mathcal{L}_{TA} + \lambda_3 \mathcal{L}_{DA}, \tag{10}$$

where λ_1, λ_2 and λ_3 are hyperparameters for adjusting the loss weight. The overall training process of our DeKD is divided into three steps. At the beginning, we fine-tune BERT on the corresponding downstream task to get the teacher model. We then distill the teacher model to obtain the teacher assistant model. After that, we run Algorithm 1 to produce the final student model.

4 Experiments

4.1 Datasets

We conduct evaluations on eight representative text classification benchmarks. (1) We choose three different NLP tasks: Paraphrase Similarity Matching (PSM), Sentiment Classification (SC), and Natural Language Inference (NLI). For the PSM tasks, we select MRPC and QQP [16]. For the SC tasks, we test on SST-2 [16] and Emotion [11]. For the NLI tasks, we evaluate on QNLI and MNLI [16]. (2) We also add two additional text classification tasks: AG News [26] and IMDb [7]. The statistics of the datasets are shown in Table 2.

Table 2. Statistics of the datasets.

Dataset	#Train	#Dev	#Test
MRPC	3,668	408	1,725
SST-2	67,349	872	1,821
QNLI	104,743	5,463	5,463
MNLI	392,702	9,832	9,847
AG News	120,000	-	7,600
QQP	363,849	40,430	390,965
Emotion	16,000	2,000	2,000
IMDb	20,000	5,000	25,000

4.2 Baselines

We choose seven representative KD methods as baselines. Moreover, we consider these methods with 3 and 6 layers of transformers as the student models. The baselines we compare are as follows:

Vanilla KD [2]: By minimizing the original KL divergence loss, the student model is trained to emulate the soft targets created by the logits of the teacher model.

BERT-PKD [13]: To fully exploit the rich knowledge contained in the deep structure of the teacher model, the patient-KD method enables the student model to patiently learn from the teacher through a multi-layer distillation process.

DFA [12]: DFA tries to learn a compact student model capable of handling the comprehensive classification issue from multiple trained teacher models, each of which specializes in a different classification problem.

CFL [6]: CFL maps the hidden representations of the teachers into a common space. The student is trained by matching the mapped features to those of the teachers, with supplemental supervision from the logits combination.

UHC [15]: The class sets of teacher models are used by UHC to divide the student logits into subsets. Each subset is trained to mimic the output of the teacher model that corresponds to it.

MUKI [3]: Based on the estimated model uncertainty, MUKI designs a model uncertainty-aware knowledge integration framework. The golden supervision is approximated by either taking the outputs of the most confident teacher or softly integrating different teacher predictions according to their relative importance.

KSM [17]: KSM proposes an actor-critic method for selecting appropriate knowledge transfer at different training steps. This optimization considers the impact of knowledge selection on future training steps.

Table 3. The performance of the student model on the test set of the benchmark datasets. The best results from each group of student models are in bold. We also report the average performance for each task in the "AVG." column.

Method	Student	MRPC	SST-2	QNLI	MNLI	AG News	QQP	Emotion	IMDb	AVG.
BERT-base	–	88.48	92.55	91.32	83.87	94.71	90.96	93.55	89.24	90.59
Vanilla KD	3-layer	75.90	79.66	78.15	71.22	60.11	81.32	53.91	79.25	72.44
BERT-PKD	3-layer	76.55	83.53	80.71	72.31	62.17	83.81	56.56	80.13	74.47
DFA	3-layer	75.47	82.61	79.33	71.01	63.23	80.57	54.63	79.36	73.28
CFL	3-layer	76.57	83.78	81.88	73.13	60.21	84.26	59.38	80.87	75.01
UHC	3-layer	78.57	84.86	82.35	73.76	75.67	84.68	64.72	81.35	78.25
MUKI	3-layer	80.99	85.67	83.05	74.37	88.19	85.81	72.91	82.93	81.74
KSM	3-layer	81.90	85.83	83.62	74.63	90.76	86.12	89.31	83.95	84.52
DeKD (Ours)	3-layer	**83.39**	**86.01**	**84.11**	**74.82**	**93.68**	**86.57**	**92.35**	**84.24**	**85.65**
Vanilla KD	6-layer	80.21	81.37	79.86	72.33	62.58	82.37	55.71	80.33	74.35
BERT-PKD	6-layer	81.42	84.16	81.57	73.67	63.57	84.28	59.23	81.27	76.15
DFA	6-layer	81.12	83.36	80.76	72.81	65.46	82.24	58.62	80.22	75.57
CFL	6-layer	82.53	84.67	82.39	74.63	64.74	85.92	62.17	82.34	77.42
UHC	6-layer	83.72	85.74	83.62	75.67	78.19	86.71	68.16	83.66	80.68
MUKI	6-layer	84.97	86.29	85.16	77.54	90.63	87.35	79.64	84.51	84.51
KSM	6-layer	87.19	89.21	86.09	79.26	92.37	88.93	91.82	85.31	87.52
DeKD (Ours)	6-layer	**87.48**	**89.91**	**87.00**	**80.17**	**94.28**	**89.17**	**93.15**	**86.12**	**88.41**

4.3 Implementation Details

We implement our method based on the HuggingFace transformers library [20]. The results of all experiments are obtained from a single NVIDIA V100 GPU.

We first fine-tune the 12-layer BERT-base as the teacher model and distill it to obtain the 6-layer and 9-layer teacher assistant models, respectively. The 6-layer teacher assistant is configured for the 3-layer student, and the 9-layer teacher assistant is configured for the 6-layer student. Then, we fine-tune the 3-layer and 6-layer students via our method to get the final student models, respectively. In our setting, we set the number of fine-tuning epochs to 3, the batch size to 32, the learning rate to 2e-5, and the distillation temperature to 5. Meanwhile, we set the loss equilibrium coefficients λ_1 as 0.5, λ_2 as 0.3, and λ_3 as 0.2. For the data selection rate r, we set it to 0.5. For the objective alignment layers M, we set M to 3 and 6 for 3-layer and 6-layer students, respectively. All experiments are repeated five times, and we report the average results over five runs with different seeds.

4.4   Evaluation Metrics

Following prior work [1], we report the F1 score for MRPC, and we use accuracy as the evaluation metric for other tasks.

4.5   Performance Comparison

Table 3 shows the performance comparison with baselines on the text classification tasks. We draw the following observations from the table:

(1) DeKD performs the best on all the text classification tasks. The reason is that DeKD not only considers efficient learning of the student at the data level, but also further improves the distillation performance from the model and objective levels. In general, the classification accuracy of DeKD is from 0.34% to 19.44% greater than the results of the best competitor. None of these baselines, with the exception of DeKD, can simultaneously stand out across all tasks.

(2) The results of traditional KD (Vanilla KD), intermediate representations-based KD (BERT-PKD), and the distillation methods considering multiple teachers (DFA, CFL, UHC, MUKI, and KSM) are all consistent with our expectations. As the conventional distillation method does not employ any additional supervision signals or intermediate representations, it performs poorly on most tasks. However, the method of combining intermediate layer information exceeds the original distillation method. The reason is that the method based on intermediate representations encourages the student model to extract knowledge from previous layers of the teacher model rather than learning parameters from the last layer of the teacher model. In fact, this demonstrates that the student can gain incremental knowledge by learning multiple intermediate layers.

(3) Compared with KSM, the performances of DFA, CFL, UHC, and MUKI are not satisfactory. Although DFA uses additional features to align objectives, it cannot achieve better results. This is due to the instability of feature-aligned supervision, and teacher features are fine-tuned to specifically target different semantic classes. CFL can learn a multitalented and lightweight student model capable of mastering the integration knowledge of heterogeneous teachers, but it suffers from the same issue as DFA in that the supervision based on feature

alignments is unstable. The performance of UHC exceeds that of DFA and CFL, but is inferior to MUKI and KSM. This demonstrates a potential supervision conflict as UHC matches the student output independently to that of teachers, thus limiting its generalizability across datasets. To summarize, the above results demonstrate that simply combining different supervision signals is ineffective. However, our DeKD designs a more efficient framework that significantly improves the distillation performance of the student model by selecting informative data, assisting the student's evolution with the teacher assistant, and dynamically selecting alignment signals.

Table 4. Ablation study on MRPC, SST-2, and IMDb tasks.

Method	MRPC	SST-2	IMDb
DeKD	87.48	89.91	86.12
w/o DS	86.84	87.84	85.52
w/o TA	86.74	87.61	85.50
w/o DOA	86.33	87.38	85.18

4.6 Ablation Study

In order to evaluate the contributions of different parts of DeKD, we design the ablation experiments. Experimental results are shown in Table 4. Due to space constraints, we show only the results on the MRPC, SST-2, and IMDb datasets. Other results are similar, so we omit them. We design three different configurations: w/o DS, w/o TA, and w/o DOA.

w/o DS. This configuration removes the entropy-based selection strategy. Compared with DeKD, the overall performance drops without DS, implying that selecting informative instances can reduce repetitive learning caused by data redundancy. This entropy-based selection strategy is capable of making better use of limited queries.

w/o TA. Without TA, the model performance declines on the three tasks. This is because the auxiliary model, i.e., the teacher assistant, becomes a boosting factor in the evolution of the student model. Therefore, the performance of this configuration is inferior to that of DeKD, demonstrating that the addition of the teacher assistant can bridge the capacity gap between the student and the teacher.

w/o DOA. When dynamic objective alignment is not taken into account, this configuration performs worse than DeKD. This implies that learning via dynamically aligning middle representations can help the student quickly understand tasks, thus improving prediction confidence.

To summarize, DeKD is superior to the first three configurations, which shows that all the components together can improve the performance of the student model.

Table 5. The mean, standard deviation, and statistically significant T-test (p-Value) of five different runs on SST-2 and IMDb. The superscripts 1, 2, and 3 respectively denote statistically significant improvements over UHC, MUKI, and KSM.

	SST-2	Mean	Stdev
UHC	85.74, 83.61, 84.22, 82.37, 85.69	84.33	1.281
MUKI	86.29, 82.16, 85.34, 84.68, 86.12	84.92	1.495
KSM	89.21, 89.01, 88.36, 89.18, 88.27	88.81	0.408
Ours[1,2,3]	89.91, 89.88, 89.65, 89.17, 88.96	89.51	0.383
	IMDb	Mean	Stdev
UHC	83.66, 81.21, 83.38, 82.93, 82.07	82.65	0.899
MUKI	84.51, 83.23, 84.47, 83.13, 82.98	83.66	0.679
KSM	85.31, 85.26, 84.93, 85.28, 84.97	85.15	0.165
Ours[1,2,3]	86.12, 86.08, 86.02, 85.85, 85.93	86.00	0.099

4.7   Model Analysis

Variance Analysis. Taking the 6-layer student setup as an example, we carry out five experiments with different seeds and calculate their mean and standard deviation (Stdev). Moreover, we also conduct a two-sided statistically significant t-test (p-value) with a threshold of 0.05 and compare the baseline methods with our DeKD method. We report the experimental results in Table 5. As shown, our method is statistically significant compared to baselines.

Data Selection Strategies. In addition to the scaled entropy-based (SE) selection strategy, we also implement three other common strategies to compute the uncertainty score U_x for each instance x:

Random, which randomly selects $N \times r$ instances as the baseline to evaluate the effectiveness of selection strategies.

Least-Confidence (LC), which indicates the uncertainty of the model to the predicted class $\hat{y} = \arg\max_y P(y \mid x)$:

$$U_x = 1 - P(\hat{y} \mid x). \tag{11}$$

Margin, which is calculated as the margin between the first and second most probable classes, y_1^* and y_2^*:

$$U_x = P(y_1^* \mid x) - P(y_2^* \mid x). \tag{12}$$

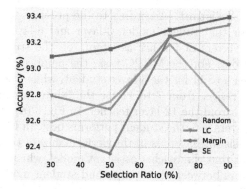

Fig. 2. The average accuracy of five experiments with four strategies under different selection ratios.

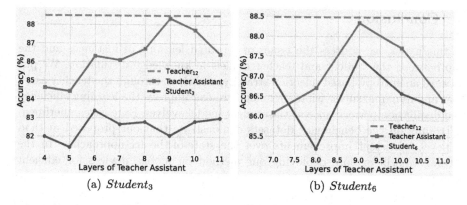

Fig. 3. The performance of the student models with 3 and 6 layers under different configurations of teacher assistants.

As shown in Fig. 2, we change the selection ratio r to check the results of different strategies on the Emotion dataset. The results on all datasets show a consistent trend, so one of them is selected for analysis.

From Fig. 2, we can observe that: (1) The selection strategy based on the entropy of student prediction can make better use of limited queries, which is better than other strategies. (2) We can use approximately 50% of the training data to achieve satisfactory performance. This shows that about 50% of the data can cover the training data well, so learning from these instances can sufficiently train the student model. It is helpful to choose informative instances for reducing repetitive learning caused by data redundancy. (3) There is a trade-off between performance and training cost, i.e., increasing the selection ratio usually improves the performance of the student model but leads to a greater training cost.

Layers of Teacher Assistant. According to the previous work [25], we make two configurations for the student model: 3-layer and 6-layer. We explore the influence of different layers of teacher assistants on the student model under these two configurations. For the MRPC task, the number of layers of teacher assistants ranges from 4 to 11 for the 3-layer student, while the number of layers of teacher assistants ranges from 7 to 11 for the 6-layer student. The teacher assistants are distilled from the 12-layer teacher. In Fig. 3 (a), when the number of teacher assistant layers is 6, the student performs best. In Fig. 3 (b), when the number of teacher assistant layers is 9, the performance of the student is at its peak. It demonstrates that the teacher assistant model, which sits in the middle of the number of layers between the teacher and student models, can solve the problem of the small student not being able to match the capacity of the large teacher.

5 Conclusion

In this paper, we address the issues of repeated learning of instances and the gap between the student and teacher models in knowledge distillation. We put forward an entropy-based selection strategy, and then, through the teacher assistant and dynamic supervision alignment, we can improve the learning efficiency and distillation performance as the student model evolves. Extensive experimental results on the benchmark datasets demonstrate that our proposed method achieves consistent improvements over the state-of-the-art approaches. In the future, we will explore deploying our method on mobile devices for efficient inference.

Acknowledgement. This work is supported by the National Key Research and Development Program of China (No. 2023YFC3303800).

References

1. Devlin, J., Chang, M., Lee, K., Toutanova, K.: BERT: pre-training of deep bidirectional transformers for language understanding. In: NAACL, pp. 4171–4186 (2019)
2. Hinton, G., Vinyals, O., Dean, J.: Distilling the knowledge in a neural network. arXiv preprint arXiv:1503.02531 (2015)
3. Li, L., et al.: From mimicking to integrating: knowledge integration for pre-trained language models. In: EMNLP, pp. 6391–6402 (2022)
4. Li, Z., Xu, X., Shen, T., Xu, C., Gu, J.C., Tao, C.: Leveraging large language models for NLG evaluation: a survey. arXiv preprint arXiv:2401.07103 (2024)
5. Liu, Y., et al.: RoBERTa: a robustly optimized BERT pretraining approach. arXiv preprint arXiv:1907.11692 (2019)
6. Luo, S., Wang, X., Fang, G., Hu, Y., Tao, D., Song, M.: Knowledge amalgamation from heterogeneous networks by common feature learning. In: IJCAI, pp. 3087–3093 (2019)
7. Maas, A.L., Daly, R.E., Pham, P.T., Huang, D., Ng, A.Y., Potts, C.: Learning word vectors for sentiment analysis. In: ACL, pp. 142–150 (2011)

8. OpenAI: GPT-4 technical report. arXiv preprint arXiv:2303.08774 (2023)

9. Raffel, C., et al.: Exploring the limits of transfer learning with a unified text-to-text transformer. J. Mach. Learn. Res. **21**, 140:1–140:67 (2020)

10. Sanh, V., Debut, L., Chaumond, J., Wolf, T.: DistilBERT, a distilled version of BERT: smaller, faster, cheaper and lighter. arXiv preprint arXiv:1910.01108 (2019)

11. Saravia, E., Liu, H.T., Huang, Y., Wu, J., Chen, Y.: CARER: contextualized affect representations for emotion recognition. In: EMNLP, pp. 3687–3697 (2018)

12. Shen, C., Wang, X., Song, J., Sun, L., Song, M.: Amalgamating knowledge towards comprehensive classification. In: AAAI, pp. 3068–3075 (2019)

13. Sun, S., Cheng, Y., Gan, Z., Liu, J.: Patient knowledge distillation for BERT model compression. In: ACL/IJCNLP, pp. 4322–4331 (2019)

14. Turc, I., Chang, M., Lee, K., Toutanova, K.: Well-read students learn better: the impact of student initialization on knowledge distillation. arXiv preprint arXiv:1908.08962 (2019)

15. Vongkulbhisal, J., Vinayavekhin, P., Scarzanella, M.V.: Unifying heterogeneous classifiers with distillation. In: CVPR, pp. 3175–3184 (2019)

16. Wang, A., Singh, A., Michael, J., Hill, F., Levy, O., Bowman, S.R.: GLUE: a multi-task benchmark and analysis platform for natural language understanding. In: ICLR (2019)

17. Wang, C., Lu, Y., Mu, Y., Hu, Y., Xiao, T., Zhu, J.: Improved knowledge distillation for pre-trained language models via knowledge selection. arXiv preprint arXiv:2302.00444 (2023)

18. Wang, W., Bao, H., Huang, S., Dong, L., Wei, F.: Minilmv2: multi-head self-attention relation distillation for compressing pretrained transformers. In: ACL/IJCNLP, pp. 2140–2151 (2021)

19. Wang, W., Wei, F., Dong, L., Bao, H., Yang, N., Zhou, M.: Minilm: deep self-attention distillation for task-agnostic compression of pre-trained transformers. In: NeurIPS (2020)

20. Wolf, T., et al.: Transformers: state-of-the-art natural language processing. In: EMNLP, pp. 38–45 (2020)

21. Xu, G., Liu, Z., Loy, C.C.: Computation-efficient knowledge distillation via uncertainty-aware mixup. Pattern Recogn. **138**, 109338 (2023)

22. Xu, Y., Yuan, F., Cao, C., Su, M., Lu, Y., Liu, Y.: A contrastive self-distillation BERT with kernel alignment-based inference. In: Mikyska, J., de Mulatier, C., Paszynski, M., Krzhizhanovskaya, V.V., Dongarra, J.J., Sloot, P.M. (eds.) ICCS 2023. LNCS, vol. 14073, pp. 553–565. Springer, Cham (2023). https://doi.org/10.1007/978-3-031-35995-8_39

23. Xu, Y., et al.: MetaBERT: collaborative meta-learning for accelerating BERT inference. In: CSCWD, pp. 119–124. IEEE (2023)

24. Yang, Z., Dai, Z., Yang, Y., Carbonell, J.G., Salakhutdinov, R., Le, Q.V.: Xlnet: generalized autoregressive pretraining for language understanding. In: NeurIPS, pp. 5754–5764 (2019)

25. Yuan, F., et al.: Reinforced multi-teacher selection for knowledge distillation. In: AAAI, pp. 14284–14291 (2021)

26. Zhang, X., Zhao, J.J., LeCun, Y.: Character-level convolutional networks for text classification. In: NeurIPS, pp. 649–657 (2015)

MPI4All: Universal Binding Generation for MPI Parallel Programming

César Piñeiro[1] , Alvaro Vazquez[2] , and Juan C. Pichel[2,3]([✉])

[1] CITIC, Universidade da Coruña, A Coruña, Spain
cesar.pomar@udc.es
[2] Electronics and Computer Science Department,
Universidade de Santiago de Compostela, Santiago de Compostela, Spain
alvaro.vazquez@usc.es
[3] CITIUS, Universidade de Santiago de Compostela, Santiago de Compostela, Spain
juancarlos.pichel@usc.es

Abstract. Message Passing Interface (MPI) is the predominant and most extensively utilized programming model in the High Performance Computing (HPC) area. The standard only provides bindings for the low-level programming languages C, C++, and Fortran. While efforts are being made to offer MPI bindings for other programming languages, the support provided may be limited, potentially resulting in functionality gaps, performance overhead, and compatibility problems. To deal with those issues, we introduce MPI4All, a novel tool aimed at simplifying the process of creating efficient MPI bindings for any programming language. MPI4All is not dependent on the MPI implementation, and adding support for new languages does not require significant effort. The current version of MPI4All includes binding generators for Java and Go programming languages. We demonstrate their good performance with respect to other state-of-the-art approaches.

Keywords: Parallel computing · MPI · Bindings · Java · Go

1 Introduction

Message Passing Interface (MPI) [13] is the most widely used and dominant programming model in High Performance Computing (HPC). In MPI, processes make explicit calls to library routines defined by the MPI standard to communicate data between two or more processes. These routines include both point-to-point (two party) and collective (many party) communications. Quality implementations can be found from prominent open-source projects like MPICH [5] and OpenMPI [6], as well as from software and hardware vendors of HPC systems (for instance, Intel MPI).

Traditionally, in the pursuit of raw performance, HPC has been closely tied to software development using low-level compiled languages such as C/C++ and Fortran. However, in contemporary science and research, high-level programming languages like Python, Go, and Julia play a crucial role. They offer researchers a

L. Franco et al. (Eds.): ICCS 2024, LNCS 14832, pp. 196–208, 2024.
https://doi.org/10.1007/978-3-031-63749-0_14

user-friendly platform for developing complex algorithms, analyzing vast datasets, and implementing sophisticated models, even for those with limited programming experience. In parallel programming, high-level languages face limitations in their support for MPI. These languages prioritize abstraction and ease of use, which can conflict with the low-level nature of MPI. While efforts exist to provide MPI bindings for these languages, support may be limited, leading to gaps in functionality, performance overhead and compatibility issues.

In this paper, we introduce MPI4All[1], a novel tool designed to simplify the creation of MPI bindings for any programming language. Adding support for a new language only requires writing a *generator* code (in any language) that maps MPI C macros, functions, and data types, automatically obtained by MPI4All, to the target language. It is important to note that unlike other approaches, MPI4All is not tied to a specific MPI implementation (such as OpenMPI or MPICH), it is compatible with all of them. Another important limitation of the existent MPI bindings is that they do not support the complete MPI API, or due to their lack of support, they only implement functions for old MPI versions. In our case, if MPI were to release a new version, running MPI4All again would suffice to generate complete API bindings for the desired MPI implementation and language. In other words, we will obtain the bindings for the new MPI version in seconds. Therefore, once there is an implementation of the *generator* code for a target language, it can be reused.

As illustrative examples, currently MPI4All provides bindings for Java and Go programming languages. We have evaluated them with respect to other state-of-the-art bindings when running different MPI applications to demonstrate the efficiency and completeness of the ones generated by our tool.

The remainder of this paper is organized as follows. Section 2 discusses related work, Sect. 3 explains the MPI4All architecture and how to create MPI bindings for a particular language using our tool. Section 4 shows the performance evaluation of the Java and Go MPI bindings built using MPI4All. Finally, the conclusions derived from this work are presented.

2   Related Work

The MPI Standard [13] only provides bindings for C and Fortran programming languages[2]. Overall, people interested in MPI features but who do not use either C or Fortran have been relying in unofficial MPI-like solutions. Languages supported by these implementations include C++, Java, C#, Go, Julia, Python, among others. However, some of these implementations are not currently maintained and only provide support for an outdated MPI version or implement a reduced set of characteristics.

Java is one of the most popular object-oriented languages and is widely used, for instance, in Big Data processing. Therefore, many efforts have been done

---

[1] It is publicly available at https://github.com/citiususc/mpi4all.

[2] C++ bindings where introduced in the MPI 2.0 specification but have been removed since the MPI 3.0 specification.

in using Java for parallel programming. Mainly, proposed Java MPI libraries subscribe to one of these three approaches:

- Relying on standalone Java APIs in order to provide a fully portable solution. Though it was followed by some MPI libraries, it finally proved out not to be practical since all of the MPI features need to be re-implemented.
- Relying on native MPI libraries for communication using the Java Native Interface (JNI), as it was adopted by solutions like mpiJava [10], and more recently the OpenMPI [16] and MVAPICH2 [7] official Java bindings. JNI allows Java programs to invoke functions and methods written in other languages including C.
- A hybrid approach, taken by MPJ Express [8] and FastMPJ [15], where message-passing libraries have custom device and network layers implemented in Java combined with JNI communication devices that call native methods.

The third approach, of implementing the MPI standard in Java aiming for high-performance communications, requires substantial development and maintenance effort. As a consequence, any change in the network and/or MPI standard version implies modifications in the code at low-level. On the other hand, the second approach allows easier development and maintenance. In order to get high-performance for Java MPI libraries, it keeps the Java layer as minimal as possible and uses JNI to invoke MPI methods implemented by native production-quality MPI libraries. The downside is that the JNI introduces a substantial amount of time overhead due to additional memory copying operations and requires recompiling the native code when porting the application to a new computer.

Currently, OpenMPI Java and FastMPJ are practically the only two well-maintained Java MPI libraries in the community. The MVAPICH2 Java implementation [7] is still in a maturation phase.

In the absence of a standard API for Java, older implementations [8,10] follow the mpiJava 1.2. API proposed by the Java Grande Forum (JGF) in late 90 s. FastMPJ [15] has support for both mpiJava 1.2 and the MPJ API, a minor upgrade to the mpiJava 1.2 API. On the other hand, the Java OpenMPI library [16] implements a custom API that is an extension of the MPJ API. Likewise, the Java MVAPICH2 bindings have adopted the OpenMPI Java API in order to facilitate end users. Though MPI4All follows the official MPI C interface style, a simple wrapper would solve the hypothetical necessity to comply with any of the proposed Java APIs.

There are other languages that include some kind of support for MPI. For example, Python supports MPI through the MPI4Py implementation [11], which underlies on the standard MPI-2 C++ bindings. The last version is compatible with both Python 2 and Python 3, and it supports various MPI-2 implementations like OpenMPI, MPICH, and Intel MPI. JuliaMPI.jl [9] is a Julia package for MPI. Though it supports up to MPI 3.1 many features are not yet available. Also, it does currently not support high performance networks such as InfiniBand, which limits its scalability to large problems. There were also several attempts to implement MPI bindings for Go programming language [1,3,4],

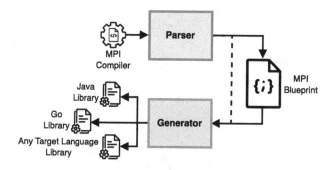

Fig. 1. Architecture of MPI4All.

but available distributions implement the MPI Standard only partially or stop keeping updating to new MPI versions.

3 MPI4All

In this section we will explain in detail the MPI4All architecture and how to proceed in order to build MPI bindings for a particular target language. MPI4All is composed of two distinct modules: the *Parser* and the *Generator*, as depicted in Fig. 1. The *Parser* takes an MPI compiler installed on the system as input. Its output, known as the *blueprint*, can be saved in JSON format. This blueprint is then utilized by the *Generator* to produce bindings for a particular programming language. The modular design allows the *Parser* and *Generator* to function independently, with the ability to exchange the blueprint file as needed.

3.1 Parser

The Parser module is in charge of collecting functions, data types and variables from an MPI implementation (for example, MPICH) and then creating a structured blueprint to organize all the extracted information. The information is obtained from the MPI compiler, which by default searches for the compiler in the system's PATH using the common names (mpicc, mpicxx, mpicpp, etc.), although the user can specify the compiler to use. Once the compiler is detected, the parsing task is carried out in two stages: extraction and typing.

The extraction phase involves retrieving the functions, types, and variables defined by MPI, that are identifiable by the prefix MPI. For example, functions like MPI_Send and MPI_Recv, data types like MPI_Comm, and variables like MPI_COMM_WORLD fall under this category. First, the preprocessor-defined macros are examined, followed by the symbols. During this initial phase, distinguishing between a variable and a data type is challenging, but the functions can be correctly detected.

```
 1  {
 2    "macros": [
 3      ...
 4      {
 5        "raw": "#define MPI_COMM_WORLD
               ↪ ((MPI_Comm)0x44000000)",
 6        "name": "MPI_COMM_WORLD",
 7        "value": "((MPI_Comm)0x44000000)",
 8        "type": "MPI_Comm",
 9        "var": true
10      },
11      ...
12    ],
13    "functions": [
14      ...
15      {
16        "header": "int MPI_Send (const void
               ↪ *, int, MPI_Datatype, int,
               ↪ int, MPI_Comm)",
17        "rtype": "int",
18        "name": "MPI_Send",
19        "args": [
20        {
21          "type": "const void *",
22          "name": "buf"
23        },
24        {
25          "type": "int",
26          "name": "count"
27        },
28        {
29          "type": "MPI_Datatype",
30          "name": "datatype"
31        },
32        {
33          "type": "int",
34          "name": "dest"
35        },
36        {
37          "type": "int",
38          "name": "tag"
39        },
40        {
41          "type": "MPI_Comm",
42          "name": "comm"
43        }
44      ]
45      },
46      ...
47    ],
48    "types": {
49      ...
50      "int": "4",
51      ...
52      "MPI_Comm": "int",
53      ...
54    },
55  }
```

Fig. 2. Example of a blueprint fragment using MPICH 4.1.

The typing phase serves to identify the type of the extracted information. This allows us to differentiate between variables and types, assigning the appropriate type to variables, as well as to the parameters and return values of each function. This process is carried out through different C and C++ tests that are compiled and executed. If a test fails to compile or compiles but fails to execute correctly, it indicates, for instance, whether we are dealing with a variable or a data type. Furthermore, this process identifies type size and aliases to gather all the necessary information, which may be required in the *Generator* module.

Figure 2 shows an example of a blueprint fragment generated with MPICH v4.1. In this example, we can see three sections: *macros*, *functions*, and *types*. The *macros* store preprocessor definitions, where we can observe the macro name, its associated type, and whether it is a variable or defines a MPI-specific type. Regarding *functions*, the most important data includes the function name, the types of arguments and return, and additional information such as parameter names or their C headers, which are useful for improving the readability of the target bindings. Finally, the *types* section contains information about all types used in macros and functions. These types are first mapped to native language types, and in the case of native types, the number of bytes they occupy is indicated. This can assist languages in finding type equivalences based on their names and sizes.

3.2 Generator

The Generator is in charge of generating the source code for a particular language bindings following a blueprint. Its goal is mapping the MPI C macros, functions, and data types included in the blueprint to the target language. The generation process is different for each programming language and employs different strategies for code generation. The current implementation of MPI4All includes generator scripts for Java and Go. However, since generators have no dependencies between them, they can be implemented as independent projects, allowing for the creation of new generators by three party users using a blueprint. It is not mandatory for the generator implementation to be in the same programming language as the target bindings. For instance, MPI4All, implemented in Python, provides generators for Go and Java programmed in that language.

As we commented previously, the implementation of a generator must take into account the interoperability of C with the target language. As illustrative examples, we will describe the design in terms of the implementation of the Java and Go generator scripts, which can be generalized to support other languages. The interoperability between C and Go is facilitated by the `cgo` tool, which allows easy calling of C functions from Go and vice versa. Likewise, in Java, the new Foreign Function Interface (FFI) allows for bidirectional interaction between Java and C, facilitating the integration of native code into Java applications. In the Go approach, interoperability is facilitated by including C headers directly into Go code, allowing seamless interaction between the two languages. This integration is achieved by compiling the combined codebase, ensuring that Go and C components work harmoniously together. In contrast, Java's FFI interacts with pre-compiled native libraries. These libraries are linked dynamically at runtime, enabling Java to access and utilize functions defined within the C code. Java communicates with these libraries using the symbols they contain, providing a bridge for the execution of native code within Java applications.

Macros and Data Types. In Go and Java, in any case, it is not possible to directly call MPI C functions because macros defined in the headers cannot be accessed. Go lacks access to compiler macros, and in Java, such information is removed after compilation. This behavior is common to other programming languages.

To deal with this issue, MPI4All uses a hybrid strategy, which requires two steps. Firstly, it generates a C auxiliary library. The primary function of this library is to convert all macros stored in the blueprint into data types or variables. For instance, while languages like Java require a separate file for this purpose, Go allows embedding C code directly within a string in the code. Nonetheless, the procedure still involves iterating over all macros in the blueprint and generating C code using the following pattern:

– Variable:

```
[type] [PREFIX][name] = [name];
```

```
1    #include<mpi.h>
2
3    MPI_Comm GO_MPI_COMM_WORLD = MPI_COMM_WORLD;
4    ...
5    MPI_Datatype GO_MPI_DOUBLE = MPI_DOUBLE;
6    ...
7    int GO_MPI_THREAD_SINGLE = MPI_THREAD_SINGLE;
8    ...
9    typedef int GO_MPI_Fint;
10   ...
```

Fig. 3. Example of auxiliary C code output from the generator script.

```
1    var MPI_COMM_WORLD = C.GO_MPI_COMM_WORLD;
2    ...
3    var MPI_DOUBLE = C.GO_MPI_DOUBLE;
4    ...
5    var MPI_THREAD_SINGLE = C.GO_MPI_THREAD_SINGLE;
6    ...
7    type MPI_Fint = C.GO_MPI_Fint;
8    ...
9    type MPI_Comm = C.MPI_Comm;
10   ...
```

Fig. 4. Example of Go macro binding output from the generator script.

– Data type:

typedef [type] [PREFIX] [name];

where the pattern names correspond to blueprint fields and PREFIX represents any chosen value defined in the generator.

Once applied to all macros in the blueprint, we would have a result similar to Fig. 3. Note that it is necessary to include the MPI header (line 1) to compile the library. Once the symbols corresponding to the macros are generated, we can use them from the target language.

The process in the second step is similar to the previous one, iterating over all the macros in the blueprint but generating code in the target language as shown in Fig. 4 for Go. Note that in the process, variables in the target language with the same name than the macro in C are assigned to the corresponding ones in the C auxiliary library. In this phase, we can also iterate over the data types defined in the blueprint and generate them along with the macros following the same procedure (line 9).

The second step for Java is slightly more complicated because it is not possible to map compiled data types in the native C library. The FFI uses a class called MemorySegment, which represents a memory address range of a known size using a MemoryLayout. Consequently, MPI4All defines MPI types as Java classes that extend a common class containing the MemorySegment that emulate different types. The MemoryLayout size of the MemorySegment are determined through the *types* section of the blueprint. The variables are defined as instances of those classes and are assigned to static attributes. In the case of classes representing primitive types, these are converted to simplify the API.

```
1   ...
2   func MPI_Send(buf unsafe.Pointer /*(const void *)*/, count C_int, datatype
        C_MPI_Datatype, dest C_int, tag C_int, comm C_MPI_Comm) error {
3       return mpi_check(C.MPI_Send(buf, count, datatype, dest, tag, comm))
4   }
5   ...
```

Fig. 5. Example of Go function binding output from the generator script.

Functions. The final task of the generator script is to map the functions as described in the blueprint to the target language. This involves generating each function using the syntax and conventions of the considered language, using the types specified in the previous step. In the function body, we call the corresponding C function, ensuring conversion of arguments and return types, and taking care of C error codes. Figure 5 presents the Go code for the MPI_Send function. Parameters are seamlessly passed to the C function without conversion, as we have defined their types as aliases of C types. Additionally, we have introduced the auxiliary function mpi_check to handle return values, converting MPI function return codes into error when they are not equal to 0.

The equivalent process in Java involves more steps. First, we need to locate the function symbol within the auxiliary C library generated previously. Then, we define the type for each parameter and return value using MemoryLayout. Finally, we can use the function definition to invoke it from Java. While primitive types are automatically converted into MemorySegment using the function layout, complex types must have their layout defined manually. Similar to Go, an auxiliary function mpiCheck is defined in Java. This function will check the return code of the MPI function and throw a RuntimeException, analogous to how Go returns an error.

4 Experimental Evaluation

Next we will evaluate the performance of the bindings for Go and Java that currently can be generated by MPI4All. With that goal in mind, we will compare their performance with respect to other state-of-the-art solutions. Since MPI4All is agnostic regarding the MPI implementation considered, unlike other approaches, we will prove that it is capable of generating bindings for OpenMPI and Intel MPI, for example.

Experiments were conducted using up to 8 computing nodes of the FinisTerrae III [2] supercomputer installed at CESGA (Spain). Each node contains two 32-core Intel Xeon Ice Lake 8352Y @2.2 GHz processors and 256 GB of memory interconnected with Infiniband HDR 100. A 100 Gb Ethernet network is also available on all nodes. It is a Linux cluster running Rocky Linux v8.4 (kernel v4.18.0).

4.1 Java

In this section we present the evaluation results carried out over three different Java MPI implementations. To provide an estimation of MPI4All performance we ran some of the NAS parallel benchmarks ported to Java described in [12]. We have introduced a few modifications in the source code in order to adapt them to the MPI4All Java bindings. The subset of the Java NAS parallel benchmarks selected for evaluation purposes is the following:

- CG – It uses a Conjugate Gradient method to compute approximations to the smallest eigenvalues of a sparse unstructured matrix.
- EP – This benchmark, short for Embarrassingly Parallel, is designed to measure the performance of a parallel application that consists of independent tasks that can be executed concurrently without any communication or synchronization between them.
- FT – It contains the computational kernel of a 3D Fast Fourier Transform (FFT).
- SP – It is a simulated Computational Fluid Dynamics (CFD) application. It solves a Scalar Pentadiagonal system of linear equations.
- MG – It uses a V-cycle Multi Grid method to compute the solution of the 3D scalar Poisson equation.

We use class D benchmarks, which correspond to considerably large problem sizes. A strong scaling test was conducted for each benchmark using up to 128 cores (in 4 nodes, 32 cores per node) of the Finisterrae III cluster and the Infiniband HDR 100 interconnection. We measured the performance using JDK 21.0.1, and Java bindings were generated for OpenMPI v4.1. Each measurement was computed as the median of five executions. Note that each new JDK release starting from version 19 has required to generate new MPI4All Java bindings. This is due to the fact that the Java FFI is currently a preview feature of the Java Platform still subject to changes. This API will be upgraded to permanent features in the next JDK 22 release, so it is expected to remain stable. Nevertheless, in our case the JDK update barely involved a change in a couple lines of code in the generation process of MPI4All Java bindings.

Also, for comparison purposes we selected and ran the same subset of the NAS parallel benchmarks over two representative MPI Java implementations: the FastMPJ library [15], and the official OpenMPI Java bindings [16]. The former uses JNI to invoke networking native library primitives (including Infiniband ones) while the later uses JNI to call MPI C primitives.

We represent in Fig. 6 the speedups obtained for both FastMPJ library and the OpenMPI Java bindings using as reference the execution times measured in the corresponding MPI4All tests.

Overall, the OpenMPI Java bindings present degraded performance figures mainly due to the overhead introduced by the JNI calls to the MPI native library. MPI4All invokes the same MPI C library functions using FFI. Therefore, in this case, the performance improvement of the MPI4All Java bindings comes from using FFI instead of JNI to minimize the cost associated with copying data from Java to C.

On the other hand, FastMPJ presents better performance numbers than the generated MPI4All Java bindings except for the MG benchmark, where MPI4All

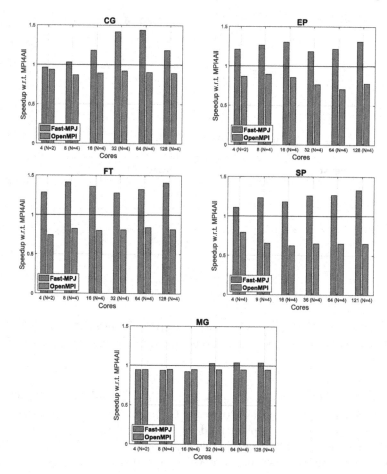

Fig. 6. Speedup of the different Java MPI implementations when executing the NAS Parallel Benchmarks (class D) using as reference the bindings for OpenMPI generated by MPI4All. N is the number of computing nodes.

is competitive with FastMPJ. FastMPJ relies on a highly efficient Java implementation of MPI-like functions aiming to provide a similar performance as native MPI implementations. However, current version only provides support for MPI-2, and upgrading or incorporating new functionalities requires a huge code programming effort.

4.2   Go

In this section, we will evaluate the performance obtained by the MPI Go bindings generated by MPI4All. In particular, we have generated bindings for Intel MPI version 2021.10.0. Go version 1.20.4 was used for compiling and deploying purposes. In the experiments we have considered Ember [14], which is a

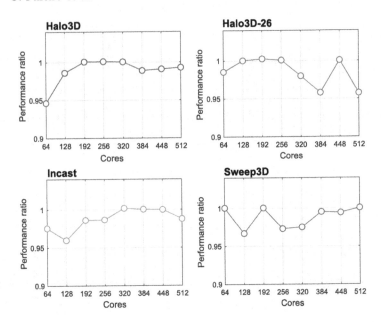

Fig. 7. Performance ratio between the original MPI C implementation and the Go one that uses the language bindings generated by MPI4All when running the Ember benchmarks.

communication pattern library developed at Sandia National Labs (USA). It is part of the Exascale Computing Project (ECP) proxy applications suite[3]. The Ember code was originally implemented in C using MPI and represents simplified communication patterns that are relevant to extreme scaled supercomputing systems. Four communication patterns were studied:

- Halo3D: It performs a structured nearest neighbor communication. In this pattern, each MPI rank communicates with ranks that are adjacent to it in each cartesian dimension. The *halo* exchanged is the data on each face.
- Halo3D-26: In this pattern, each MPI rank communicates with ranks along each cartesian face, as well as each edge of the local grid and each vertex. It represents a typical unstructured nearest neighbor communication.
- Incast: The purpose of this benchmark is to represent small collections of nodes which attempt to simultaneously send messages to the same remote node, similar to some parallel I/O systems.
- Sweep3: There are many scientific applications that have a strong level of dependencies, which affect their communication patterns. This benchmark attempts to mimic that behavior by decomposing a 3D data domain over a 2D array of MPI ranks.

Our experimentation focuses on comparing the performance of the original C Ember benchmark with our Go port implementation, which utilizes MPI4All

---

[3] http://proxyapps.exascaleproject.org/ecp-proxy-apps-suite.

bindings (see Fig. 7). Through this comparison, we aim to evaluate the efficacy of our ported solution against the established C implementation. The experiment was conducted using up to 8 nodes of the Finisterrae III cluster (comprising a total of 512 cores, 64 per node). The benchmarking process involved running each of the four communication patterns, both in C and Go, a total of five times. Subsequently, the median of the values obtained in each run for each pattern and language was taken.

Figure 7 shows that the MPI4All-generated Go bindings achieve similar performance comparable to the native Intel C MPI library for all four communication patterns since the ratio remained very close to 1 regardless of the number of nodes used. It also indicates that in terms of performance, Go and C are highly similar. One notable distinction between both languages is Go's garbage collector. However, given the absence of memory creation and destruction during the communication process, it should not significantly impact the measurements. The seamless integration of C into Go has enabled MPI4All to produce a binding library with performance nearly identical to that of a C implementation.

On the other hand, as we pointed out in Sect. 2, there are some other MPI Go bindings proposals [1,3,4]. However, to the best of our knowledge, none of them can implement the Ember benchmark in Go. This limitation arises from both the absence of necessary functionalities such as asynchronous communications and the lack of support for Intel MPI.

5   Conclusions

In this paper, we introduced MPI4All[4], an innovative tool aimed at facilitating the creation of MPI bindings for any programming language. Adding support for new languages merely requires the development of an script code that maps MPI C macros, functions, and data types to the target language. It is noteworthy that unlike other approaches, MPI4All is not bound to a specific MPI implementation; it is compatible with all of them. Once the script for some programming language is available, generating bindings for the complete API of any MPI implementation (and version) takes seconds. This assures completeness and avoids maintenance problems.

MPI4All includes scripts for generating MPI bindings for Go and Java programming languages. We evaluated them in terms of performance compared to other state-of-the-art solutions. The MPI4All Java bindings for OpenMPI clearly outperform the official Java OpenMPI bindings when running several NAS benchmarks. Although their performance is lower compared to FastMPJ, it is important to note that FastMPJ only supports MPI-2 routines. Regarding Go, the MPI4All bindings for Intel MPI achieve very similar performance to the native Intel C MPI library when running the Ember benchmarks.

---

[4] It is publicly available at https://github.com/citiususc/mpi4all.

Acknowledgments. The authors would like to thank Guillermo López Taboada and Roberto Expósito for providing access to the FastMPJ library. This work was supported by Xunta de Galicia [ED431G 2019/04, ED431F 2020/08, ED431C 2022/16]; MICINN [PLEC2021-007662, PID2022-137061OB-C2, PID2022-141027NB-C22]; and European Regional Development Fund (ERDF). Authors also wish to thank CESGA (Galicia, Spain) for providing access to their supercomputing facilities.

Disclosure of Interests. The authors have no competing interests to declare that are relevant to the content of this article.

References

1. A Golang Wrapper for MPI. https://github.com/yoo/go-mpi. Accessed 26 Feb 2024
2. CESGA (Galician Supercomputing Center) - Computing Infrastructures. https://www.cesga.es/en/infrastructures/computing/. Accessed 26 Feb 2024
3. GoMPI: Message Passing Interface for Parallel Computing. https://github.com/sbromberger/gompi. Accessed 26 Feb 2024
4. MPI-binding package for Golang. https://github.com/marcusthierfelder/mpi. Accessed 26 Feb 2024
5. MPICH. https://www.mpich.org. Accessed 26 Feb 2024
6. Open MPI. https://www.open-mpi.org/. Accessed 26 Feb 2024
7. Al-Attar, K., Shafi, A., Subramoni, H., Panda, D.K.: Towards Java-based HPC using the MVAPICH2 library: early experiences. In: IEEE International Parallel and Distributed Processing Symposium Workshops (IPDPSW), pp. 510–519 (2022)
8. Baker, M., Carpenter, B., Shafi, A.: MPJ express: towards thread safe Java HPC. In: 2006 IEEE International Conference on Cluster Computing, pp. 1–10 (2006)
9. Byrne, S., Wilcox, L.C., Churavy, V.: MPI. jl: Julia bindings for the message passing interface. In: Proceedings of the JuliaCon Conferences **1**(1), 68 (2021)
10. Carpenter, B., Getov, V., Judd, G., Skjellum, A., Fox, G.: MPJ: MPI-like message passing for Java. Concurr. Pract. Exp. **12**(11), 1019–1038 (2000)
11. Dalcin, L., Fang, Y.L.L.: mpi4py: Status update after 12 years of development. Comput. Sci. Eng. **23**(4), 47–54 (2021)
12. Mallón, D.A., Taboada, G.L., Touriño, J., Doallo, R.: NPB-MPJ: NAS parallel benchmarks implementation for message-passing in Java. In: 2009 17th Euromicro International Conference on Parallel, Distributed and Network-based Processing, pp. 181–190 (2009)
13. Message Passing Interface Forum: MPI: a message-passing interface standard version 4.1 (2023). https://www.mpi-forum.org/docs/mpi-4.1/mpi41-report.pdf. Accessed 26 Feb 2024
14. Sandia National Laboratories: Ember Communication Pattern Library (2018). https://github.com/sstsimulator/ember. Accessed 26 Feb 2024
15. Taboada, G.L., Touriño, J., Doallo, R.: F-MPJ: scalable Java message-passing communications on parallel systems. J. Supercomput. **60**, 117–140 (2012)
16. Vega-Gisbert, O., Roman, J.E., Squyres, J.M.: Design and implementation of Java bindings in Open MPI. Parallel Comput. **59**, 1–20 (2016)

Time Series Predictions Based on PCA and LSTM Networks: A Framework for Predicting Brownian Rotary Diffusion of Cellulose Nanofibrils

Federica Bragone$^{(\boxtimes)}$ ⓘ, Kateryna Morozovska ⓘ, Tomas Rosén ⓘ,
Daniel Söderberg ⓘ, and Stefano Markidis ⓘ

KTH Royal Institute of Technology, 100-44 Stockholm, Sweden
{bragone,kmor,trosen,dansod,markidis}@kth.se

Abstract. As the quest for more sustainable and environmentally friendly materials has increased in the last decades, cellulose nanofibrils (CNFs), abundant in nature, have proven their capabilities as building blocks to create strong and stiff filaments. Experiments have been conducted to characterize CNFs with a rheo-optical flow-stop technique to study the Brownian dynamics through the CNFs' birefringence decay after stop. This paper aims to predict the initial relaxation of birefringence using Principal Component Analysis (PCA) and Long Short-Term Memory (LSTM) networks. By reducing the dimensionality of the data frame features, we can plot the principal components (PCs) that retain most of the information and treat them as time series. We employ LSTM by training with the data before the flow stops and predicting the behavior afterward. Consequently, we reconstruct the data frames from the obtained predictions and compare them to the original data.

Keywords: Principal Component Analysis · Long Short-Term Memory · Time Series · Cellulose Nanofibrils

1 Introduction

Cellulose nanofibrils (CNFs), the fundamental building block of all plants and trees, have extraordinary mechanical properties in terms of strength and stiffness and are considered one of the major materials in terms of providing sustainable options to many advanced materials used today [16]. Through controlled alignment and assembly of dispersed CNFs, very strong and stiff filaments can be created by means of flow-focusing wet spinning, as described in [20]. Successful spinning relies on a delicate balance of timescales and, in particular, the competing effects from hydrodynamic forcing (causing alignment through shear and extensional flow) and Brownian rotary diffusion (causing de-alignment) [21].

Supported by VINNOVA.

These effects, in turn, depend heavily on how the CNFs are extracted from biomass and the raw material source. Even the same raw material and extraction protocol can yield completely different CNF behavior in flowing processes. Therefore, developing methodologies for quick quality determination of CNF dispersions is crucial to assess their suitability in material processes. A potential method is the rheo-optical flow-stop experiment described by Rosén et al. [22]. A flow cell is used to align the dispersed CNFs by means of flow-focusing (see Fig. 1). When placed between two cross-polarized linear polarization filters, the transmitted intensity of light through the filters and flow cell will be a measurement of the CNF alignment as the system becomes birefringent. When the flow is instantly stopped, the decay of birefringence will thus be a measurement of the Brownian rotary diffusion of the CNFs. Both the behavior during flow, as well as the timescales of decay after stop, become a unique fingerprint of the CNF dispersion that can be classified and used to determine the CNF quality. In this work, we will explore the possibility of predicting the behavior after stopping the flow by training our model on the information before the flow stop and for some frames after the flow stop. This would allow us to monitor a material process *in operando* and perform classification without actually stopping the flow.

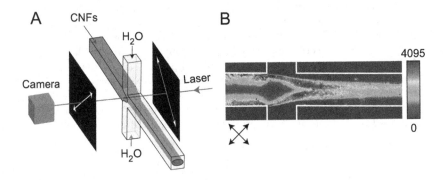

Fig. 1. Illustration of the rheo-optical flow-stop experiment as described by Rosén *et al.* [22]; (A) a core flow of dispersed CNFs (flow rate 23.4 ml/h) is focused by two sheath flows of water (each flow rate 13.5); the setup is illuminated by laser light and placed between cross-polarized linear polarization filters; the transmitted light intensity, corresponding to CNF alignment, is recorded with a camera; channel width is 1 mm; (B) a typical camera image, measuring intensities from 0 to 4095 (12 bit)

These flow-stop experiments produce large amounts of datasets. In particular, the datasets comprise data frames with specific intensities. One dataset consists of the flow of one CNF material at a particular concentration. Several materials and concentrations are tried out, resulting in large quantities of data. Dimensionality reduction techniques can then help analyze the data with smaller amounts of features. In particular, Principal Component Analysis (PCA) [12,30] is one of the major techniques to reduce data dimensions in the area of Machine

Learning (ML). PCA transforms the original dataset into a new set of orthogonal features, a linear combination of the original variables defined as principal components (PCs). The transformation accounts for the maximum variance of the data stored in the first PC, and consequently, the other PCs add up cumulatively in descending order.

As these PCs create a new set of coordinates retaining most of the variance, they can be visualized as time series over the data frames of our dataset. Working with these time series, we aim to create an automatic model that predicts the behavior of the CNFs dynamics after the flow is stopped. If successful, it would allow the creation and running of a monitoring system at the production line without involving any physical stop. Consequently, we would have an artificial stop capable of predicting the Brownian dynamics of the different CNFs, given their behavior, before the flow is stopped. To achieve an automatic prediction of the lower dimensional PCs, we adopt an ML technique capable of working with sequences and time series: Long Short-Term Memory (LSTM) networks [11]. LSTM is a type of Recurrent Neural Network (RNN) [19,29] designed to overcome the vanishing gradient problem that RNNs often encounter [1]. The class of RNNs is distinguished from classical Artificial Neural Networks (ANNs) as they present recurrent connections in their hidden layers, meaning that their looping constraints capture the sequential information stored in the data. LSTM are more complex structures that can learn complicated and long patterns, given their capability of selectively remembering or forgetting significant information. For this reason, LSTM networks are suitable for time series predictions [23].

Our work aims to implement an LSTM for each time series created by PCs: each network is trained on the time steps before the flow-stop, including a few data frames after the stop, and then predicts the following behavior. Collecting all the results, we can reconstruct the data frames by inverting the PCA transformation and finally compare the predictions with the original data. This study aims to create an automatic method that can capture the dynamic behavior of the CNFs after the flow is stopped. In return, it can help further studies on the characterization of the materials and concentrations of CNFs. Moreover, it could simulate other materials using the proposed method rather than physically setting up and running the experiment. The overall goal is to create a model that simulates an artificial flow stopping capable of predicting the Brownian dynamics of any CNF after being trained on the flow before the flow stop on different CNF samples and concentrations. In this way, the physical stopping experiments could be avoided and implemented easily in the production line without waiting a long time to produce several experiments in the labs. This paper is the first step towards the broader goal and has the following contributions:

- We propose a method to automatically predict the CNF's behavior after the flow is stopped in the rheo-optical flow-stop technique. The model will possibly simulate and characterize further combinations of materials and concentrations without requiring physical experiments.

- Our model combines a dimensionality reduction technique, like PCA, and an ML method, like LSTM, to predict sequential data created by the PCs. The results show promising behavior predictions, especially after the flow stop.
- The reconstruction from the predicted PCs is already accurate; however, PCA did not capture certain particle-like CNF clusters completely.

The paper is structured as follows. Section 2 presents related work using similar methods and techniques for time series predictions. In Sect. 3, the methods used for our model are introduced. Section 4 describes the data utilized and the simulation's architecture and details. The results are shown in Sect. 5. Finally, Sect. 6 includes a discussion of the results and closing conclusions.

2 Related Work

The first RNN was introduced in 1989 [29], and consequently, partially connected RNNs were developed in [7,13]. They focus on time series by discovering and modeling their relations and information. Several works afterward extended these preliminary implementations, applying them to several problems. The main drawback of RNNs is that they suffer from the vanishing gradient problem. Therefore, RNNs cannot fully capture the non-stationary dependencies over long periods and multiple time dependencies [18]. Gating mechanisms are introduced to substitute the classical activation functions to overcome this problem. LSTM is one of the models that, with its three gates, can update a cell state by capturing the long-term dependencies [11]. In 2014, the Gated Recurrent Unit (GRU) was introduced [4], a variant of the LSTM network. This mechanism improves short-term information integration but also predicts long-term dependencies. GRU comprises a gating system with a simplified cell structure compared to the LSTM. In [5], an evaluation of gated recurrent neural networks on sequencing models is presented. In particular, a comparison between LSTM and GRU networks is given, and their predictions provide related results. A large-scale analysis on several network architectures is performed in [3], where LSTM outperformed GRU networks. Several LSTM architectures were developed, including Bidirectional LSTM [10], hierarchical and attention-based LSTM [27,32], Convolutional LSTM [15], LSTM autoencoder [9], Grid LSTM [14], and cross-modal [26] and associative LSTM [6]. Sequence-to-Sequence (Seq2Seq) networks, introduced in [24], also work on using an input sequence to predict output sequences. More findings and comparisons of the network and architecture variants to predict nonlinear time series are presented in [18].

More recently, transformers have been considered for time series analysis and predictions, and several studies have been made to investigate their effectiveness [28,33]. Transformers are a machine learning architecture based on multi-head self-attention mechanisms that work with sequential data [25]. They became popular in applications like natural language processing (NLP), speech recognition, and computer vision. In [17], a temporal-fusion transformer (TFT) is introduced to forecast multi-horizon and multivariate time series. TFT can interpret the temporal dependencies with different time scales by merging the LSTM Sequence-to-Sequence and the transformers' self-attention mechanism.

In computational science, many problem domains produce large amounts of data, which is not always essential for analysis. For this reason, reduced order models and dimensionality reduction techniques become powerful tools for reducing the number of features while retaining most information. Several works utilize PCA to apply their models then. In [8], the authors apply PCA and LSTM for the short-term power system load forecast. In the context of time series predictions, the work in [31] presents a PCA-LSTM model for anomaly detection and prediction of time series power data. Stock prices rely on time series forecasting, and in [2], the authors first use PCA to reduce the data features and then apply RNNs to predict stock prices. In [34], it also works on stock price prediction but applies a PCA-LSTM model instead.

To the best of our knowledge, our model is the first to consider the time dependencies of the CNF flow by considering the principal components and applying LSTM networks to predict the behavior after the flow is stopped.

3   Methods

3.1   Dimensionality Reduction

Dimensionality reduction techniques are fundamental methods that reduce the number of attributes of a dataset while preserving most of the variation of the original dataset. One of these techniques is Principal Component Analysis (PCA), which transforms the data into a new set of uncorrelated variables called principal components (PCs). The first PCs contain most of the variance present in the dataset. The first step of performing PCA involves the standardization of the dataset; each variable is scaled using its corresponding mean and standard deviation values. The covariance matrix of the variables is computed, identifying the correlations between features. The following step involves computing the eigenvectors and the eigenvalues of the covariance matrix to identify the PCs. Finally, the PCs are created and arranged according to their eigenvalues, from the highest to the lowest.

3.2   Long Short-Term Memory

Recurrent Neural Networks (RNNs) are artificial neural networks that capture sequential data dependencies. Long Short-Term Memory (LSTM) networks [11] are a type of RNN capable of dealing with long-term dependencies. The LSTM network comprises the input gate i, the forget gate f, and the output gate o. The input gate gives information on the new inputs loaded in the cell state C_t. The forget gate highlights the information that should not be kept in the cell state. Finally, the output gate gives the final output of the LSTM block, h_t, at time step t. The equations representing the gates at time step t are as follows:

$$f_t = \sigma(W_f x_t + U_f h_{t-1} + b_f) \tag{1}$$
$$i_t = \sigma(W_i x_t + U_i h_{t-1} + b_i) \tag{2}$$
$$o_t = \sigma(W_o x_t + U_o h_{t-1} + b_o) \tag{3}$$

where σ represents the sigmoid function; W_f, W_i, and W_o are the input weights for the forget, input, and output gates, respectively; U_f, U_i, and U_o are the recurrent weights for the forget, input, and output gates, respectively; h_{t-1} is the output of the previous LSTM block at time step $t-1$; x_t is the input at the current time step t; and b_f, b_i, and b_o are the biases of the input, forget and output gates, respectively. The following equations represent the cell state C_t, the candidate cell state $\tilde{C}_t$ and the final output of the LSTM block h_t:

$$\tilde{C}_t = \tanh(W_c x_t + U_c h_{t-1} + b_c) \tag{4}$$

$$C_t = f_t \odot C_{t-1} + i_t \odot \tilde{C}_t \tag{5}$$

$$h_t = o_t \odot \tanh(C_t) \tag{6}$$

where W_c and U_c represent the candidate cell's input and recurrent weights, respectively; the operator $\odot$ represents the element-wise product, and the tanh is the hyperbolic tangent activation function. The cell state C_t includes both the information that needs to be forgotten from the previous cell state C_{t-1} and the information that needs to be looked at in the current time step from the candidate cell state $\tilde{C}_t$. Finally, we define h_t as the output of the current LSTM block, which includes the current cell state passed through the activation function to decide the block's output.

4   Input Data and Experiment Description

The data used in this study is generated from the cellulose spinning process and subsequent flow-stop experiments designed to capture changes in the process. We consider one experiment generated by one carboxymethylated (CM) CNF at a concentration of 3.0 g/l. The choice of the sample used for this study was purely arbitrary. The dataset contains 15000 frames of size 640×100 at 1000 fps recorded by a 12-bit camera with intensity values between 0 and 4095. The experiment shows the flow-stop experiment, meaning the flow is stopped around frames 5000–6000. For our work, we consider the data frames around the flow stop. In particular, we consider a total of 5000 frames: 3000 and 2000 frames before and after the flow-stop, respectively. We are interested in finding an automatic model that can predict what happens after the flow is stopped without actually stopping the flow. In this way, the physical experiments would not need to be stopped to characterize the materials, which is fundamental given the complexity of the experiment. Despite considering a smaller amount of data frames, the dimension of the data is still significant. For this reason, we can apply PCA, a dimensionality reduction technique: by keeping most of the information available in the dataset, we reduce the amount of variables in the dataset. First, the data is normalized in the range of (0, 1), then PCA is performed. Only 10 PCs are considered for this work, which already retains 85.19%. The first PC keeps most of the information, precisely 82.24%. 10 PCs can already identify the flow well: using fewer PCs, certain particle-like CNF clusters would be missed; using more

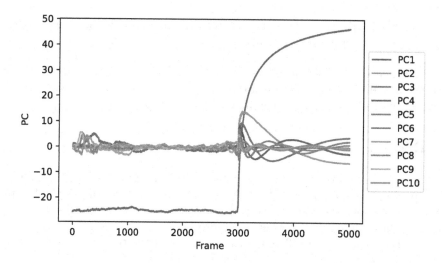

Fig. 2. Plot of the 10 PCs

PCs, we risk introducing more noise. If we plot the resulting PCs over the frames, we can observe that they create some time series. It is also visible in the plots where the flow is stopped, as shown in Fig. 2.

Working on the lower-dimensional space created by the PCs, we could treat these components as time series, train our model on the first part, and predict when the flow is stopped.

4.1 LSTM Architecture

The LSTM model is created using the open-source libraries Keras and Tensor-Flow in Python. The model consists of 10 LSTM networks, one for each PC. For a better generalization, each network has the same hyperparameters and structure. The only difference arises when using `EarlyStopping` for the number of the argument `patience` for a few PCs.

First, we split the time series into 67% training and 33% test for each PC, giving 3350 data frames for training and 1650 data frames for testing. The network predicts the time series using a window. Considering the data between $t - w$ and t, for some value w, we would like to predict one time step forward, i.e., $t + 1$. The value w defines the size of the window, specifying how many previous time steps we are looking to predict the future value. In particular, it is defined as the look-back period. For our model, we chose a look-back period of three time steps, considering three previous values to predict the following time step.

The network has an input layer with one input, two hidden layers with 20 LSTM blocks each, and an output layer with a single-value prediction. The activation function used is the rectified linear unit (ReLU). The model is compiled using the mean squared error (MSE) loss function and the Adam optimizer with a default learning rate of 10^{-3}. The model is fitted and trained for 100 epochs, and an `EarlyStopping` is placed to monitor the test loss with a `patience` value of 50. This function stops the training when the test loss no longer improves, specifically for 50 consecutive epochs. For two cases, in particular PCs 2 and 3, the `patience` value is set up to be 5 and 10, respectively, since it was noticed that the model was starting to overfit at early epochs. Finally, the model is used to generate predictions for the training and the test datasets.

5   Results

This section reports the results of training an LSTM network for each PC. The corresponding predictions are then stacked and reversed to the original data format to compare the results. We train each network for 100 epochs, possibly stopping the training earlier when no improvement in the minimization of the test loss is observed. Figure 3 shows the evolution of the training loss, the blue line, and the test loss, the orange line, for the first PC.

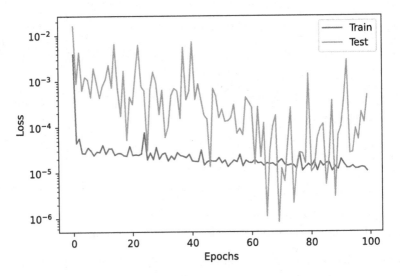

Fig. 3. Loss functions for training (blue) and test (orange) for the first PC (Color figure online)

Figure 4 shows the training and test predictions of the LSTM networks for the 10 PCs. In particular, the light blue line is the original data, the dashed darker blue line is the LSTM training predictions, and the dashed red line is the LSTM test predictions. The plots show that the LSTM model can capture the component values evolving over the data frames. In particular, the test score of the first PC is of the most interest, as it captures most of the information stored in the original dataset. As for the second and third components, we can notice some discrepancies between the original data and the test predictions. It

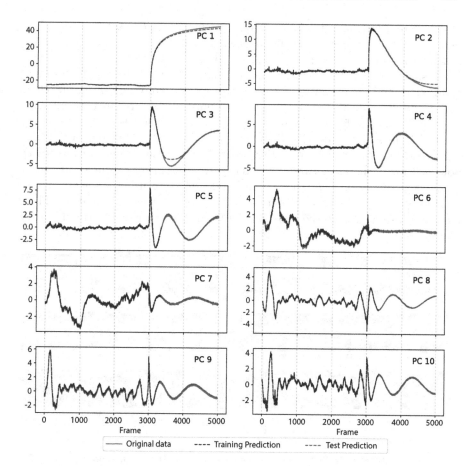

Fig. 4. LSTM predictions on the 10 PCs

might be due to more noise in these components, hence more abrupt changes in the data. The train and test scores for the 10 PCs are presented in Table 1. The scores capture the root mean squared error (RMSE) between the original data and the predictions.

All the results were tested multiple times with different random seeds to assess the robustness of the model in terms of initialization of the network. For the first PC, a standard deviation of the scale of 10^{-2} for training and 10^{-1} for test was noticed between different runs. While for the other PCs, a scale of 10^{-3} and 10^{-1} for training and test, respectively, was observed.

Table 1. RMSE scores for each PC for training and test datasets of LSTM.

PC	1	2	3	4	5	6	7	8	9	10
Train score	0.26	0.15	0.14	0.12	0.14	0.14	0.11	0.10	0.19	0.10
Test score	1.61	0.62	0.70	0.12	0.16	0.07	0.06	0.04	0.17	0.05

218   F. Bragone et al.

The predicted data is then reversed using the inverse_transform function available for the sklearn.decomposition.PCA from the sklearn library. Finally, the reconstructed predictions are compared to the original data frames. Fig. 5 shows the snapshots of the original data, the predicted data using PCA and LSTM, and the pointwise absolute error for certain data frames. The choice of the specific data frames to display was purely arbitrary. In particular, the plots in Fig. 5a and 5b display some results before the flow is stopped, precisely for data frames 1100 and 2980. They represent the predicted results of the training set used for the model. We noticed that the predictions already captured the flow well for both cases. However, some parts are poorly captured, as seen in their respective error evaluations. This could relate to using a smaller dataset created by only 10 PCs, which retains 85.19% of the total information. Therefore, it is assumed that the considered PCs might not have captured entirely

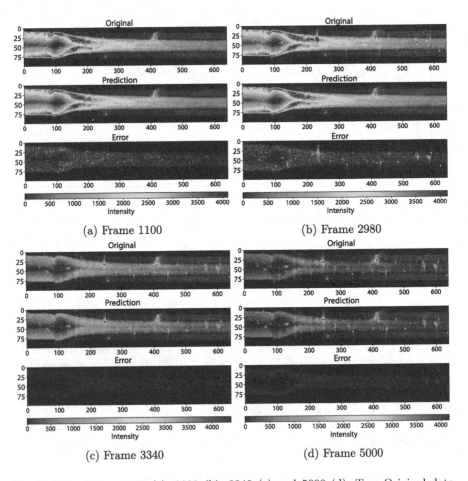

Fig. 5. Data frames 1100 (a), 2980 (b), 3340 (c) and 5000 (d). Top: Original data. Middle: LSTM predictions. Bottom: Pointwise absolute error between the original and predicted snapshots. Each frame has a .size of 640 × 100 pixels

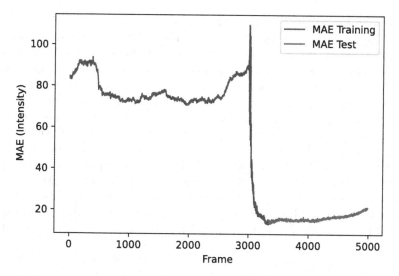

Fig. 6. MAE between the original data and the LSTM predictions for each frame (Color figure online)

some CNFs. Instead, the plots in Fig. 5c and 5d represent the flow after the stop. Snapshot 3340 refers to the training dataset; however, snapshot 5000 is part of the test data, where LSTM networks validated the model. By looking at the error plots for data frames 3340 and 5000, the results improved compared to the data frames before the flow stop. The intensity of the errors is lower, meaning that both PCA and LSTM can capture most of the CNFs flow in each data frame. Figure 6 shows the mean absolute error (MAE) over the intensity of the images between the original data and the LSTM predictions for each frame. It can be noticed that the predictions for the test set (red line) are more accurate than the predictions for the training set (blue line).

Regarding the computational time, evaluating each LSTM network requires, on average, 280 s per network, amounting to around 46 min for the whole model. The model was run using an Intel Core i7-1185G7 processor at 3.00 GHz × 8.

6 Discussion and Conclusion

CNFs have been proven to be essential building blocks for creating more sustainable materials in the future. Their mechanical strength and stiffness are fundamental properties for achieving environmentally friendly alternatives.

In our work, we propose an ML model that involves dimensionality reduction of the data through PCA and LSTM predictions of the time series created by the principal components. First, we scale down the number of variables of our data frames, keeping 10 PCs for the analysis. The 10 PCs already explain 85.19% of the total variance. The first PC is the most critical as it already retains 82.24% of the information. Representing 85% of the variance, we can already identify the flow well. Using fewer PCs would lead to neglect of some particle-like CNF clusters, especially before the flow stops. On the other hand, using more PCs would introduce more noise. Further analysis should be considered to distinguish the noise from the particle-like CNF clusters, as we would like to retain as much information as possible about the flow. By reducing the dimensionality of the workspace, we are left with new coordinates that, if plotted over the time steps of the data frames, produce ten time series. In the flow-stop experiments, we are interested in analyzing the Brownian dynamics for the given CNF materials and concentrations that emerge after the flow is stopped. The goal is to make this analysis automatically without performing physical experiments by stopping the flow. For this reason, we propose a time series prediction of the PCs after the flow is stopped. First, by reducing the dimensionality, we reduce the large amount of data available and the computational costs. Secondly, we can exploit ML models' capabilities in predicting and forecasting sequential data and time series.

Classic time series forecast methods include Autoregressive Moving Average (ARMA), Autoregressive Integrated Moving Average (ARIMA), and Seasonal Autoregressive Integrated Moving Average (SARIMA) models. They combine the autoregressive model to make predictions based on the data observed in the previous steps and the moving average model, which regulates the model by monitoring the average prediction errors of earlier predictions. While ARMA is the base model, ARIMA addresses time series trends, and SARIMA focuses on seasonality. Given that we are considering time series that do not have proper trends and seasonality and are looking at longer time frames, these models were not considered appropriate for our problem. Machine learning models are identified to be suitable for the kind of problem we are addressing. Especially given the number of data frames we are taking into account. Our choice of LSTM is because we can use a relatively more straightforward model with accurate results. Several networks and models are mentioned in Sect. 2. However, the basic LSTM network is a good candidate for the work without using more complex architectures that would require more parameter tuning and computational power. The trade-off between accuracy and simplicity makes us choose a model that stands in between. Without giving up on the efficiency of the predictions, we can build a model that uses few layers and neurons but can guarantee a precise result within minutes of training. A choice of transformer architecture would have possibly increased the accuracy of the predictions. However, the improvement would not have been so high as to disregard the simplicity of the LSTM that was just implemented. Moreover, when we want to achieve a broader goal using more datasets simultaneously, we believe that the computational time and power will increase, leading to the choice of a model like LSTM. However, variants of LSTM networks like Bidirectional LSTM and GRU were tried without gaining any benefit in accuracy and computational execution.

The results obtained using LSTM are already satisfactory. First, comparing the LSTM model to the PCA result for each PC shows that LSTM performs accurately both for the training and test. The most critical PC is the first, and the corresponding RMSE test score is 1.61 over the intensity. Considering that it is the component retaining most of the information, it follows that it requires more hyperparameter tuning. By looking at the accuracy of the LSTM predictions, the results are particularly accurate for the data after the flow stop. In particular, analyzing both the pointwise absolute error of the data frames and the mean absolute error shows that the accuracy in the predictions for the test dataset is higher than in the training dataset. This follows from the fact that for the data frames before the flow stops, higher intensities are present, showing the flow of CNFs. PCA has already struggled to capture a few of the particle-like CNF clusters, transferring some errors to the LSTM predictions. Therefore, the predictions for the data frames before the flow stop are less accurate. However, as the flow is stopped and less intensity is present in the images, PCA is capable of capturing most of the particle-like CNF clusters. As a consequence, the LSTM model can predict more precise results. The model takes about 45 min to run on a laptop. However, as we would like to implement this approach on a real-time production line, we will consider different computational approaches for future work, like running the code on GPUs and parallel computing, since the outcome of one network does not affect the others.

The work presented is the first step that allows for extending more experiments. It helps to focus and understand the flow behaviors in a reduced dimensional space created by PCA. In particular, we distinguish the PCA values before the flow stop and the PCA values after the flow stop. Using a machine learning method, we can predict the dynamics after the flow is stopped with reasonable accuracy. However, further steps will allow us to have a more robust model. The following steps will include training the model on the principal components before the stop of repeated experiments of the same material and same concentration and predicting the behavior after the flow is stopped of an experiment that the model did not train on (but maintaining the same material and same concentration). Further, we could train on the same sample but different concentrations and analyze the model's predictions on an unseen concentration of the same material. Finally, the ultimate goal would involve predicting the Brownian dynamics after the flow stop of any CNF material and concentration. Therefore, the preliminary work presented in this paper allows us to identify the type of models that can be used to achieve this goal and proves the feasibility of the predictions. If we can reach the final step, we could run a monitoring system at the production line without performing the actual stop in the experiment. We could, therefore, achieve an artificial stopping that will not need to stop the flow and will need extra manual work, which will benefit the industry.

References

1. Bengio, Y., Simard, P., Frasconi, P.: Learning long-term dependencies with gradient descent is difficult. IEEE Trans. Neural Netw. **5**(2), 157–166 (1994)
2. Berradi, Z., Lazaar, M.: Integration of principal component analysis and recurrent neural network to forecast the stock price of Casablanca stock exchange. Procedia Comput. Sci. **148**, 55–61 (2019)

3. Britz, D., Goldie, A., Luong, M.T., Le, Q.: Massive exploration of neural machine translation architectures. arXiv preprint arXiv:1703.03906 (2017)
4. Cho, K., Van Merriënboer, B., Bahdanau, D., Bengio, Y.: On the properties of neural machine translation: encoder-decoder approaches. arXiv preprint arXiv:1409.1259 (2014)
5. Chung, J., Gulcehre, C., Cho, K., Bengio, Y.: Empirical evaluation of gated recurrent neural networks on sequence modeling. arXiv preprint arXiv:1412.3555 (2014)
6. Danihelka, I., Wayne, G., Uria, B., Kalchbrenner, N., Graves, A.: Associative long short-term memory. In: International Conference on Machine Learning, pp. 1986–1994. PMLR (2016)
7. Elman, J.L.: Finding structure in time. Cogn. Sci. **14**(2), 179–211 (1990)
8. Fang, Q., Zhong, Y., Xie, C., Zhang, H., Li, S.: Research on PCA-LSTM-based short-term load forecasting method. In: IOP Conference Series: Earth and Environmental Science, vol. 495, p. 012015. IOP Publishing (2020)
9. Gensler, A., Henze, J., Sick, B., Raabe, N.: Deep learning for solar power forecasting-an approach using autoencoder and LSTM neural networks. In: 2016 IEEE International Conference on Systems, Man, and Cybernetics (SMC), pp. 002858–002865. IEEE (2016)
10. Graves, A., Fernández, S., Schmidhuber, J.: Bidirectional LSTM networks for improved phoneme classification and recognition. In: Duch, W., Kacprzyk, J., Oja, E., Zadrożny, S. (eds.) ICANN 2005. LNCS, vol. 3697, pp. 799–804. Springer, Heidelberg (2005). https://doi.org/10.1007/11550907_126
11. Hochreiter, S., Schmidhuber, J.: Long short-term memory. Neural Comput. **9**(8), 1735–1780 (1997)
12. Jolliffe, I.T., Cadima, J.: Principal component analysis: a review and recent developments. Philos. Trans. Roy. Soc. A: Math. Phys. Eng. Sci. **374**(2065), 20150202 (2016)
13. Jordan, M.: Attractor dynamics and parallelism in a connectionist sequential machine. In: Eighth Annual Conference of the Cognitive Science Society, 1986, pp. 513–546 (1986)
14. Kalchbrenner, N., Danihelka, I., Graves, A.: Grid long short-term memory. arXiv preprint arXiv:1507.01526 (2015)
15. Kim, T.Y., Cho, S.B.: Predicting residential energy consumption using CNN-LSTM neural networks. Energy **182**, 72–81 (2019)
16. Li, T., et al.: Developing fibrillated cellulose as a sustainable technological material. Nature **590**(7844), 47–56 (2021)
17. Lim, B., Arık, S.Ö., Loeff, N., Pfister, T.: Temporal fusion transformers for interpretable multi-horizon time series forecasting. Int. J. Forecast. **37**(4), 1748–1764 (2021)
18. Lindemann, B., Müller, T., Vietz, H., Jazdi, N., Weyrich, M.: A survey on long short-term memory networks for time series prediction. Procedia CIRP **99**, 650–655 (2021)
19. Medsker, L.R., Jain, L.: Recurrent neural networks. Des. Appl. **5**(64–67), 2 (2001)
20. Mittal, N., et al.: Multiscale control of nanocellulose assembly: transferring remarkable nanoscale fibril mechanics to macroscale fibers. ACS Nano **12**(7), 6378–6388 (2018)
21. Rosén, T., Hsiao, B.S., Söderberg, L.D.: Elucidating the opportunities and challenges for nanocellulose spinning. Adv. Mater. **33**(28), 2001238 (2021)
22. Rosén, T., Mittal, N., Roth, S.V., Zhang, P., Lundell, F., Söderberg, L.D.: Flow fields control nanostructural organization in semiflexible networks. Soft Matter **16**(23), 5439–5449 (2020)

23. Song, X., et al.: Time-series well performance prediction based on long short-term memory (LSTM) neural network model. J. Petrol. Sci. Eng. **186**, 106682 (2020)
24. Sutskever, I., Vinyals, O., Le, Q.V.: Sequence to sequence learning with neural networks. In: Advances in Neural Information Processing Systems, vol. 27 (2014)
25. Vaswani, A., et al.: Attention is all you need. In: Advances in Neural Information Processing Systems, vol. 30 (2017)
26. Veličković, P., et al.: Cross-modal recurrent models for weight objective prediction from multimodal time-series data. In: Proceedings of the 12th EAI International Conference on Pervasive Computing Technologies for Healthcare, pp. 178–186 (2018)
27. Villegas, R., Yang, J., Zou, Y., Sohn, S., Lin, X., Lee, H.: Learning to generate long-term future via hierarchical prediction. In: International Conference on Machine Learning, pp. 3560–3569. PMLR (2017)
28. Wen, Q., et al.: Transformers in time series: a survey. arXiv preprint arXiv:2202.07125 (2022)
29. Williams, R.J., Zipser, D.: A learning algorithm for continually running fully recurrent neural networks. Neural Comput. **1**(2), 270–280 (1989)
30. Wold, S., Esbensen, K., Geladi, P.: Principal component analysis. Chemom. Intell. Lab. Syst. **2**(1–3), 37–52 (1987)
31. Xie, W., et al.: PCA-LSTM anomaly detection and prediction method based on time series power data. In: 2022 China Automation Congress (CAC), pp. 5537–5542. IEEE (2022)
32. Xue, H., Huynh, D.Q., Reynolds, M.: SS-LSTM: a hierarchical LSTM model for pedestrian trajectory prediction. In: 2018 IEEE Winter Conference on Applications of Computer Vision (WACV), pp. 1186–1194. IEEE (2018)
33. Zeng, A., Chen, M., Zhang, L., Xu, Q.: Are transformers effective for time series forecasting? In: Proceedings of the AAAI Conference on Artificial Intelligence, vol. 37, pp. 11121–11128 (2023)
34. Zheng, X., Xiong, N.: Stock price prediction based on PCA-LSTM model. In: Proceedings of the 2022 5th International Conference on Mathematics and Statistics, pp. 79–83 (2022)

Cost-Effective Defense Timing Selection for Moving Target Defense in Satellite Computing Systems

Lin Zhang[1,2,3], Yunchuan Guo[1,2,3], Siyuan Leng[1,2,3], Xiaogang Cao[1,2,3], Fenghua Li[1,2,3], and Liang Fang[1,3(✉)]

[1] Institute of Information Engineering, Chinese Academy of Sciences, Beijing, China
{zhanglin1716,guoyunchuan,lengsiyuan,caoxiaogang,lifenghua,
fangliang}@iie.ac.cn
[2] School of Cyber Security, University of Chinese Academy of Sciences,
Beijing, China
[3] Key Laboratory of Cyberspace Security Defense, Beijing, China

Abstract. Satellite computing system (SCS), with its huge economic value, is suffering from increasing attacks. Moving Target Defense (MTD) can create the asymmetric situation between attacks and defenses by changing the attack surface. As SCS's limited defense resources, current MTD defense timing selection methods are not suitable for SCS. This paper proposes a Markov Game based Defense Timing Selection (MGDTS) approach for MTD in SCS. MGDTS formulates attack-defense adversarial relationship as a Markov game with incomplete information, and explicit costs are used to define the resource consumption of a defender. For defense timing decision, MGDTS uses a Markov decision process to construct the defense timing decision equation, and a real-time dynamic programming to solve the equation. Experimental results show that compared with other MTDs, MGDTS can improve the security of MTD while reducing its costs.

Keywords: Satellite computing system · Moving target defense · Defense timing selection · Network scanning

1 Introduction

Satellite Computing System (SCS), which virtualizes the computing resources of high-performance satellites, offers large-scale computing and communication services to space users (e.g., remote sensing satellites). Large aerospace companies are trying their best to develop SCSs. For example, Starlink constellation has more than 30000 Linux nodes (and more than 6000 microcontrollers) in space now and provides extra computing power [24]. Undoubtedly, SCSs own huge economic value. According to The Wall Street Journal, Starlink brought in $1.4 billion in revenue in 2022 [12]. However, due to its huge economic value, SCS is suffering from increasing attacks. For example, Viasat's KA-SAT satellite network was attacked during the Russia-Ukraine war, as a result, communication services in Ukraine and Europe were interrupted [18].

L. Franco et al. (Eds.): ICCS 2024, LNCS 14832, pp. 224–239, 2024.
https://doi.org/10.1007/978-3-031-63749-0_16

To effectively prevent attacks, active defense has been proposed in academia and industry, including honeypots, mimic defense [11] and Moving Target Defense (MTD) [6], where MTD constantly changes systems' attack surfaces and creates the asymmetric situation between attacks and defenses in cyber-security, thus proactively defending against network attacks. In MTD, defense timing selection, which is used to determine when to transform the attack surface, is crucial to increase defense capability and decrease defense cost.

From the perspective of defense timing selection, existing MTD can be divided into three categories: time-driven MTD, event-driven MTD, and hybrid-driven MTD, where a time-driven MTD changes the attack surface in a fixed or variable time period. Similarly, in an event-driven MTD, the attack surface is changed only after specific events are detected. The hybrid-driven scheme combines the above approaches and provides more proactive and immediate defense against attacks. However, existing hybrid-driven schemes cannot be directly applied to SCSs for the following reasons.

(1) **Defense resources in SCS are limited.** Although SCS owns huge communication resources, most resources are reserved for business and only negligible resources are for security defense. However, existing hybrid-driven MTDs consume a large number of communication resources because SCS has to execute multiple round communication interaction with users in MTD.

(2) **Existing hybrid-driven MTDs do not distinguish internal attacks from external attacks.** In SCS, internal attacks can succeed with a high probability and bring more serious consequences, but the probability of internal attacks is relatively low. Oppositely, external attacks against SCS happen with a high probability and succeed with a relatively low probability. As a result, if we do not distinguish these two types of attacks, we either consume more defense resources, or obtain poor defense effects.

(3) **Existing hybrid-driven MTDs rely on historical experience to select defense timing.** This scheme lacks quantitative analysis of the MTD effectiveness under different system security states. As a result, the selected defense timing may be unsuitable.

To address the above challenges, considering that network scanning is the first stage of network attacks, in the paper, a Markov Game based Defense Timing Selection (MGDTS) approach is proposed for MTD to guide the VM IP shuffling in SCS to resist network scanning. Our main contributions are as follows.

(1) Considering that defenders cannot accurately differentiate between interior attacks and exterior attacks, we formulate attack-defense adversarial relationship as a Markov game with incomplete information. In the game, to decrease the consumption of communication resources, explicit costs are used to define the communication resource consumption of a defender, and guide the defender's timing selection.

(2) We construct the defense timing decision equation using a Markov Decision Process (MDP), where the MTD effectiveness is quantified using Bellman

equations. Further, Real-Time Dynamic Programming (RTDP) is designed to solve the decision equation and decide the defense timing of MTD.

(3) A series of experiments are conducted and experimental results show that, compared with existing MTDs, MGDTS can effectively reduce the communication resource consumption of MTD while enhancing its security effect.

2   Related Work

MTD can be classified into time-driven, event-driven and hybrid time- and event-driven from the perspective of defense timing selection.

2.1   Time-Driven MTD

Time-driven MTD employs a fixed or variable time interval between adjacent system configuration changes. Fixed-interval MTD uses game theory [10,17,26], Stochastic Petri Net [1,27,28], and quantitative analysis models [2,14] to determine optimal time intervals. For example, Connell et al. [2] proposed a method to quantify the MTD security and system performance, and determined the optimal reconfiguration period by balancing the two factors. Variable-interval MTD adjusts time intervals based on attack severity [31] or system overhead [15], such as Zangeneh et al. [31] tuning the movement period inversely with the changing of the adversarial severity. Time-driven MTD has good proactivity, and can effectively defend against covert attacks, such as network infiltration [29]. But it cannot respond immediately and adaptively to detected attack behaviors.

2.2   Event-Driven MTD

Event-driven MTD takes security alerts [22] and system events (like system calls [13], system errors [7], security level changes [32] and Quality of Service (QoS) variations [8]) as triggers for system configuration changing. For example, Smith et al. [22] proposed an MTD triggered by intrusion alerts to mitigate denial of service, employing Neuro-Evolution of Augmented Topologies (NEAT) to construct an intrusion detector and initiating IP hopping on alerts. Khan et al. [7] introduced an MTD triggered by system errors to resist ransomware, immediately changing file extensions on detection of system errors related to ransomware. Event-driven MTD responds immediately and adaptively to detected attack behaviors, but lags in reacting to network attacks, and missed detections of security events can reduce its effectiveness.

2.3   Hybrid-Driven MTD

Hybrid-driven MTD combines the triggering conditions above, with the movement of the attack surface being initiated by time and events. For example, Potteiger et al. [19,20] use the temporal schedule and attack alerts to trigger address space randomization, thereby mitigating memory corruption attacks.

Huang et al. [4] rotate virtual servers based on a schedule and anomaly detection results to resist Web attacks. To counter scanning attacks, Prakash [21] and Xu [30] initiate MTD when pre-setting timers expire or attack behaviors are detected. Hybrid-driven MTD can synthesize the advantages of time-driven and event-driven MTD, offering proactive defense while responding immediately and adaptively to attacks. However, defense resources in SCS are limited, and existing hybrid-driven MTDs do not distinguish internal attacks from external attacks and rely on historical experience to select defense timing. As a result, existing hybrid-driven MTDs cannot be directly applied to SCS.

3   System and Threat Model

SCS enables sharing of computing satellite resources via resource pooling and virtualization, generating m VMs, each of which has a unique IP address assigned from the IP address pool Γ at any given moment (Fig. 1).

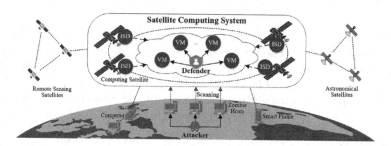

Fig. 1. System and threat model

To prepare for deeper active attacks, the ground-based attacker, who is either internal or external, uses zombie hosts and consumes resources such as money to scan VM IPs. The scanning policy is cyclic non-repetitive scanning, which involves iteratively scanning all IP addresses in Γ in a random order without repetition. To render the IP information obtained by the attacker invalid, the defender on SCS consumes communication resources to execute VM IP shuffling, which also affects the network QoS. When the IP Scanning Detector (ISD) on the computing satellite detects the zombie host IP scan, it informs the defender and blacklists the zombie host's IP. The defender selects shuffling timing based on the detection results from ISDs and time, and the attacker eliminates ISDs' interference by changing the zombie host IP.

Assume that whether the attacker is internal or external is his private information, while other information is public (including probabilities of the attacker's two identities). Both the rational attacker and defender make decisions based on all the information they possess. Assuming negligible VM IP switching time, the maximum transmission delays of the new IP address and the scanning packet determine the time required for each shuffling and each scan, respectively, and their maximum transmission delay is τ.

4   Markov Game

MGDTS views the entire game process as a time sequence $\{\tau_1, \tau_2, \ldots\}$ composed of time steps. In each time step τ, each zombie host and each VM can perform IP scan and VM IP shuffling at most once, respectively. The state of each VM at each time step is defined based on the security indexes that the defender can actually measure.

Definition 1. The state s of a VM is a tuple $<k, t>$. $k \in \{0, 1\}$ indicates whether ISDs have detected ($k = 1$) or not detected ($k = 0$) a scan for the VM's IP since the last VM IP shuffling. $t \in \{0, 1, \ldots, t_{max}\}$ represents the number of time steps since the last VM IP shuffling, where t_{max} is the maximum validity period of IP. When $t = t_{max}$, VM requests a new IP address. The state of a VM at time step τ_i is denoted as $s_{\tau_i} = <k_{\tau_i}, t_{\tau_i}>$, where $i \in \{1, 2, \ldots\}$.

MGDTS defines the players of the game as the attacker and the defender. Whether the attacker is internal or external, which affects his benefit, is not the defender's knowledge. Thus the game is an incomplete information game.

Definition 2. The type θ of the player refers to all the private information possessed by the player. The type θ_1 of the attacker has two possible values: θ_{11} for the attacker being internal and θ_{12} for the attacker being external. As all the defender information needed by the attacker is public, the type θ_2 of the defender has only one possible value. The probabilities of the attacker's type being θ_{11} and θ_{12}, denoted as $p(\theta_{11})$ and $p(\theta_{12})$ respectively, satisfy: $p(\theta_{11}) + p(\theta_{12}) = 1$.

Each player has a policy that guides their actions at each time step.

Definition 3. The action a of the player refers to the decision variable of the player at a certain time point. The action $a_{1,\tau_i}(\theta_1)$ of the attacker represents that the attacker of type θ_1 uses $n_{\tau_i}(\theta_1)$ zombie hosts to execute IP scanning at time step τ_i, where $\theta_1 \in \{\theta_{11}, \theta_{12}\}$, $n_{\tau_i}(\theta_1) \in \{0, 1, \ldots, n_{max}\}$, and n_{max} represents the total number of zombie hosts owned by the attacker. According to the linear model [7], the relationship between the defender's scanning detection success rate α and $n_{\tau_i}(\theta_1)$ is: $\alpha = min(dn_{\tau_i}(\theta_1), 1)$, where d represents the change rate of α relative to $n_{\tau_i}(\theta_1)$. The action $a_2(s) \in \{es, ns\}$ of the defender represents whether the defender executes (es) or does not execute (ns) an IP address shuffling for the VM in state s.

Definition 4. The policy π of the player specifies the action that the player selects at each time point. The policy π_1 of the attacker specifies the number of zombie hosts used by different types of attackers at each time step. The policy π_2 of the defender specifies the action taken by the defender in each state.

The attacker expends resources to launch scans and cope with the defender's interference, preparing for deeper active attacks on SCS. The defender protects SCS's assets through VM IP shuffling, which consumes communication resources and impacts network QoS, such as increased network latency. The costs, benefits and utilities of the players are defined below:

Definition 5. The cost of the attacker's action $a_{1,\tau_i}(\theta_1)$ includes the scanning cost and the cost of changing zombie host IPs. The scanning cost is the product of the scanning times and the single scan cost $c_{11}(\theta_1)$. The cost $c_{11}(\theta_1)$ represents the total resources expended by the attacker of type θ_1 on executing a scan. The cost of changing zombie host IPs is the product of the blacklisted IP count and the single IP changing cost $c_{12}(\theta_1)$. The cost $c_{12}(\theta_1)$ represents the total resources expended by the attacker of type θ_1 on changing a zombie host IP. The cost of the defender's action $a_2(s)$ includes explicit and implicit costs. The explicit cost $c_{21}(a_2(s))$ represents the total resources expended by the defender on executing $a_2(s)$. The implicit cost $c_{22}(a_2(s))$ represents the reduced network QoS when executing $a_2(s)$.

Definition 6. The benefit of the attacker's action $a_{1,\tau_i}(\theta_1)$ is the product of the scanning times and the single scan benefit $e_1(\theta_1)$. The benefit $e_1(\theta_1)$ represents the increase in SCS's asset risk resulting from the attacker of type θ_1 executing a scan. The asset risk is determined by the success rate of the subsequent active attack and the asset value λ. If an active attack is successful, SCS loses all its assets. The success rate of the active attack is proportional to the number of VM IPs obtained by the attacker. When the attacker of type θ_1 obtains all VM IPs, the success rate reaches its maximum value $\mu(\theta_1)$. The benefit $e_2(s, a_2(s))$ of the defender's action $a_2(s)$ on a VM in state s represents the reduction of SCS's asset risk resulting from executing $a_2(s)$.

Definition 7. The utility $v_{1,\pi_1,\tau_i}(\theta_1)$ of the attacker of type θ_1 represents the difference between his benefit and cost at time step τ_i under policy π_1. As the defender's benefit is related to the state of VM, the utility $v_{2,\pi_2}(s)$ of the defender represents the sum of his instantaneous utility and subsequent utility at a VM of state s under policy π_2. The instantaneous utility represents the difference between the defender's benefit and cost for executing his action, and the subsequent utility represents the expected utility in future states resulting from the defender's action.

As the attacker's utility is independent of VMs' state, the number of zombie hosts chosen by the attacker of type θ_1 at different time steps is the same, which can be denoted as $n(\theta_1)$. Thus the game is a Markov game associated with the process of VM state transition, where the attacker aims to decide $n(\theta_1)$ according to his utility and the defender aims to decide $a_2(s)$ in each state s.

5   Defense Timing Decision

5.1   Decision Equations of Players

Decision Equation of the Attacker. The rational attacker and defender aim to maximize utilities in decision-making. The attacker's decision variable is his policy, represented by $\pi_1 = <n(\theta_{11}), n(\theta_{12})>$, and the optimal policy is denoted as $\pi_1^* = <n^*(\theta_{11}), n^*(\theta_{12})>$. The attacker's objective functions are the utilities of all types of attackers. According to Definitions 5–7, the utility of the attacker of type θ_1 in τ under the policy π_1 is:

$$v_{1,\pi_1,\tau}(\theta_1) = n(\theta_1)\left(\frac{\lambda\mu(\theta_1)}{|\Gamma|} - c_{11}(\theta_1) - \alpha c_{12}(\theta_1)\right) \tag{1}$$

The attacker's constraints are the ranges of $n(\theta_{11})$ and $n(\theta_{12})$. Thus, the attacker's decision equation can be defined as:

$$\begin{cases} \forall \theta_1 \in \{\theta_{11}, \theta_{12}\}, max\,(v_{1,\pi_1,\tau}(\theta_1)) \\ s.t.\ \forall \theta_1 \in \{\theta_{11}, \theta_{12}\}, n(\theta_1) \in \{0, 1, \ldots, n_{max}\} \end{cases} \tag{2}$$

Decision Equation of the Defender. As the defender's utility is related to the VM's state, the VM's state transition is analyzed before defining the defender's decision equation. When the attacker of type θ_1 and the defender execute actions on the VM in state $s_{\tau_i} = <k_{\tau_i}, t_{\tau_i}>$ at time step τ_i, the VM's state changes based on the following rules:

(1) $a_2(s_{\tau_i}) = es$: The VM's state at the next time step changes to $s_{\tau_{i+1}} = <0, 0>$.
(2) $a_2(s_{\tau_i}) = ns$ and $k_{\tau_i} = 0$: Given the attacker's cyclic non-repetitive scanning policy and the condition $k_{\tau_i} = 0$, the probability $p(0|k_{\tau_i} = 0)$ that the attacker has scanned the VM's IP 0 times in the current round of non-repetitive scanning can be calculated using Bayes' theorem:

$$p(0|k_{\tau_i} = 0) = \frac{p(0) \times p(k_{\tau_i} = 0|0)}{\sum_{x \in \{0,1\}} p(x) \times p(k_{\tau_i} = 0|x)} \tag{3}$$
$$= \frac{|\Gamma| - n(\theta_1)t_{\tau_i} + \left\lfloor \frac{n(\theta_1)t_{\tau_i}}{|\Gamma|} \right\rfloor |\Gamma|}{|\Gamma| - n(\theta_1)t_{\tau_i}\alpha + \left\lfloor \frac{n(\theta_1)t_{\tau_i}}{|\Gamma|} \right\rfloor |\Gamma|\alpha}$$

where $p(x)$ represents the probability that the attacker has scanned the VM's IP x times in the current round of non-repetitive scanning, $p(k_{\tau_i} = 0|x)$ represents the probability of $k_{\tau_i} = 0$ given that the attacker has scanned the VM's IP x times in the current round of non-repetitive scanning. Therefore, when $a_2(s_{\tau_i}) = ns$ and $k_{\tau_i} = 0$, the probability that the VM's state changes to $s_{\tau_{i+1}} = <k_{\tau_i} + 1, t_{\tau_i} + 1>$ in the next time step is:

$$\beta_{t_{\tau_i}}(\theta_1) = p(0|k_{\tau_i} = 0) \times \frac{n(\theta_1)}{|\Gamma| - n(\theta_1)t_{\tau_i} + \left\lfloor \frac{n(\theta_1)t_{\tau_i}}{|\Gamma|} \right\rfloor |\Gamma|} \times \alpha$$
$$= \frac{n(\theta_1)\alpha}{|\Gamma| - n(\theta_1)t_{\tau_i}\alpha + \left\lfloor \frac{n(\theta_1)t_{\tau_i}}{|\Gamma|} \right\rfloor |\Gamma|\alpha} \tag{4}$$

and the probability that the VM's state changes to $s_{\tau_{i+1}} = <k_{\tau_i}, t_{\tau_i} + 1>$ in the next time step is $1 - \beta_{t_{\tau_i}}(\theta_1)$.
(3) $a_2(s_{\tau_i}) = ns$ and $k_{\tau_i} = 1$: The VM's state at the next time step changes to $s_{\tau_{i+1}} = <k_{\tau_i}, t_{\tau_i} + 1>$.

When $t_{max} = 2$, the state transition process of VM is shown in Fig. 2. As the VM state is finite and the state transition is random and Markovian, MGDTS constructs the defender's decision equation based on MDP.

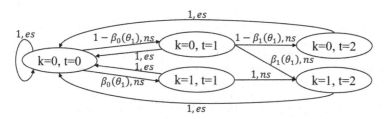

Fig. 2. Example of the state transition process

The defender's MDP is a six-tuple $<S, A, p\left(s^{'}|s, a_2(s), \theta_1\right), \pi_2, v_{2,\pi_2}(s), \gamma>$:

(1) $S = \{s = <k, t>|k \in \{0, 1\}, t \in \{0, 1, \ldots, t_{max}\}, k \le t\}$ is a set of states.

(2) $A = \{es, ns\}$ is a set of actions. The defender can only choose one action $a_2(s)$ from A to execute for VM in state $s \in S$.

(3) $p\left(s^{'}|s, a_2(s), \theta_1\right)$ is the probability that VM changes to the new state $s^{'}$ given the current state s, the taken action $a_2(s)$, and the attacker type θ_1.

(4) $\pi_2 = <a_2(s_1), a_2(s_2), \ldots, a_2(s_{|S|})>$ is the defender's policy.

(5) $v_{2,\pi_2}(s)$ is the defender's utility obtained in state s under policy π_2.

(6) $\gamma \in [0, 1]$ is the discount factor representing the importance of future utilities.

The defender's decision variable is the defender's policy, where the optimal policy is denoted as $\pi_2^* = <a_2^*(s_1), a_2^*(s_2), \ldots, a_2^*(s_{|S|})>$.

The defender's objective functions are the defender's utilities in all possible VM states. According to Definition 7 and MDP, the defender's utility in state s under policy π_2 is defined using the Bellman equation:

$$v_{2,\pi_2}(s) = \sum_{\theta_1 \in \{\theta_{11}, \theta_{12}\}} p(\theta_1) v_{2,\pi_2}(s, \theta_1) \tag{5}$$

$$v_{2,\pi_2}(s, \theta_1) = e_2(s, a_2(s)) - c_{21}(a_2(s)) - c_{22}(a_2(s))$$
$$+ \gamma \sum_{s' \in S} p\left(s^{'}|s, a_2(s), \theta_1\right) v_{2,\pi_2}\left(s^{'}, \theta_1\right) \tag{6}$$

where $v_{2,\pi_2}(s, \theta_1)$ is the defender's utility given the policy π_2, state s, and attacker type θ_1. According to Definition 6, the defender's benefit $e_2(s, a_2(s))$ is calculated as follows:

$$e_2(s, a_2(s)) = \sum_{s' \in S} p\left(s^{'}|s, a_2(s), \theta_1\right) \left(r(\theta_1, s) - r\left(\theta_1, s^{'}\right)\right) \tag{7}$$

where $r(\theta_1, s)$ is the risk value brought to SCS by VM in state s given the attacker type θ_1. Assuming that all VMs are identical in terms of configuration and software, except for their IP addresses, the increased risk value after VM IP leakage is $\frac{\lambda\mu(\theta_1)}{m}$. Then, the risk value $r(\theta_1, s)$ is calculated as follows:

$$r(\theta_1, s) = \eta(\theta_1, s) \times \frac{\lambda\mu(\theta_1)}{m} \qquad (8)$$

where $\eta(\theta_1, s)$ is the probability of VM IP leakage in state s given the attacker type θ_1, and is calculated based on the elements k and t of state s:

$$\eta(\theta_1, s) = \begin{cases} \frac{n(\theta_1)t - n(\theta_1)t\alpha}{|\Gamma| - n(\theta_1)t\alpha} & k = 0, t < \frac{|\Gamma|}{n(\theta_1)} \\ 1 & other \end{cases} \qquad (9)$$

The defender's constraints include the ranges of $a_2(s_1), a_2(s_2), ..., a_2(s_{|S|})$, as well as $n(\theta_{11}) = n^*(\theta_{11})$ and $n(\theta_{12}) = n^*(\theta_{12})$.

Based on the decision variable, objective functions and constraints, the defender's decision equation can be defined as:

$$\begin{cases} \forall s \in S, max\,(v_{2,\pi_2}(s)) \\ s.t.\; \forall s \in \{<k,t>|<k,t> \in S, t \neq t_{max}\}, a_2(s) \in A \\ \forall s \in \{<k,t>|<k,t> \in S, t = t_{max}\}, a_2(s) \in \{es\} \\ n(\theta_{11}) = n^*(\theta_{11}), n(\theta_{12}) = n^*(\theta_{12}) \end{cases} \qquad (10)$$

5.2   Solutions of Decision Equations

Solution of the Attacker. From Formula (2), the attacker's decision equation can be decomposed into 2 independent integer programming problems, and each problem corresponds to an objective function. The integer programming problem can be solved by the branch and bound [9] method, which can reduce the computational complexity of the solution process through pruning. The solution of the decision equation is obtained by integrating the solutions of all integer programming problems.

Solution of the Defender. The size of the defender's policy space is exponentially related to the state space, rendering Exhaustive Enumeration (EE) [5] computationally intractable. Synchronous dynamic programming algorithms, such as Value Iteration (VI) [3], update the value function or policy for all states simultaneously, which are inefficient when the state space is large. Asynchronous dynamic programming algorithms, such as RTDP, are more efficient by selectively updating states. Therefore, MGDTS uses RTDP to find the defender's optimal policy, and the process is shown in Algorithm 1.

6   Experiments

6.1   Experimental Environment and Settings

To evaluate the performance and effect of MGDTS, a SCS is constructed using the OMNET++ simulation software, as shown in Fig. 3.

The SCS consists of 3 computing satellites ($ComSat0 \sim ComSat2$) and the attacker is *Attacker*. According to references [16, 23, 25, 33], the parameter values of the system, attacker and defender are listed in Table 1:

The single scan cost is determined by the satellite tariff of the scan packet. The implicit and explicit costs of each VM IP shuffling are determined by the satellite tariffs of the IP notification packet and affected user service traffic, respectively.

Algorithm 1. Solution of the defender

Input: the defender's decision variable π_2, objective functions: $\forall s \in S, v_{2,\pi_2}(s)$, constraints, state transition probability $p\left(s^{'}|s, a_2(s), \theta_1\right)$, threshold of objective function change th_φ, maximum validity period of IP address t_{max}
Output: the optimal policy π_2^*

1:   $\forall s \in S, v_{2,\pi_2}(s) = 0, \pi_2^* =< es, es, \ldots, es >$
2:   Initialize the maximum change of objective functions: $\varphi = 0$, initialize the variables: $k = 0, t = 0, flag = false$
3: **while** $k \neq 0$ or $t \neq 0$ or $!flag$ **do**
4:      **if** $k = 0$ and $t = 0$ **then**
5:         $flag = true$
6:      Generate a new episode with the initial state $< k_{\tau_1}, t_{\tau_1} >=< k, t >$
7:      Set the episode's current time step τ_i to τ_1
8:      **while** $k_{\tau_i}! = 0$ or $t_{\tau_i}! = 0$ or $\tau_i = \tau_1$ **do**
9:         Based on the current values of objective functions of all states, calculate the utilities of performing all possible actions in state $< k_{\tau_i}, t_{\tau_i} >$
10:        Update $v_{2,\pi_2} (< k_{\tau_i}, t_{\tau_i} >)$ with the larger utility, and update π_2^* with the action that generates the larger utility
11:        Randomly select the next time step state $< k_{\tau_{i+1}}, t_{\tau_{i+1}} >$ based on the above action and the state transition probability
12:        $k_{\tau_i} = k_{\tau_{i+1}}, t_{\tau_i} = t_{\tau_{i+1}}, \tau_i = \tau_{i+1}$
13:      Calculate φ in the episode
14:      **if** $\varphi < th_\varphi$ **then**
15:         $t = \left(t + \left\lfloor \frac{k+1}{min(2, t+1)} \right\rfloor\right) \% (t_{max} + 1), k = (k+1) \% min(2, t+1)$
16:      **else**
17:         $flag = false$
18: **return** π_2^*

Fig. 3. Simulation construction of SCS

Table 1. Parameters of the system, attacker and defender

	Parameter	Value
SCS	VM IP address space	192.168.0.1–192.168.15.255
	User bandwidth	200 Mbps
	Average user service traffic in a VM	6.14 MB/s
	Packet transmission delay	≤200 ms
	Satellite tariff	0.25¥/MB
Attacker	Probabilities of θ_{11} & θ_{12}	0.3 & 0.7
	Max. active attack success rate of θ_{11} & θ_{12}	0.04 & 0.02
	Total number of zombie hosts	100
	Scanning packet size	78 B
	Cost of changing a single zombie host IP	0.2¥
Defender	Asset value	18000
	Change rate of α relative to $n(\theta_1)$	0.1
	IP notification packet size	40 KB
	Discount factor	0.9

6.2 Performance Evaluation

To evaluate RTDP's performance in solving MDP, this experiment compares the solutions and the update counts of objective functions of EE [5], VI [3] and RTDP in state sets with different sizes (by changing t_{max}). Table 2 shows that under different sizes of state sets, the solutions obtained by the 3 algorithms are consistent with the exact solution. Additionally, Fig. 4 reveals that under different sizes of state sets, the number of updates in RTDP is far lower than that in EE and VI. The above results indicate that RTDP can efficiently obtain solutions that are consistent with the exact solutions.

Table 2. Comparison of the solutions

$\|S\|$	Is the same as the exact solution?		
	EE	VI	RTDP
20481	Yes	Yes	Yes
40961	Yes	Yes	Yes
61441	Yes	Yes	Yes
81921	Yes	Yes	Yes
102401	Yes	Yes	Yes

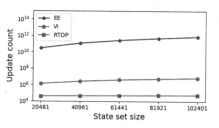

Fig. 4. Comparison of update counts

6.3 Defense Effect Evaluation

To evaluate MGDTS's defensive effect against IP scanning attacks, this experiment compares it with the typical time-driven, event-driven and hybrid-driven MTD methods, which are Quantitative Analytic Model based MTD (QAMMTD) [2], NEAT Detector based MTD (NDMTD) [22] and Attack Detector and Timer based MTD (ADTMTD) [30], respectively. All 4 MTD methods randomly select the new IP address of VM. The evaluation indexes include: defender utility, IP leakage time, bandwidth consumption and network QoS. The following results are the average of 10 simulation runs. Each run lasts 3600 s.

Defender Utility. Figure. 5 shows the defender utility obtained by the 4 MTD methods for each VM state $<k,t>$. Figure 5(a) and Fig. 5(b) demonstrate the trend of defender utility changing with t when $k = 0$, and Fig. 5(c) and Fig. 5(d) demonstrate the trend when $k = 1$, with $t_{max} = 10240$. According to Fig. 5, MGDTS achieves higher defender utilities compared to the other 3 methods for all states, especially when MGDTS takes different actions from the other 3 methods, such as when $k = 1$ and $0 \leq t < 65$ for QAMMTD, $k = 0$ and $90 < t < 10240$ for NDMTD, and $k = 0$ and $90 < t < 100$ for ADTMTD.

IP Leakage Time. The leakage time of a VM IP address is the duration from the moment an attacker successfully scans the address to the time when the address is changed. This experiment's IP leakage time is the sum of the leakage time of all VMs' used IP addresses. The change of IP leakage time in the simulation time under 4 MTD methods is shown in Fig. 6. Overall, the IP leakage time of MGDTS is less than that of the other 3 methods. This indicates that when the VM IP is leaked, MGDTS can replace the secure IP for VM more immediately, making the attacker's information invalid earlier, thereby improving the security of SCS.

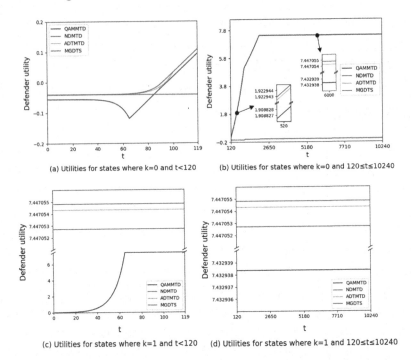

(a) Utilities for states where k=0 and t<120   (b) Utilities for states where k=0 and 120≤t≤10240

(c) Utilities for states where k=1 and t<120   (d) Utilities for states where k=1 and 120≤t≤10240

Fig. 5. Defender utilities obtained by 4 MTD methods

Bandwidth Consumption and Network QoS. The change of the cumulative bandwidth consumption and QoS impact of 4 MTD methods in the simulation time is shown in Fig. 7. Overall, the cumulative bandwidth consumption and QoS impact of MGDTS are lower than those of the other 3 methods. Combined with the experimental results of IP leakage time, it can be concluded that MGDTS provides better security for the system while reducing the explicit and implicit costs of MTD. Therefore, MGDTS is a cost-effective timing selection method.

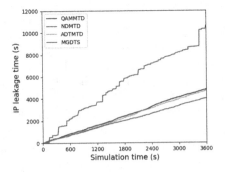

Fig. 6. IP leakage time under 4 MTD methods

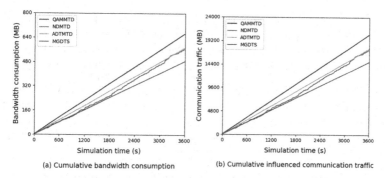

Fig. 7. Cumulative bandwidth consumption and QoS impact of 4 MTD methods

7   Conclusion

This paper introduces MGDTS, an approach for selecting the defense timing of MTD. It disrupts network attacks during the network scanning phase by guiding VM IP shuffling in SCS. MGDTS uses a Markov game with incomplete information to formulate attack-defense confrontation, and explicit costs are used to define the resource consumption of a defender. For defense timing decision, MGDTS uses a MDP and a RTDP to construct and solve the decision equation. Experiments show that MGDTS is efficient. It has higher defender utilities than existing MTDs in all VM states. Compared with these MTDs, MGDTS can better enhance the security of SCS while reducing the explicit and implicit costs of MTD. Consequently, it is cost-effective in defense timing selection.

Future work can consider combining VM IP shuffling with other types of MTD, such as VM migration.

Acknowledgment. This work is supported by the National Key Research and Development Program of China (No. 2023YFB3107605), the National Natural Science Foundation of China (No. 62202463), the Youth Innovation Promotion Association CAS (No. 2021154) and the Pandeng Project of IIE CAS.

References

1. Chen, Z., Chang, X., Han, Z., Yang, Y.: Numerical evaluation of job finish time under MTD environment. IEEE Access **8**, 11437–11446 (2020)
2. Connell, W., Menasce, D.A., Albanese, M.: Performance modeling of moving target defenses with reconfiguration limits. IEEE Trans. Depend. Secure Comput. (2018). https://doi.org/10.1109/TDSC.2018.2882825
3. Farahmand, A., Ghavamzadeh, M.: Pid accelerated value iteration algorithm. In: International Conference on Machine Learning, pp. 3143–3153. PMLR (2021)
4. Huang, Y., Ghosh, A.K.: Introducing diversity and uncertainty to create moving attack surfaces for web services. In: Jajodia, S., Ghosh, A., Swarup, V., Wang, C., Wang, X. (eds.) Moving Target Defense. Advances in Information Security, vol. 54, pp. 131–151. Springer, New York (2011). https://doi.org/10.1007/978-1-4614-0977-9_8

5. Ingels, F., Azaïs, R.: Enumeration of irredundant forests. Theor. Comput. Sci. **922**, 312–334 (2022). https://doi.org/10.1016/j.tcs.2022.04.033

6. Jajodia, S., Ghosh, A.K., Swarup, V., Wang, C., Wang, X.S.: Moving Target Defense: Creating Asymmetric Uncertainty for Cyber Threats, vol. 54. Springer, New York (2011). https://doi.org/10.1007/978-1-4614-0977-9

7. Khan, M.M., Hyder, M.F., Khan, S.M., Arshad, J., Khan, M.M.: Ransomware prevention using moving target defense based approach. Concurr. Comput. Pract. Exp. **35**(7), e7592 (2023)

8. Kim, D.S., Kim, M., Cho, J.H., Lim, H., Moore, T.J., Nelson, F.F.: Design and performance analysis of software defined networking based web services adopting moving target defense. In: DSN-S, pp. 43–44. IEEE (2020)

9. Lan, J., Brückner, B., Lomuscio, A.: A semidefinite relaxation based branch-and-bound method for tight neural network verification. In: AAAI Conference on Artificial Intelligence (2023). https://doi.org/10.1609/aaai.v37i12.26745

10. Lei, C., Zhang, H.Q., Wan, L.M., Liu, L., Ma, D.: Incomplete information Markov game theoretic approach to strategy generation for moving target defense. Comput. Commun. **116**, 184–199 (2018)

11. Lingshu, L., Jiangxing, W., Hongchao, H., Wenyan, L., Zehua, G.: Secure cloud architecture for 5G core network. Chin. J. Electron. (2021)

12. Maidenberg, M., Winkler, R.: Starlink surges but is still far short of spacex's goals, documents show (2023). https://www.wsj.com/tech/spacexs-starlink-demonstrates-its-power-but-still-needs-growth-9906c5b0?mod=hp_lead_pos5

13. Masumoto, T., Oo, W.K.K., Koide, H.: MTD: run-time system call mapping randomization. In: ISCSIC, pp. 257–263. IEEE (2021)

14. Mendonça, J., et al.: Performability analysis of services in a software-defined networking adopting time-based moving target defense mechanisms. In: Annual ACM Symposium on Applied Computing, pp. 1180–1189 (2020)

15. Mercado-Velázquez, A.A., Escamilla-Ambrosio, P.J., Ortiz-Rodriguez, F.: A moving target defense strategy for internet of things cybersecurity. IEEE Access **9**, 118406–118418 (2021). https://doi.org/10.1109/ACCESS.2021.3107403

16. Michel, F., Trevisan, M., Giordano, D., Bonaventure, O.: A first look at starlink performance. In: ACM Internet Measurement Conference, pp. 130–136 (2022)

17. Osei, A.B., Yeginati, S.R., Al Mtawa, Y., Halabi, T.: Optimized moving target defense against DDOS attacks in IoT networks: When to adapt? In: GLOBECOM 2022-2022 IEEE Global Communications Conference, pp. 2782–2787. IEEE (2022)

18. Pearson, J., Satter, R., Bing, C.J.: Exclusive: U.S. spy agency probes sabotage of satellite internet during Russian invasion, sources say (2022). https://www.reuters.com/world/europe/exclusive-us-spy-agency-probes-sabotage-satellite-internet-during-russian-2022-03-11/

19. Potteiger, B., Cai, F., Dubey, A., Koutsoukos, X., Zhang, Z.: Security in mixed time and event triggered cyber-physical systems using moving target defense. In: International Symposium on Real-Time Distributed Computing. IEEE (2020)

20. Potteiger, B., Dubey, A., Cai, F., Koutsoukos, X., Zhang, Z.: Moving target defense for the security and resilience of mixed time and event triggered cyber-physical systems. J. Syst. Architect. **125**, 102420 (2022)

21. Prakash, A., Wellman, M.P.: Empirical game-theoretic analysis for moving target defense. In: ACM Workshop on Moving Target Defense, pp. 57–65 (2015).https://doi.org/10.1145/2808475.2808483

22. Smith, R.J., Zincir-Heywood, A.N., Heywood, M.I., Jacobs, J.T.: Initiating a moving target network defense with a real-time neuro-evolutionary detector. In: Genetic and Evolutionary Computation Conference Companion, pp. 1095–1102 (2016)

23. SpaceX: Starlink (2024). https://www.starlink.com/
24. SpaceX: We are the spacex software team, ask us anything (2024). https://old.
    reddit.com/r/spacex/comments/gxb7j1/we_are_the_spacex_software_team_ask_us_
    anything/?limit=500
25. StormProxies: Stormproxies (2024). https://www.stormproxies.cn/
26. Tan, J., Zhang, H., Zhang, H., Hu, H., Lei, C., Qin, Z.: Optimal temporospatial
    strategy selection approach to moving target defense: a flipit differential game
    model. Comput. Secur. **108**, 102342 (2021)
27. Torquato, M., Maciel, P., Vieira, M.: Pymtdevaluator: a tool for time-based moving
    target defense evaluation: tool description paper. In: International Symposium on
    Software Reliability Engineering, pp. 357–366. IEEE (2021). https://doi.org/10.
    1109/ISSRE52982.2021.00045
28. Torquato, M., Maciel, P., Vieira, M.: Software rejuvenation meets moving target
    defense: modeling of time-based virtual machine migration approach. In: Interna-
    tional Symposium on Software Reliability Engineering, pp. 205–216. IEEE (2022)
29. Venkatesan, S., Albanese, M., Cybenko, G., Jajodia, S.: A moving target defense
    approach to disrupting stealthy botnets. In: Proceedings of the 2016 ACM Work-
    shop on Moving Target Defense, pp. 37–46 (2016)
30. Xu, X., Hu, H., Liu, Y., Zhang, H., Chang, D.: An adaptive IP hopping approach
    for moving target defense using a light-weight CNN detector. Secur. Commun.
    Netw. **2021**, 1–17 (2021). https://doi.org/10.1155/2021/8848473
31. Zangeneh, V., Shajari, M.: A cost-sensitive move selection strategy for moving
    target defense. Comput. Secur. **75**, 72–91 (2018)
32. Zheng, J., Siami Namin, A.: A Markov decision process to determine optimal poli-
    cies in moving target. In: ACM SIGSAC Conference on Computer and Communi-
    cations Security (2018). https://doi.org/10.1145/3243734.3278489
33. Zhou, Y., Cheng, G., Jiang, S., Zhao, Y., Chen, Z.: Cost-effective moving target
    defense against DDOS attacks using trilateral game and multi-objective Markov
    decision processes. Comput. Secur. **97**, 101976 (2020)

Energy- and Resource-Aware Graph-Based Microservices Placement in the Cloud-Fog-Edge Continuum

Imane Taleb[1]([✉]), Jean-Loup Guillaume[1], and Benjamin Duthil[2]

[1] L3i, La Rochelle University, La Rochelle, France
{imane.taleb,jean-loup.guillaume}@univ-lr.fr
[2] EIGSI, La Rochelle, France

Abstract. The development of Cloud-Fog-Edge computing infrastructures in response to the rapid advance of IoT technologies requires applications to be positioned closer to users at the edge of the network. Characterised by a geographically distributed configuration with numerous heterogeneous nodes, these infrastructures face challenges such as node failures, mobility constraints, resource limitations and network congestion. To address these issues, the adoption of microservices-based application architectures has been encouraged. However, the interdependencies and function calls between services require careful optimisation, as each has unique resource requirements. In this paper, we propose a new model and heuristic for the placement of microservices in the Cloud-Fog-Edge continuum, based on community detection and a greedy algorithm to optimise energy use while taking into account the resource constraints and ensuring that response time is acceptable. This method aims to reduce energy consumption and network load, thereby improving the efficiency and sustainability of the infrastructure. Results have been compared with different scenarios and show that our approach can significantly reduce energy consumption and make efficient use of resources.

Keywords: Energy efficiency · Microservices placement · Sustainability · Cloud-Fog-Edge Continuum · Optimization · Graph partitioning

1 Introduction

The development of smart technologies, particularly those associated with the Internet of Things (IoT), has played a central role in the ongoing transformation of the internet in response to user behaviour and emerging needs [1]. This evolution has led to the emergence of new applications, services and network infrastructures. Given the significant increase in data exchanges and the need for rapid response times, these applications need to be deployed as close as possible to users. The resulting reduction in latency and improved responsiveness of the services guarantee more efficient interaction and a better user experience.

L. Franco et al. (Eds.): ICCS 2024, LNCS 14832, pp. 240–255, 2024.
https://doi.org/10.1007/978-3-031-63749-0_17

To meet these requirements, the Cloud-Fog-Edge continuum is emerging as a key architectural solution, offering a strategic distribution of computing, storage resources and capacities closer to end users [2]. However, this infrastructure is also characterised by a geographically distributed configuration with numerous heterogeneous nodes facing challenges such as node mobility, failures, resource limitations and network congestion.

These constraints have encouraged the adoption of application architectures based on microservices, which are lightweight, flexible and modular modules, unlike traditional monolithic architectures. According to [3] microservices are defined as a series of small services that operate independently, offering modularity, ease of deployment and scalability. These are key advantages in heterogeneous and resource-constrained environments. However, microservices architectures involve dependencies and function calls between services, each of which has specific resource requirements and needs to be optimised. Inadequate positioning of microservices can cause high latency, over-consumption of energy and a higher environmental impact, increased costs and, more generally, impact on service availability and quality. This highlights the importance of adopting energy-efficient management and deployment strategies, particularly in the dynamic resource-constrained environments of the Cloud-Fog-Edge continuum.

Many researchers have focused on creating placement algorithms for distributed applications in Fog environments [4–8], focusing on metrics such as latency and quality of service. However, the specific placement of Fog applications based on microservices remains widely underexplored, particularly with regard to optimising energy consumption. For these reasons and since this is a complex multi-criteria optimisation problem, we propose a model and a heuristic for placing microservices in the Cloud-Fog-Edge continuum, based on community detection and a greedy algorithm. This approach aims to optimise energy consumption by minimising the communication distances between services, while taking into account node resource constraints and respecting response times. By strategically adapting microservices according to the interaction groups detected, we are able to reduce energy consumption and the load on the network.

This paper is structured as follows: Sect. 2 reviews recent research on microservice placement and load balancing. Section 3 describes the proposed model, and Sect. 4 details the heuristic and its constituent algorithms. Section 5 presents a comprehensive evaluation of the heuristic through various tests and analyses. Finally, the paper summarizes the main contributions in Sect. 6.

2   Background

Microservices placement methods mainly use meta-heuristics or graph-based approaches. Meta-heuristics encompass different approaches such as particle swarm optimisation (PSO) in its various forms, and different load balancing mechanisms to ensure efficient resource allocation. Graph-based methods are also used since they provide a systematic approach, using network topologies to refine placement decisions, offering a structured contrast to the adaptive and often probabilistic nature of meta-heuristic solutions.

Pallewatta et al. [4] introduce a method focusing on resource balancing and service efficiency utilizing Multi-Objective PSO (MOPSO) for optimal microservice instance creation, coupled with Load-Balancing Request Routing (LBRR) and Load-Balancing Instance Placement (LBIP) for equitable service request allocation and CPU utilization. The authors [6] approach microservices placement as a bi-criteria optimization problem. They also use a PSO-based metaheuristic to establish instances and identify the Pareto front, alongside techniques for evenly distributing requests akin to an unbounded bin-packing problem. Djemai et al. [9] formulated the placement problem as an optimization task, focusing on minimizing energy consumption and reducing delays in IoT applications. The approach uses a Discrete Particle Swarm Optimization (DPSO) algorithm, which operates with real-valued velocities to find optimal service locations. The physical setup is mapped as a graph, while IoT applications are represented as Directed Acyclic Graphs (DAGs). Mortazavi et al. [10] introduce a custom cuckoo search algorithm for service placement (CSA-SP) to optimize the positioning of services in fog nodes, aiming to minimize energy consumption while considering data transfer constraints and resource availability. The article [11] outlines a self-managing approach using the Whale Optimization Algorithm (WOA) for IoT services deployment across a three-tiered fog architecture, focusing on Quality of Service (QoS) for enhanced throughput.

Saboor et al. [12] propose a mathematical framework for the dynamic placement of container microservices based on rank matrix optimization, utilizing the stochastic matrix from microservices call graphs. They use eigenvector centrality to identify the most central microservices of an application, which will be grouped together in energy-efficient containers. Samani et al. [13] introduce a multilayer partitioning algorithm for Fog computing, tackling devices diversity by depicting resources as a four-layered graph, each reflecting similarities in network, CPU, memory, and storage among Fog nodes. Utilizing the Louvain algorithm [14], nodes are grouped by attributes, leading to a condensed graph of averaged clusters, thereby enabling application placement in similar resource environments. In [8] the authors propose a two-phase strategy for Fog computing: first, using community detection to map and partition nodes, placing applications based on deadlines; second, allocating services within Fog communities, using first-fit. Selimi et al. [15] propose a streamlined service deployment method for services in community micro-clouds, using a set number of clusters and network state information. Their Bandwidth and Availability-aware Service Placement algorithm involves node clustering with K-means, cluster head selection for optimal bandwidth, and dynamic service placement.

Authors in [7] introduce DECA, a method for energy and carbon-efficient placement in Edge Clouds, optimizing setup and migration to cut costs and emissions. Utilizing graph models and an A* algorithm, DECA considers geographical data and CO_2 outputs, aiming for a balance between energy use and expense. This strategic placement enhances both environmental sustainability and operational efficiency. In [5] authors present a fully decentralized placement strategy designed for Fog-Edge environments, utilizing Markov approximation to optimize communication among microservices, starting with random placement on Fog nodes and adjusting based on a Markov Chain representation.

In light of this state of the art, existing methods focus primarily on optimising availability, resource allocation, minimising delays and improving service quality. However, these approaches often neglect the critical dimension of energy efficiency, which is paramount given growing environmental concerns and the escalating costs associated with energy consumption in large-scale deployments. Our approach aims to fill this gap by specifically targeting the reduction of energy consumption in microservices placement. By incorporating energy-sensitive measures into our placement strategy, we aim not only to improve the efficiency of resource use and communication within the network, but also to significantly reduce the overall energy footprint of microservices deployment.

3 Model Formulation

This section presents our model. We first describe our network topology and microservices applications graphs. Then we expand upon this model to discuss the microservices placement problem.

3.1 Network Infrastructure

We propose a network model for the Cloud-Fog-Edge continuum. Each computing node is responsible for processing application placement requests. These nodes, spread across various locations, are heterogeneous in terms of geographical placement and available resources, including processing power and memory. The ability to adapt to these dynamic conditions is crucial to maintaining network functionality and ensuring continuous operation despite the inherent challenges posed by outages, node movement and varying levels of network congestion.

In our work, we consider three layers in the network architecture: Cloud layer, Fog layer and Edge/IoT layer. The devices of each level have specific characteristics and have resources to host and execute services.

- Cloud layer: refers to data centers that possess high-performance computing resources and resources high enough to be considered unlimited.
- Fog layer: is situated between the Cloud and the end-users. These nodes offer computational and storage services in close proximity to the end-users.
- Edge layer, or client layer: corresponds to a set of user devices from which placement requests originate.

We model the physical topology of Cloud-Fog-Edge architecture as a connected undirected graph $G_P = (V_P, E_P)$, where vertices represent physical execution nodes and edges are network links between these devices. We denote V_P the set of physical nodes. Each of these nodes $n_i \in V_P$ has the following characteristics: a speed of processing capacity cpu_i in MIPS, a memory size ram_i in GB and a power consumption ranging from p_i^{idle} when the device in on but not in use to p_i^{max} when used to the maximum, with a linear increase between

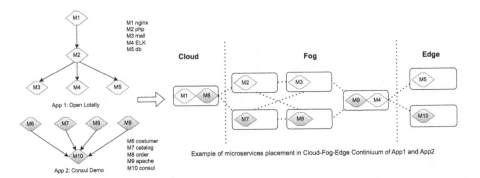

Fig. 1. Deployment and placement example of two microservices applications provided from the dataset [16]

these two values. Each physical link $l = \{n_i, n_j\} \in E_P$ is identified by the two nodes n_i and n_j it connects and is characterized by a bandwidth $bw_l = bw_{i,j}$ and a propagation delay $Pr_l = Pr_{i,j}$. Following this, we can build a logical link l'. We assume that if a node can communicate with another non adjacent node in the network, we can add a logical link between these two nodes. This link represents the shortest path P in terms of propagation delay composed of a sequence of physical links between the two nodes. $bw_{l'} = min_{l \in P}(bw_l)$ and $Pr_{l'} = \sum_{l \in P}(Pr_l)$, i.e. the minimum bandwidth on the path and the sum of the propagation delay on the path. Therefore, the network connections represent a complete graph with the consideration of physical and logical links (Fig. 1).

3.2 Microservices Applications and Function Paths

In our research, applications are a collection of microservices. Each of them is implemented as a standalone container, with specified resource needs. These microservices communicate, inter operate with each other and exchange data via function calls (API) to accomplish a specific application task. A micro-services application can be modeled as a directed acyclic graph (DAG) $G_S = (V_S, E_S)$ where the nodes represent the services $V_S = \{s_1, s_2, ..., s_t)$ and the edges are the dependencies and the function calls between the services. Each service $s_k \in V_S$ requires some resources consumption: a requested CPU cpu_k in million instructions and a requested RAM ram_k in GB. Each directed link $(s_k, s_l) \in E_S$ that connects s_k to s_l represents the request need and the data dependencies between these two microservices. It is characterised by a message size $data_{k,l}$.

We define three distinct categories of services, depending on the origin of the service request: firstly, sensors and actuators, for which requests come only from users, these components act as source nodes within the DAG application. The second category comprises internal services, responsible for processing operations requested exclusively by other microservices. The third category represents the final microservices, which constitute the sinks in the DAG.

Within the described microservices graph framework, we introduce the concept of Microservice Function Paths, or MFP, denoting an organized sequence of dependencies designed to execute a particular task. Essentially, an MFP encapsulates a microservice source and all its direct and indirect successors, extending down to the sinks. An MFP therefore consists of one and only one source, but may have several sinks as shown in Fig. 2. This graph has 4 sources and is therefore made up of 4 MFP. In microservices architecture, source nodes serve as the main entry points for user requests. The descendants of these nodes, and their connections, represent the dependency network and the trajectory of function calls throughout the system. We design the process in such a way that each action is initiated by a source node and traverses the network, ultimately ending at terminal nodes, which correspond to the final microservices where transactions or processes are completed.

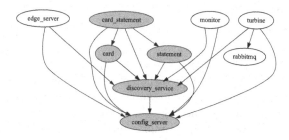

Fig. 2. Example of microservices dependencies graph of an IT blog called the card statement application taken from the dataset [16]. One MFP is highlighted

3.3 Energy Efficiency Placement Problem

In the context of Cloud-Fog-Edge computing, effective management of energy is crucial to maintain an optimal use of resources. Our goal is to place microservices in the network graph in a way that minimises energy consumption. We consider that the total energy consumption is the sum of energy from node execution and energy from network communication. Energy used by node n_i, see Eq. (1), is composed of a fixed term p_i^{idle} if there is at least one microservice placed on n_i plus the fraction of CPU used by each microservice s_k placed on node n_i. Similar to the works of [9–11], we consider a linear correlation between resources utilisation and energy consumption and therefore each microservice use a fraction of $\left(p_i^{max} - p_i^{idle}\right)$. Similarly, a communication between two microservices s_k and s_l uses some energy on both end nodes n_i and n_j involved for the total duration (transmission time plus propagation delay Pr) of the transmission, see Eq. (2). If two microservices are placed on the same physical node there is no communication (except memory transfers that have very low consumption) and we therefore consider the energy to be zero. Also, if two microservices are not placed on physically adjacent nodes but use a logical link, all intermediate nodes are switched on even if they are not used for computation.

$$
\begin{cases}
\text{Minimize } E_{tot} = \sum_{n_i \in V_P} E_{node}(n_i) + \sum_{(n_i,n_j) \in V_P^2} E_{com}(i,j) \\
\text{such that all capacity constraints are respected}
\end{cases}
$$

$$
E_{node}(n_i) = \overbrace{p_i^{idle}\delta(\_,n_i)}^{\text{Node}_i \text{ is on}} + \sum_{s_k \in V_S} \delta(s_k,n_i) \underbrace{\frac{cpu_k}{cpu_i}\left(p_i^{max} - p_i^{idle}\right)s_k}_{\text{Fraction of CPU for MS}} \tag{1}
$$

$$
E_{com}(n_i,n_j) = \sum_{(s_k,s_l) \in V_S^2} \delta(s_k,n_i)\delta(s_l,n_j)\overbrace{\left(\frac{data_{k,l}}{bw_{i,j}} + Pr_{i,j}\right)}^{\text{Transmission time}} \underbrace{\left[p_i^{max} + p_j^{max} - p_i^{idle} - p_j^{idle}\right]}_{\text{Energy used on both ends}} \tag{2}
$$

where $\delta(s_k,n_i)$ is equal to 1 if microservice s_k is placed on node n_i and 0 otherwise, and $\delta(\_,n_i)$ is equal to 1 if there is at least one microservice s_k such that $\delta(s_k,n_i) = 1$. Note that the communication energy is only described for physical links but the same applies for all the nodes alongside a logical link.

We focus solely on RAM as the primary constraint for our analysis. However, it is important to note that other constraints, such as storage limitations or other resources, could also be added in the microservices constraints:

$$
\sum_{s_k \in V_S} ram_k.\delta(s_k,n_i) \le ram_i, \quad \forall n_i \in V_P \tag{3}
$$

3.4   Minimum Execution Time of an MFP

The execution time of an MFP is a composite of two primary components: the processing time at each node and the time taken to traverse the network's edges when microservices on different network nodes must communicate with each other. If there is a direct or indirect dependency between two microservices, they cannot be run at the same time since one needs the other and conversely if there is no dependency then microservices can be executed in parallel.

To formalize this, we use the notion of depth in directed acyclic graph. In a DAG, the depth of a node measures its distance from the source nodes (i.e., nodes without predecessors) which are at depth 0. A node u is at depth k if the longest path from a source node to that node u has k edges. For example, the microservices in Fig. 2 are organized according to their depth levels. In this way, each depth represents an additional step in the graph's sequence of dependencies. With this definition, two microservices at the same depth can be run in parallel (no dependency at all) and two microservices that are not at the same depth can be interdependent and must not be run in parallel. In practice, two microservices can be independent even with a different depth if they are on different paths of the DAG. This notion of depth is extended to DAGs and MFP by considering it to be the maximum depth of their nodes.

From this, we define the minimum time execution of any MFP as its depth. Indeed, we consider that if the entire MFP is on the same server, i.e. without any network transmission, then all microservices on the longest path from a source to a sink will have to be executed one after the other. If microservices are not placed in the same network node, their execution will generate additional network communications and the overall execution time will increase.

It should be noted that, in practice, when a microservice s_k queries another microservice s_l, s_k will perform calculations, send a request to s_l, which will in turn perform calculations and possibly query other services, then s_l will send back a response to s_k, which will perform final calculations. The return time will just double the transmission times, but in a homogeneous way. We therefore disregard it and consider that there is no response.

4   Proposed Solution for Energy Efficiency Microservices Placement

To solve the placement problem, we propose a heuristic placement approach based on community detection that aims to optimize the allocation of applications in communities of devices by considering available resources and their requirements. In a nutshell, our approach consists of two main phases: firstly, we identify a community of network nodes in which to place the application, and secondly, we select the network nodes within the selected community.

4.1   Community Based Selection

In a graph, communities are groups of nodes that are densely connected. That means, there is a lot of edges within a community, and few or less edges between communities. In our context, a community therefore represents a group of network nodes which are strongly connected: if communications between microservices are required, it would be preferable to place them in an area where communication possibilities are numerous. We make the assumption that the communication between microservices that are not placed in the same network nodes consumes the most energy and that the power electric cost is lower within communities compared to outside communities. Once the network communities are determined, we select all the MFP from the microservices dependency graph. We aim to allocate entire MFP within a single community to minimize inter-community exchanges.

The first phase therefore starts by calculating communities of devices using Louvain algorithm [14] which is fast and gives good results. We assign a score to each community, determined by its available resources and its size. We identify communities that are both resource-rich and have a large number of nodes. The score gives priority to communities that are both resource-rich and large in size, in which we begin to place the largest MFP as follows:

$$\text{Community\_score} = \text{total\_cpu} \times \text{total\_ram} \times \text{num\_Node} \qquad (4)$$

These communities are then ranked in ascending order according to their respective scores. Similarly, the applications are assigned scores based on their resource requirements and sorted in descending order as follows:

$$\text{MFP\_score} = \text{total\_cpu} \times \text{total\_ram} \times \text{num\_MS.} \qquad (5)$$

The algorithm iterates through the MFP with the aim of placing the largest MFP in the smallest community that can fully accommodate it. If a community meets the requirements, the MFP is placed in that community, and the resource usages are updated accordingly. However, if no community has enough resources, the algorithm divides the MFP, by placing as many as possible of the partitions in the community with the highest score and the remaining partitions in another adjacent community identified by the shortest distance between them i.e. the lowest edge weight between two pairs of nodes belonging to the two communities. This iterative process ensures efficient allocation of applications in device communities, considering both available resources and application requirements.

Algorithm 1. Community detection based placement algorithm for MFP

Input Microservices applications DAG, Network graph
Output placement communities for all the applications
1: $C \leftarrow$ calculate device communities and assign a score to each
2: $OC \leftarrow$ order communities C in ascending score
3: $MFP \leftarrow$ extract the MFP from the DAGs App and assign a score to each
4: $OMFP \leftarrow$ order applications by descending score
5: $MFPlacement \leftarrow \emptyset$
6: **for** MF in OMFP **do**
7:     **for** Comm in OC **do**
8:         **if** Comm has enough resources **then**
9:             MFPlacement(Comm,MF)
10:             updateResourceUsages(Comm,MF)
11:             RecalculateScore(Comm)
12:             ReorderCommunitiesByScores(C)
13:         **end if**
14:     **end for**
15:     Divide MFP, place partitions in highest-scored and the remainder in adjacent communities
16: **end for**

4.2   Node Inside Community Selection

The second phase partitions MFP using the Kernighan-Lin algorithm [17] to obtain clusters of microservices. The algorithm is a graph partitioning method used to divide a graph into two parts while minimizing the number of edges between the two parts. It is a bipartitioning algorithm that offers granular control and predictability over partitioning, enabling a minimal but efficient division, in alignment with our goal of grouping as many microservices as possible. We start by dividing the MFP into two partitions and then applying the algorithm to each partition until we have 1 node per partition to obtain a complete (binary) dendrogram. We then calculate a fitness function for each node within the community, considering both available resources and the node betweenness centrality (node_BetCent), a measure that quantifies the number of times a node acts as a bridge on the shortest path between two other nodes. We assign a factor of 0.7 to available resources and a factor of 0.3 to intermediate centrality as we consider that available resources are more critical than intermediate centrality. More in-depth study should be performed to evaluate the impact of these 0.7/0.3 factors. In the context of microservices placement networks, leveraging this measure helps in identifying strategic nodes, thereby optimizing energy consumption through more efficient communication paths and reduced transmission distances:

$$Node\_fit = 0.7 \times \left(\frac{MFP\_resources\_requirement}{node\_available\_resources} \right) + 0.3 \times (node\_BetCent) \quad (6)$$

We then select the node with the highest fitness and we place as much microservices as possible on this node, respecting the dendrogram (i.e. placing MS in order one by one). Once the node is full, the other MS are placed on the next node and so on. This method ensures that microservices are distributed in a manner that optimizes resource utilization and maintains the logical grouping determined by the KL algorithm.

5   Evaluation and Results

Direct comparison of our heuristic with existing state-of-the-art approaches presents difficulties, mainly because few studies focus on microservices placement taking into account specific architectural dependencies and inter-microservices communications. Furthermore, the existing literature does not address the energy efficiency aspect of microservices placement. Therefore, to evaluate the effectiveness of our microservices placement heuristic, we carried out extensive validations to guarantee that (1) all the microservices of each MFP are deployed, (2) energy consumption is reduced, (3) resources consumption are respected, and finally (4) deployment time does not exceed twice the minimum execution time. To evaluate our heuristic approach, we developed three baseline scenarios:

- **No overload** where we use the same heuristic but limit network node resource usage at 70 %. This prevent overloading and ensure redundancy.

Algorithm 2. Node inside community-based application placement

---

1: $KLMFP \leftarrow$ generate the KL partitions
2: $ON \leftarrow$ order devices in community by the fitness
3: **for** MFPpartition in KLSFC **do**
4:     **for** partition in MFPpartition **do**
5:         **for** each N in ON **do**
6:             **if** N has enough resources **then**
7:                 **for** each service-id in partition **do**
8:                     **if** service-id not already placed **then**
9:                         SelectNode(N, service-id)
10:                        Update resource usages(N)
11:                        ReorderNodesByFitness(ON)
12:                    **else**
13:                        Continue to next service-id if already placed
14:                    **end if**
15:                **end for**
16:            **else**
17:                Continue to next node if insufficient resources
18:            **end if**
19:        **end for**
20:    **end for**
21: **end for**

- **Without community** where we do not make use of communities for MFP placement. Nevertheless, the algorithm assigns scores to nodes based on resources and attempts to rank MFP from largest to smallest, similar to the proposed heuristic.
- **Random** implements random placement, this algorithm starts by selecting a node at random and places as many MFP as possible in that vertex. Then another node is chosen, again at random, and this process is repeated until all the MFP have been placed.

All our tests were carried out on an Intel(R) Core(TM) i7-9850H CPU @ 2.60GHz 2.59 GHz computer, with 32 GB of RAM running on the Windows 10 Professional Education system. Simulations were performed using Python 3.9.13. We used the Networkx library for the graph algorithms used in our heuristic.

5.1 Experimental Setup

Network. We used the topology of Oteglobe, obtained from The Internet Zoo topology library [18]. Oteglobe is a European operator known for providing telecommunications services to network operators. We divided this topology into three sets of nodes to create a single Cloud node, which represents the most central node determined by its betweenness centrality, this measure quantifies the centrality of a node as a key bridge in the shortest paths between two other nodes. Of the remaining nodes, 50% are designated as Fog nodes (nodes with the highest betweenness except the Cloud one). Finally, the remaining nodes

Table 1. Parameters (network and microservices) used in the experiments

Parameter (unit)	Value or range
Network	Zoo Topology Dataset Oteglobe [18]
Number of nodes/links	81/103
Type of nodes	1 Cloud, 40 Fog, 40 Edge
CPU(GIPS)	[250-500] (Cloud), [1.5-4] (Fog), [1-1.5] (Edge)
Ram(GB)	[100-250] (Cloud), [2.5-5] (Fog), [1-2] (Edge)
p_{idle} (J)	145 (Cloud), 45 (Fog), 30 (Edge)
p_{max} (J)	320 (Cloud), 169 (Fog), 90 (Edge)
Bandwidth (KB/ms)	75
Link Propagation delay (ms)	1
Applications	Microservices dataset [16]
Microservices number	[5-25] depending on the application
CPU (MI)	[300-800]
Ram (MB)	[100-600]
Message size (KB)	[1500-4500]

are classified as Edge nodes. This partitioning strategy is aimed at modeling a diverse network infrastructure. For the simulation of resource characteristics (such as the number of cores, CPU speed and memory) of each Fog device, we employed a uniform random distribution similar to [13,19]. In addition, the bandwidth and latency configurations in the Fog network were defined in accordance with the methodologies used in previous studies [9,13]. We have additionally assigned weights to the topology graph based on the maximum and minimum electrical power values. All the specific details are provided in Table 1.

Microservices Applications. We used the microservices dependency graph dataset taken from [16]. This dataset contains 20 distinct application architectures, each application consisting of a variable number of microservices ranging from 5 to 25. Each is represented by an acyclic directed graph where the nodes represent the microservices and the edges the function calls and dependencies between them. We used a uniform random distribution to assign values for various computing resources (such as cpu_i speed, memory size ram_i and also to specify the message size $data_{i,j}$ between microservices. To obtain a realistic and more challenging placement environment, we have duplicated the MFP obtained from these applications.

5.2   Results Analysis

To analyse and evaluate the performance of our heuristic and compare it with the different scenarios mentioned above, we have used a number of metrics: (1) the

energy consumption of the placement by measuring intra-nodes energy in nodes
and inter-nodes communication energy; (2) the difference between the observed
execution time and the minimal execution time; (3) the number of nodes used
for the deployment of all MFP, considering that a node is used if at least one
microservice is placed on it or if it is used for transmission; and (4) the number
of links used as a measure of the dispersion of microservices.

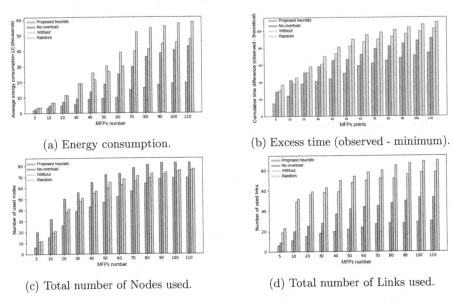

(a) Energy consumption.

(b) Excess time (observed - minimum).

(c) Total number of Nodes used.

(d) Total number of Links used.

Fig. 3. Comparison of the different microservice placement scenarios

Our heuristic has a clear advantage in terms of optimizing energy consumption over the other scenarios, as our evaluation shows in Fig. 3a Our approach reduces energy consumption in comparison with the strategy that limits node utilization to only 70% of capacity. This constraint restricts the efficient use of resources, since we are forced to place microservices in a higher number of nodes, generating more communications and therefore more energy between nodes. The comparison becomes even broader if we compare it to the version with a similar placement, but without the communities. This version, lacking community-based strategic allocation, does not place microservices that often communicate closely. Finally, the random strategy is not at all effective as expected. This disparity underlines the effectiveness of the heuristic we propose, not only in exploiting the full potential of the nodes, but also in ensuring a more coherent community-oriented deployment of services, which significantly improves the overall performance and energy consumption of the system.

Observed Time of MFP Execution. We compared the observed execution time obtained through the placement with the minimum execution times that depends on the depth of the MFP. Figure 3b shows the cumulative time difference between the observed values and the minimum values (where each MFP is placed on a single node), This allows us to assess the extent to which the deployment times observed correspond to theoretical expectations. The proposed heuristic shows lower cumulative time differences than the other methods, indicating closer alignment with theoretical expectations. By contrast, the "No Overload" approach shows slightly higher time differences, followed by the "Without Communities" and "Random" approaches, suggesting that these approaches deviate more from expected deployment times.

Nodes and Links Number. The analysis demonstrates that our method surpasses other approaches in efficiency by utilizing fewer links and nodes as the quantity of MFP to be placed increases (see Figs. 3c and 3d). This indicates a more effective optimization of communications and resource distribution. While the "No overload" strategy consumes more network nodes, its efficiency in link usage remains high due to strategic node selection within the same community, minimizing distance despite underutilized capacities because of imposed limitations. Conversely, the "Without community" and "Random" strategies, despite their relatively efficient employment of network nodes, exhibit significant inefficiencies in network communications. The "Random" strategy, in particular, is noted for its potential to select nodes that are exceedingly distant for placing microservices of the same MFP, leading to suboptimal communication paths.

6   Conclusion and Future Work

In this study, we developed a heuristic for the placement of microservices in IoT environments, based on community detection to optimise energy consumption in heterogeneous Cloud-Fog-Edge nodes with limited resources. Our proposed heuristic partitions the network into communities to identify nearby, strongly connected network nodes in order to prioritize the placement of MFP within these communities. We then used a best-fit approach to place the largest MFP in the smallest communities that could accommodate them. Our heuristic provides better results in terms of node and link utilization, and consequently a lower energy consumption than three other strategies.

In our future research, we aim to refine the placement of microservices in the Cloud-Fog-Edge continuum by including user nodes in the network topology, improving the management of microservice instances for energy efficiency, and implementing dynamic placement strategies that respond to real-time user needs and resource changes. In addition, we plan to integrate a wider range of environmental metrics, such as embodied energy and greenhouse gas emissions, to assess and optimise the environmental footprint of our deployment strategies. These efforts aim to optimise content delivery and minimise energy consumption, helping to create more sustainable and efficient IT environments.

References

1. Madakam, S., Lake, V., Lake, V., Lake, V., et al.: Internet of things (IoT): a literature review. J. Comput. Commun. **3**(05), 164 (2015)
2. Bittencourt, L., et al.: The internet of things, fog and cloud continuum: integration and challenges. Internet Things **3**, 134–155 (2018)
3. Lewis, J., Fowler, M.: Microservices: a definition of this new architectural term. MartinFowler. com **25**(14–26), 12 (2014)
4. Pallewatta, S., Kostakos, V., Buyya, R.: Microservices-based IoT application placement within heterogeneous and resource constrained fog computing environments. In: Proceedings of the 12th IEEE/ACM International Conference on Utility and Cloud Computing, pp. 71–81 (2019)
5. Kayal, P., Liebeherr, J.: Autonomic service placement in fog computing. In: 2019 IEEE 20th International Symposium on A World of Wireless, Mobile and Multimedia Networks (WoWMoM), pp. 1–9. IEEE (2019)
6. Yu, Y., Yang, J., Guo, C., Zheng, H., He, J.: Joint optimization of service request routing and instance placement in the microservice system. J. Netw. Comput. Appl. **147**, 102441 (2019)
7. Ahvar, E., Ahvar, S., Mann, Z.Ã., Crespi, N., Glitho, R., Garcia-Alfaro, J.: DECA: a dynamic energy cost and carbon emission-efficient application placement method for edge clouds. IEEE Access **9**, 70192–70213 (2021)
8. Lera, I., Guerrero, C., Juiz, C.: Availability-aware service placement policy in fog computing based on graph partitions. IEEE Internet Things J. **6**(2), 3641–3651 (2018)
9. Djemai, T., Stolf, P., Monteil, T., Pierson, J.-M.: A discrete particle swarm optimization approach for energy-efficient IoT services placement over fog infrastructures. In: 2019 18th International Symposium on Parallel and Distributed Computing (ISPDC), pp. 32–40. IEEE (2019)
10. Mortazavi, M.G., Shirvani, M.H., Dana, A.: A discrete cuckoo search algorithm for reliability-aware energy-efficient IoT applications multi-service deployment in fog environment. In: 2022 International Conference on Electrical, Computer and Energy Technologies (ICECET), pp. 1–6. IEEE (2022)
11. Ghobaei-Arani, M., Shahidinejad, A.: A cost-efficient IoT service placement approach using whale optimization algorithm in fog computing environment. Expert Syst. Appl. **200**, 117012 (2022)
12. Saboor, A., Mahmood, A.K., Omar, A.H., Hassan, M.F., Shah, S.N.M., Ahmadian, A.: Enabling rank-based distribution of microservices among containers for green cloud computing environment. Peer-to-Peer Netw. Appl. **15**(1), 77–91 (2022)
13. Samani, Z.N., Saurabh, N., Prodan, R.: Multilayer resource-aware partitioning for fog application placement. In: 2021 IEEE 5th International Conference on Fog and Edge Computing (ICFEC), pp. 9–18. IEEE (2021)
14. Blondel, V.D., Guillaume, J.-L., Lambiotte, R., Lefebvre, E.: Fast unfolding of communities in large networks. J. Stat. Mech: Theory Exp. **2008**(10), P10008 (2008)
15. Selimi, M., Cerdà-Alabern, L., Sánchez-Artigas, M., Freitag, F., Veiga, L.: Practical service placement approach for microservices architecture. In: 2017 17th IEEE/ACM International Symposium on Cluster, Cloud and Grid Computing (CCGRID), pp. 401–410. IEEE (2017)
16. Rahman, M.I., Taibi, D.: A curated dataset of microservices-based systems. In: Joint Proceedings of the Summer School on Software Maintenance and Evolution. CEUR-WS (2019)

17. Hendrickson, B., Leland, R.W., et al.: A multi-level algorithm for partitioning graphs. SC **95**(28), 1–14 (1995)
18. Knight, S., Nguyen, H.X., Falkner, N., Bowden, R., Roughan, M.: The internet topology zoo. IEEE J. Sel. Areas Commun. **29**(9), 1765–1775 (2011)
19. Guerrero, C., Lera, I., Juiz, C.: A lightweight decentralized service placement policy for performance optimization in fog computing. J. Ambient. Intell. Humaniz. Comput. **10**(6), 2435–2452 (2019)

Fast and Layout-Oblivious Tensor-Matrix Multiplication with BLAS

Cem Savaş Başsoy[(✉)]

Hamburg University of Technology, Schwarzenbergstrasse 95, Hamburg, Germany
cem.bassoy@gmail.com

Abstract. The tensor-matrix multiplication is a basic tensor operation required by various tensor methods such as the ALS and the HOSVD. This paper presents flexible high-performance algorithms that compute the tensor-matrix product according to the Loops-over-GEMM (LoG) approach. Our algorithms can process dense tensors with any linear tensor layout, arbitrary tensor order and dimensions all of which can be runtime variable. We discuss different tensor slicing methods with parallelization strategies and propose six algorithm versions that call BLAS with subtensors or tensor slices. Their performance is quantified on a set of tensors with various shapes and tensor orders. Our best performing version attains a median performance of 1.37 double precision Tflops on an Intel Xeon Gold 6248R processor using Intel's MKL. We show that the tensor layout does not affect the performance significantly. Our fastest implementation is on average at least 14.05% and up to 3.79x faster than other state-of-the-art approaches and actively developed libraries like Libtorch and Eigen.

Keywords: Tensor computation · Tensor contraction · Tensor-matrix multiplication · High-performance computing

1 Introduction

Tensor computations are found in many scientific fields such as computational neuroscience, pattern recognition, signal processing and data mining [6,14]. These computations use basic tensor operations as building blocks for decomposing and analyzing multidimensional data which are represented by tensors [7,9]. Tensor contractions are an important subset of basic operations that need to be fast for efficiently solving tensor methods.

There are three main approaches for implementing tensor contractions. The Transpose-Transpose-GEMM-Transpose (TGGT) approach reorganizes (flattens) tensors in order to perform a tensor contraction using optimized General Matrix Multiplication (GEMM) implementations [1,18]. Implementations of the GEMM-like Tensor-Tensor multiplication (GETT) method have macro-kernels that are similar to the ones used in fast GEMM implementations [12,19]. The

© The Author(s), under exclusive license to Springer Nature Switzerland AG 2024
L. Franco et al. (Eds.): ICCS 2024, LNCS 14832, pp. 256–271, 2024.
https://doi.org/10.1007/978-3-031-63749-0_18

third method is the Loops-over-GEMM (LoG) approach in which BLAS are utilized with multiple tensor slices or subtensors if possible [2,10,13,17]. Implementations of the LoG and TTGT approaches are in general easier to maintain and faster to port than GETT implementations which might need to adapt vector instructions or blocking parameters according to a processor's microarchitecture.

In this work, we present high-performance algorithms for the tensor-matrix multiplication which is used in many numerical methods such as the alternating least squares method [7,9]. It is a compute-bound tensor operation and has the same arithmetic intensity as a matrix-matrix multiplication which can almost reach the practical peak performance of a computing machine.

To our best knowledge, we are the first to combine the LoG approach described in [2,16] for tensor-vector multiplications with the findings on tensor slicing for the tensor-matrix multiplication in [10]. Our algorithms support dense tensors with any order, dimensions and any linear tensor layout including the first- and the last-order storage formats for any contraction mode all of which can be runtime variable. They compute the tensor-matrix product in parallel using efficient GEMM or batched GEMM without transposing or flattening tensors. Despite their high performance, all algorithms are layout-oblivious and provide a sustained performance independent of the tensor layout and without tuning.

Moreover, every proposed algorithm can be implemented with less than 150 lines of C++ code where the algorithmic complexity is reduced by the BLAS implementation and the corresponding selection of subtensors or tensor slices. We have provided an open-source C++ implementation of all algorithms and a python interface for convenience. While Intel's MKL is used for our benchmarks, the user is free to select any other library that provides the BLAS interface and even integrate it's own implementation to be library independent.

The analysis in this work quantifies the impact of the tensor layout, the tensor slicing method and parallel execution of slice-matrix multiplications with varying contraction modes. The runtime measurements of our implementations are compared with state-of-the-art approaches discussed in [12,15,19] including Libtorch and Eigen. In summary, the main findings of our work are:

- A tensor-matrix multiplication can be implemented by an in-place algorithm with 1 GEMV and 7 GEMM calls, supporting all combinations of contraction mode, tensor order and dimensions for any linear tensor layout.
- Our fastest algorithm with tensor slices is on average 17% faster than Intel's batched GEMM implementation when the contraction and leading dimensions of the tensors are greater than 256.
- The proposed algorithms are layout-oblivious. Their performance does not vary significantly for different tensor layouts if the contraction conditions remain the same.
- Our fastest algorithm computes the tensor-matrix multiplication on average, by at least 14.05% and up to a factor of 3.79 faster than other state-of-the art library implementations, including LibTorch and Eigen.

The remainder of the paper is organized as follows. Section 2 presents related work. Section 3 introduces some notation on tensors and defines the tensor-matrix multiplication. Algorithm design and methods for slicing and parallel execution are discussed in Sect. 4. Section 5 describes the test setup. Benchmark results are presented in Sect. 6. Conclusions are drawn in Sect. 7.

2   Related Work

Springer et al. [19] present a tensor-contraction generator TCCG and the GETT approach for dense tensor contractions that is inspired from the design of a high-performance GEMM. Their unified code generator selects implementations from generated GETT, LoG and TTGT candidates. Their findings show that among 48 different contractions 15% of LoG-based implementations are the fastest.

Matthews [12] presents a runtime flexible tensor contraction library that uses GETT approach as well. He describes block-scatter-matrix algorithm which uses a special layout for the tensor contraction. The proposed algorithm yields results that feature a similar runtime behavior to those presented in [19].

Li et al. [10] introduce InTensLi, a framework that generates in-place tensor-matrix multiplication according to the LOG approach. The authors discusses optimization and tuning techniques for slicing and parallelizing the operation. With optimized tuning parameters, they report a speedup of up to 4x over the TTGT-based MATLAB tensor toolbox library discussed in [1].

Başsoy [2] presents LoG-based algorithms that compute the tensor-vector product. They support dense tensors with linear tensor layouts, arbitrary dimensions and tensor order. The presented approach is to divide into eight cases calling GEMV and DOT. He reports average speedups of 6.1x and 4.0x compared to implementations that use the TTGT and GETT approach, respectively.

Pawlowski et al. [16] propose morton-ordered blocked layout for a mode-oblivious performance of the tensor-vector multiplication. Their algorithm iterate over blocked tensors and perform tensor-vector multiplications on blocked tensors. They are able to achieve high performance and mode-oblivious computations.

3   Background

Notation. An order-p tensor is a p-dimensional array [11] where tensor elements are contiguously stored in memory. We write a, $\mathbf{a}$, $\mathbf{A}$ and $\underline{\mathbf{A}}$ in order to denote scalars, vectors, matrices and tensors. If not otherwise mentioned, we assume $\underline{\mathbf{A}}$ to have order $p > 2$. The p-tuple $\mathbf{n} = (n_1, n_2, \ldots, n_p)$ will be referred to as a dimension tuple with $n_r > 1$. We will use round brackets $\underline{\mathbf{A}}(i_1, i_2, \ldots, i_p)$ or $\underline{\mathbf{A}}(\mathbf{i})$ to denote a tensor element where $\mathbf{i} = (i_1, i_2, \ldots, i_p)$ is a multi-index. A subtensor is denoted by $\underline{\mathbf{A}}'$ and references elements of a tensor $\underline{\mathbf{A}}$. They are specified by a selection grid consisting of p index ranges. The index range in this work shall either address all indices of a given mode or a by a single index i_r with $1 \leq r \leq p$. Elements n'_r of a subtensor's dimension tuple $\mathbf{n}'$ are $n'_r = n_r$ if all indices of mode r are selected or $n_r' = 1$. We will annotate subtensors using only their non-unit modes such as $\mathbf{A}_{u,v,w}'$ where $n_u > 1, n_v > 1$ and $n_w > 1$ and $1 \leq u \neq v \neq w \leq p$. The remaining single indices of a selection grid correspond to the loop induction variables. A subtensor is called a slice $\mathbf{A}_{u,v}'$ if only two modes of $\mathbf{A}$ is selected with a full range. A fiber $\mathbf{A}_u'$ is a tensor slice with only one dimension greater than 1.

Linear Tensor Layouts. We use a layout tuple $\boldsymbol{\pi} \in \mathbb{N}^p$ to encode all linear tensor layouts including the first-order or last-order layout. They contain permuted tensor modes whose priority is given by their index. For instance, the general k-order tensor layout for an order-p tensor is given by the layout tuple $\boldsymbol{\pi}$ with $\pi_r = k - r + 1$ for $1 < r \leq k$ and r for $k < r \leq p$. The first- and last-order storage formats are given by $\boldsymbol{\pi}_F = (1, 2, \ldots, p)$ and $\boldsymbol{\pi}_L = (p, p-1, \ldots, 1)$. An inverse layout tuple $\boldsymbol{\pi}^{-1}$ is defined by $\boldsymbol{\pi}^{-1}(\boldsymbol{\pi}(k)) = k$. Given a layout tuple $\boldsymbol{\pi}$ with p modes, the π_r-th element of a stride tuple is given by $w_{\pi_r} = \prod_{k=1}^{r-1} n_{\pi_k}$ for $1 < r \leq p$ and $w_{\pi_1} = 1$. Tensor elements of the π_1-th mode are contiguously stored in memory. The location of tensor elements is determined by the tensor layout and the layout function. For a given tensor layout and stride tuple, a layout function $\lambda_\mathbf{w}$ maps a multi-index to a scalar index with $\lambda_\mathbf{w}(\mathbf{i}) = \sum_{r=1}^p w_r(i_r - 1)$, see [3, 16].

Non-modifying Flattening and Reshaping. The flattening operation $\varphi_{r,q}$ transforms an order-p tensor $\mathbf{A}$ to another order-p' view $\mathbf{B}$ that has different a shape $\mathbf{m}$ and layout $\boldsymbol{\tau}$ tuple of length p' with $p' = p - q + r$ and $1 \leq r < q \leq p$. It is related to the tensor unfolding operation as defined in [7, p.459] but neither changes the element ordering nor copies tensor elements. Given a layout tuple $\boldsymbol{\pi}$ of $\mathbf{A}$, the flattening operation $\varphi_{r,q}$ is defined for contiguous modes $\hat{\boldsymbol{\pi}} = (\pi_r, \pi_{r+1}, \ldots, \pi_q)$ of $\boldsymbol{\pi}$. Let $j = 0$ if $k \leq r$ and $j = q - r$ otherwise for $1 \leq k \leq p'$. Then the resulting layout tuple $\boldsymbol{\tau} = (\tau_1, \ldots, \tau_{p'})$ of $\mathbf{B}$ is given by $\tau_r = \min(\boldsymbol{\pi}_{r,q})$ and $\tau_k = \pi_{k+j} - s_k$ if $k \neq r$ where $s_k = |\{\pi_i \mid \pi_{k+j} > \pi_i \wedge \pi_i \neq \min(\hat{\boldsymbol{\pi}}) \wedge r \leq i \leq p\}|$. Elements of the shape tuple $\mathbf{m}$ are defined by $m_{\tau_r} = \prod_{k=r}^q n_{\pi_k}$ and $m_{\tau_k} = n_{\pi_{k+j}}$ if $k \neq r$. Reshaping ρ transforms an order-p tensor $\mathbf{A}$ to another order-p tensor $\mathbf{B}$ with the shape tuple $\mathbf{m}$ and layout tuple $\boldsymbol{\tau}$ tuples, both of length p. In this work, it permutes the shape and layout tuple simultaneously without changing the element ordering and without copying tensor elements. The operation ρ is defined by a permutation tuple $\boldsymbol{\rho} = (\rho_1, \ldots, \rho_p)$ that defines elements of $\mathbf{m}$ and $\boldsymbol{\tau}$ with $m_r = n_{\rho_r}$ and $\tau_r = \pi_{\rho_r}$, respectively.

Tensor-Matrix Multiplication. Let $\mathbf{A}$ and $\mathbf{C}$ be order-p tensors with shapes $\mathbf{n}_a = (n_1, \ldots, n_q, \ldots, n_p)$ and $\mathbf{n}_c = (n_1, \ldots, n_{q-1}, m, n_{q+1}, \ldots, n_p)$. Let $\mathbf{B}$ be a matrix of shape $\mathbf{n}_b = (m, n_q)$. A mode-q tensor-matrix product is denoted by $\mathbf{C} = \mathbf{A} \times_q \mathbf{B}$. An element of $\mathbf{C}$ is defined by

$$\mathbf{C}(i_1, \ldots, i_{q-1}, j, i_{q+1}, \ldots, i_p) = \sum_{i_q=1}^{n_q} \mathbf{A}(i_1, \ldots, i_q, \ldots, i_p) \cdot \mathbf{B}(j, i_q) \qquad (1)$$

with $1 \leq i_r \leq n_r$ and $1 \leq j \leq m$ [7,10]. Mode q is called the contraction mode with $1 \leq q \leq p$. The tensor-matrix multiplication generalizes the computational aspect of the two-dimensional case $\mathbf{C} = \mathbf{B} \cdot \mathbf{A}$ if $p = 2$ and $q = 1$. Its arithmetic intensity is equal to that of a matrix-matrix multiplication and is not memory-bound. In the following, we assume that the tensors $\underline{\mathbf{A}}$ and $\underline{\mathbf{C}}$ have the same tensor layout $\boldsymbol{\pi}$. Elements of matrix $\mathbf{B}$ can be stored either in the column-major or row-major format. Without loss of generality, we assume $\underline{\mathbf{B}}$ to have the row-major storage format. Also note that the following approach can be applied, if indices j and i_q of matrix $\mathbf{B}$ are swapped.

4   Algorithm Design

4.1   Baseline Algorithm with Contiguous Memory Access

Equation 1 can be implemented with one sequential `C++` function which consists of a nested recursion which is described in [3]. The algorithm consists of two `if` statements with an `else` branch that computes a fiber-matrix product with two loops. The outer loop iterates with j over dimension m of $\underline{\mathbf{C}}$ and $\mathbf{B}$. The inner loop iterates with i_q over dimension n_q of $\underline{\mathbf{A}}$ and $\mathbf{B}$, computing an inner product. However, elements of $\underline{\mathbf{A}}$ and $\underline{\mathbf{C}}$ are accessed non-contiguously if $\pi_1 \neq q$. Matrix $\mathbf{B}$ is contiguously accessed if i_q or j is incremented with unit-strides depending on the storage format of $\underline{\mathbf{B}}$.

The above algorithm can be modified such that tensor elements are accessed according to the tensor layout that is specified by layout tuple $\boldsymbol{\pi}$. The resulting baseline algorithm is given in Algorithm 1 which contiguously accesses memory for $\pi_1 \neq q$ and $p > 1$. In line number 5, one multi-index element i_{π_r} is incremented with a stride w_{π_r}. With increasing recursion level and decreasing r, indices are incremented with smaller strides as $w_{\pi_r} \leq w_{\pi_{r+1}}$. The second `if` statement in line number 4 allows the loop over dimension π_1 to be placed into the base case which contains three loops performing a slice-matrix multiplication. The inner-most loop increments i_{π_1} and contiguously accesses tensor elements of $\underline{\mathbf{A}}$ and $\underline{\mathbf{C}}$. The second loop increments i_q with which elements of $\mathbf{B}$ are contiguously accessed if $\mathbf{B}$ is stored in the row-major format. The third loop increments j and could be placed as the second loop if $\mathbf{B}$ is stored in the column-major format.

While spatial data locality is improved by adjusting the loop ordering, slices $\underline{\mathbf{A}}'_{\pi_1, q}$, fibers $\underline{\mathbf{C}}'_{\pi_1}$ and elements $\mathbf{B}(j, i_q)$ are accessed m, n_q and n_{π_1} times, respectively. While the specified fiber of $\underline{\mathbf{C}}$ can fit into first or second level cache, slice elements of $\underline{\mathbf{A}}$ are unlikely to fit in the local caches if the slice size $n_{\pi_1} \times n_q$ is large leading to higher cache misses and suboptimal performance. Instead of optimizing for better temporal data locality, we use existing high-performance BLAS implementations for the base case.

4.2   BLAS-Based Algorithms with Tensor Slices

Algorithm 1 is the starting point for BLAS-based algorithms. It computes the mode-q tensor-matrix product by recursively multiplying tensor slices with the

```
1  tensor_times_matrix(A, B, C, n, i, m, q, q̂, r)
2  │ if r = q̂ then
3  │ │  tensor_times_matrix(A, B, C, n, i, m, q, q̂, r − 1)
4  │ else if r > 1 then
5  │ │ for i_{π_r} ← 1 to n_{π_r} do
6  │ │ └  tensor_times_matrix(A, B, C, n, i, m, q, q̂, r − 1)
7  │ else
8  │ │ for j ← 1 to m do
9  │ │ │ for i_q ← 1 to n_q do
10 │ │ │ │ for i_{π_1} ← 1 to n_{π_1} do
11 │ │ │ │ └  C(i_1, ..., i_{q−1}, j, i_{q+1}, ..., i_p) += A(i_1, ..., i_q, ..., i_p) · B(j, i_q)
```

Algorithm 1: Modified baseline algorithm with contiguous memory access for the tensor-matrix multiplication. The tensor order p must be greater than 1 and the contraction mode q must satisfy $1 \leq q \leq p$ and $\pi_1 \neq q$. The initial call must happen with $r = p$ where **n** is the shape tuple of $\underline{\mathbf{A}}$ and m is the q-th dimension of $\underline{\mathbf{C}}$.

matrix for $q \neq \pi_1$. Instead of optimizing the multiplication, it is possible to insert a gemm routine in the base case and to compute slice-matrix products. Additionally, there are seven other (corner) cases where a single gemv or gemm call suffices. All eight cases are listed in table 1. The arguments of gemv or gemm are chosen depending on the tensor order p, tensor layout $\boldsymbol{\pi}$ and contraction mode q except for parameter CBLAS_ORDER which is set to CblasRowMajor. The CblasColMajor format can be used as well if the following case descriptions are changed accordingly. The parameter arguments are given in our C++ library. Note that with table 1 all linear tensor layout are supported with no limitations on tensor order and contraction mode.

Case 1: If $p = 1$, The tensor-vector product $\underline{\mathbf{A}} \times_1 \mathbf{B}$ can be computed with a gemv operation where $\underline{\mathbf{A}}$ is an order-1 tensor $\mathbf{a}$ of length n_1 such that $\mathbf{a}^T \cdot \mathbf{B}$.

Case 2-5: If $p = 2$, $\underline{\mathbf{A}}$ and $\underline{\mathbf{C}}$ are order-2 tensors with dimensions n_1 and n_2. In this case the tensor-matrix product can be computed with a single gemm. If $\mathbf{A}$ and $\mathbf{C}$ have the column-major format with $\boldsymbol{\pi} = (1, 2)$, gemm either executes $\mathbf{C} = \mathbf{A} \cdot \mathbf{B}^T$ for $q = 1$ or $\mathbf{C} = \mathbf{B} \cdot \mathbf{A}$ for $q = 2$. Reshaping both matrices using ρ with $\boldsymbol{\rho} = (2, 1)$, gemm interprets $\mathbf{C}$ and $\mathbf{A}$ as matrices in row-major format although both are stored column-wise. If $\mathbf{A}$ and $\mathbf{C}$ have the row-major format with $\boldsymbol{\pi} = (2, 1)$, gemm either executes $\mathbf{C} = \mathbf{B} \cdot \mathbf{A}$ for $q = 1$ or $\mathbf{C} = \mathbf{A} \cdot \mathbf{B}^T$ for $q = 2$. The transposition of $\mathbf{B}$ is necessary for the cases 2 and 5 which is independent of the chosen layout.

Case 6-7: If $p > 2$ and if $q = \pi_1$(case 6), a single gemm with the corresponding arguments executes $\mathbf{C} = \mathbf{A} \cdot \mathbf{B}^T$ and computes a tensor-matrix product $\underline{\mathbf{C}} = \underline{\mathbf{A}} \times_{\pi_1} \mathbf{B}$. Tensors $\underline{\mathbf{A}}$ and $\underline{\mathbf{C}}$ are flattened with $\varphi_{2,p}$ to row-major matrices $\mathbf{A}$ and $\mathbf{C}$. Matrix $\mathbf{A}$ has $\bar{n}_{\pi_1} = \bar{n}/n_{\pi_1}$ rows and n_{π_1} columns while matrix $\mathbf{C}$ has the

Table 1. Eight cases with gemv and gemm for the mode-q tensor-matrix multiplication. Arguments T, M, N, etc. of the BLAS are chosen with respect to the tensor order p, layout $\boldsymbol{\pi}$ and contraction mode q where T specifies if **B** is transposed. gemm* denotes multiple gemm calls with different tensor slices. Argument $\bar{n}_q$ for case 6 and 7 is given by $\bar{n}_q = 1/n_q \prod_r^p n_r$. Matrix **B** has the row-major format.

Case	Order p	Layout $\boldsymbol{\pi}$	Mode q	Routine	T	M	N	K	A	LDA	B	LDB	LDC
1	1	-	1	gemv	-	m	n_1	-	**B**	n_1	$\underline{\mathbf{A}}$	-	-
2	2	$(1,2)$	1	gemm	B	n_2	m	n_1	$\underline{\mathbf{A}}$	n_1	**B**	n_1	m
3	2	$(1,2)$	2	gemm	-	m	n_1	n_2	**B**	n_2	$\underline{\mathbf{A}}$	n_1	n_1
4	2	$(2,1)$	1	gemm	-	m	n_2	n_1	**B**	n_1	$\underline{\mathbf{A}}$	n_2	n_2
5	2	$(2,1)$	2	gemm	B	n_1	m	n_2	$\underline{\mathbf{A}}$	n_2	**B**	n_2	m
6	> 2	any	π_1	gemm	B	$\bar{n}_q$	m	n_q	$\underline{\mathbf{A}}$	n_q	**B**	n_q	m
7	> 2	any	π_p	gemm	-	m	$\bar{n}_q$	n_q	**B**	n_q	$\underline{\mathbf{A}}$	$\bar{n}_q$	$\bar{n}_q$
8	> 2	any	$\pi_2,..,\pi_{p-1}$	gemm*	-	m	n_{π_1}	n_q	**B**	n_q	$\underline{\mathbf{A}}$	w_q	w_q

same number of rows and m columns. If $\pi_p = q$ (case 7), $\underline{\mathbf{A}}$ and $\underline{\mathbf{C}}$ are flattened with $\varphi_{1,p-1}$ to column-major matrices **A** and **C**. Matrix **A** has n_{π_p} rows and $\bar{n}_{\pi_p} = \bar{n}/n_{\pi_p}$ columns while **C** has m rows and the same number of columns. In this case, a single gemm executes $\mathbf{C} = \mathbf{B} \cdot \mathbf{A}$ and computes $\underline{\mathbf{C}} = \underline{\mathbf{A}} \times_{\pi_p} \mathbf{B}$. Noticeably, the desired contraction are performed without copy operations, see Subsect. 3.

Case 8 ($p > 2$): If the tensor order is greater than 2 with $\pi_1 \neq q$ and $\pi_p \neq q$, the modified baseline algorithm 1 is used to successively call $\bar{n}/(n_q \cdot n_{\pi_1})$ times gemm with different tensor slices of $\underline{\mathbf{C}}$ and $\underline{\mathbf{A}}$. Each gemm computes one slice $\underline{\mathbf{C}}'_{\pi_1,q}$ of the tensor-matrix product $\underline{\mathbf{C}}$ using the corresponding tensor slices $\underline{\mathbf{A}}'_{\pi_1,q}$ and the matrix **B**. The matrix-matrix product $\mathbf{C} = \mathbf{B} \cdot \mathbf{A}$ is performed by interpreting both tensor slices as row-major matrices **A** and **C** which have the dimensions (n_q, n_{π_1}) and (m, n_{π_1}), respectively. Please note that Algorithm 2 in [10] suggests to transpose matrix **B**.

4.3   BLAS-Based Algorithms with Subtensors

The eighth case can be further optimized by slicing larger subtensors and use additional dimensions for the slice-matrix multiplication. The selected dimensions must adhere to flatten the subtensor into a matrix without reordering or copying elements, see lemma 4.1 in [10]. The number of additional modes is $\hat{q} - 1$ with $\hat{q} = \boldsymbol{\pi}^{-1}(q)$ and the corresponding modes are $\pi_1, \pi_2, \ldots, \pi_{\hat{q}-1}$. Applying flattening $\varphi_{1,\hat{q}-1}$ and reshaping ρ with $\boldsymbol{\rho} = (2,1)$ on a subtensor of $\underline{\mathbf{A}}$ yields a row-major matrix **A** with shape $(n_q, \prod_{r=1}^{\hat{q}-1} n_{\pi_r})$. Analogously, tensor $\underline{\mathbf{C}}$ becomes a row-major matrix with the shape $(m, \prod_{r=1}^{\hat{q}-1} n_{\pi_r})$. This description supports all linear tensor layouts and generalizes lemma 4.2 in [10].

Algorithm 1 needs a minor modification so that gemm can be used with flattened subtensors instead of tensor slices. The non-base case of the modified algorithm only iterates over dimensions with indices that are larger than $\hat{q}$, omitting the first $\hat{q}$ modes $\boldsymbol{\pi}_{1,\hat{q}} = (\pi_1, \ldots, \pi_{\hat{q}})$ with $\pi_{\hat{q}} = q$. The conditions in line 2 and 4 are changed to $1 < r \leq \hat{q}$ and $\hat{q} < r$, respectively. The single indices of the subtensors $\underline{\mathbf{A}}'_{\boldsymbol{\pi}_{1,\hat{q}}}$ and $\underline{\mathbf{C}}'_{\boldsymbol{\pi}_{1,\hat{q}}}$ are given by the loop induction variables that belong to the π_r-th loop with $\hat{q} + 1 \leq r \leq p$.

4.4 Parallel BLAS-Based Algorithms

Next, three parallel approaches for the eighth case. Note that cases 1 to 7 already call a multi-threaded `gemm`.

Sequential Loops and Multithreaded Matrix Multiplication. A simple approach is to not modify algorithm 1 and sequentially call a multi-threaded `gemm` in the base case as described in Subsect. 4.2. This is beneficial if $q = \pi_{p-1}$, the inner dimensions $n_{\pi_1}, \ldots, n_q$ are large or if the outer-most dimension n_{π_p} is smaller than the available processor cores. However, when the above conditions are not met, the algorithm executes multi-threaded `gemm` with small subtensors. This might lead to a low utilization of available computational resources. This algorithm version will be referred to as `<seq-loops,par-gemm>`.

Parallel Loops and Multithreaded Matrix Multiplication. A more advanced version of the above algorithm executes a single-threaded `gemm` in parallel with all available (free) modes. The number of free modes depends on the tensor slicing. If subtensors are used, all $\pi_{\hat{q}+1}, \ldots, \pi_p$ modes are free and can be used for parallel execution. In case of tensor slices, only dimensions with indices π_1 and $\pi_{\hat{q}}$ are free.

Using tensor slices for the multiplication, $\underline{\mathbf{A}}$ and $\underline{\mathbf{C}}$ are flattened twice with $\varphi_{\pi_{\hat{q}+1},\pi_p}$ and $\varphi_{\pi_2,\pi_{\hat{q}-1}}$. The flattened tensors are of order 4 with dimensions n_{π_1}, $\hat{n}_{\pi_2}$, n_q or m, $\hat{n}_{\pi_4}$ where $\hat{n}_{\pi_2} = \prod_{r=2}^{\hat{q}-1} n_{\pi_r}$ and $\hat{n}_{\pi_4} = \prod_{r=\hat{q}+1}^{p} n_{\pi_r}$. This approach transforms the tree-recursion into two loops. The outer loop iterates over $\hat{n}_{\pi_4}$ while the inner loop iterates over $\hat{n}_{\pi_2}$ calling `gemm` with slices $\underline{\mathbf{A}}'_{\boldsymbol{\pi}_{1,q}}$ and $\underline{\mathbf{C}}'_{\boldsymbol{\pi}_{1,q}}$. Both loops are parallelized using `omp parallel for` together with the `collapse(2)` and the `num_threads` clause which specifies the thread number.

If subtensors are used, both tensors are flattened twice with $\varphi_{\pi_{\hat{q}+1},\pi_p}$ and $\varphi_{\pi_1,\pi_{\hat{q}-1}}$. The flattened tensors are of order 3 with dimensions $\hat{n}_{\pi_1}$, n_q or m, $\hat{n}_{\pi_4}$ where $\hat{n}_{\pi_1} = \prod_{r=1}^{\hat{q}-1} n_{\pi_r}$ and $\hat{n}_{\pi_4} = \prod_{r=\hat{q}+1}^{p} n_{\pi_r}$. The corresponding algorithm consists of one loops which iterates over $\hat{n}_{\pi_4}$ calling single-threaded `gemm` with multiple subtensors $\underline{\mathbf{A}}'_{\boldsymbol{\pi}',q}$ and $\underline{\mathbf{C}}'_{\boldsymbol{\pi}',q}$ with $\boldsymbol{\pi}' = (\pi_1, \ldots, \pi_{\hat{q}-1})$.

Both algorithm variants will be referred to as `<par-loops,seq-gemm>` which can be used with subtensors or tensor slices. Note that `<seq-loops,par-gemm>` and `<par-loops,seq-gemm>` are opposing versions where either `gemm` or the free loops are performed in parallel. The all-parallel version `<par-loops,par-gemm>` executes available loops in parallel where each loop thread executes a multi-threaded `gemm` with either subtensors or tensor slices.

Multithreaded Batched Matrix Multiplication. The next version of the base algorithm is a modified version of the general subtensor-matrix approach that calls a single batched `gemm` for the eighth case. The subtensor dimensions and remaining `gemm` arguments remain the same. The library implementation is responsible how subtensor-matrix multiplications are executed and if subtensors are further divided into smaller subtensors or tensor slices. This version will be referred to as the `<gemm_batch>` variant.

5 Experimental Setup

Computing System. The experiments have been carried out on an Intel Xeon Gold 6248R processor with a Cascade micro-architecture. The processor consists of 24 cores operating at a base frequency of 3 GHz. With 24 cores and a peak AVX-512 boost frequency of 2.5 GHz, the processor achieves a theoretical data throughput of ca. 1.92 double precision Tflops. We measured a peak performance of 1.78 double precision Tflops using the likwid performance tool.

We have used the GNU compiler v10.2 with the highest optimization level -O3 and -march=native, -pthread and -fopenmp. Loops within for the eighth case have been parallelized using GCC's OpenMP v4.5 implementation. We have used the `gemv` and `gemm` implementation of the 2020.4 Intel MKL and its own threading library `mkl_intel_thread` together with the threading runtime library `libiomp5`.

If not otherwise mentioned, both tensors $\underline{\mathbf{A}}$ and $\underline{\mathbf{C}}$ are stored according to the first-order tensor layout. Matrix **B** has the row-major storage format.

Tensor Shapes. We have used asymmetrically and symmetrically shaped tensors in order to cover many use cases. The dimension tuples of both shape types are organized within two three-dimensional arrays with which tensors are initialized. The dimension array for the first shape type contains $720 = 9 \times 8 \times 10$ dimension tuples where the row number is the tensor order ranging from 2 to 10. For each tensor order, 8 tensor instances with increasing tensor size is generated. A special feature of this test set is that the contraction dimension and the leading dimension are disproportionately large. The second set consists of $336 = 6 \times 8 \times 7$ dimensions tuples where the tensor order ranges from 2 to 7 and has 8 dimension tuples for each order. Each tensor dimension within the second set is 2^{12}, 2^8, 2^6, 2^5, 2^4 and 2^3. A detailed explanation of the tensor shape setup is given in [2,3].

6 Results and Discussion

Slicing Methods. The next paragraphs analyze the two proposed slicing methods and discuss runtime results of `<par-loops,seq-gemm>` and `<gemm-batch>` using asymmetrically and symmetrically shaped tensors. Figure 1 contains six contour plots (performance maps) in which `<par-loops,seq-gemm>` either uses subtensors

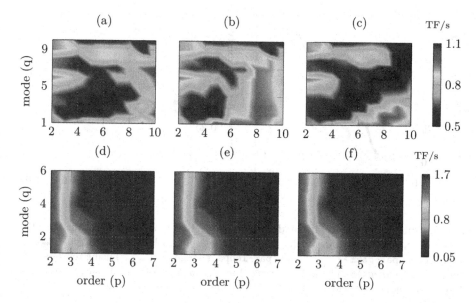

Fig. 1. Performance maps in double-precision Tflops of the proposed algorithms with varying tensor orders p and contraction modes q. Tensors are asymmetrically shaped on the top plots and symmetrically shaped on the bottom plots. In (a) and (d) function `<gemm_batch>` is executed, in (b) and (e) `<par-loops,seq-gemm>` with tensor slices, in (c) and (f) `<par-loops,seq-gemm>` with subtensors.

or tensor slices and `<gemm-batch>` loops over subtensors only. Every performance value within the maps represent a mean value that has been averaged over tensor sizes for a tensor order[1].

For asymmetrically shaped tensors, function `<par-loops,seq-gemm>` with tensor slices performs on average 18% better than with subtensors and is on average 11% faster than Intel's `gemm_batch` routine. It reaches almost 1.1 Tflops for non-edge cases with $q > 2$ and $p > 6$. This suggests that the Intel's implementation does not divide subtensors into smaller blocks.

With symmetrically shaped tensors, `<par-loops,seq-gemm>` with tensor slices and `<gemm-batch>` almost show the same runtime behavior, reaching 221.52 Gflops and 236.21 Gflops, respectively. Moreover, the slicing method seems to have only little affect on the performance of `<par-loops,seq-gemm>`. In contrast to the performance maps with asymmetrically shaped tensors, all functions almost reach the attainable peak performance of 1.7 Tflops when $p = 2$. This can by the fact that both dimensions are equal or larger than 4096 enabling `gemm` to operate under optimal conditions.

Parallelization Methods. The contour plots in Fig. 1 contain performance data of all cases except for 4 and 5, see Table 1. The effects of the presented slic-

---

[1] Note that Fig. 2 suggests that the contraction mode q can be greater than p which is not possible. Our profiling program sets $q = p$ in such cases.

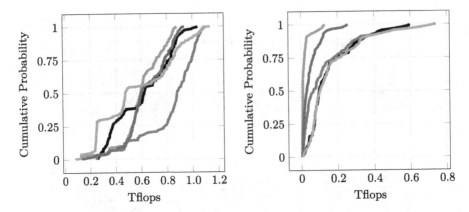

Fig. 2. Cumulative performance distributions of the proposed algorithms for the eighth case. Each distribution line belongs to one algorithm: `<gemm_batch>` ━━ , `<seq-loops,par-gemm>` (▬▬) and `<par-loops,seq-gemm>` (▬▬) using tensor slices, `<seq-loops,par-gemm>` (▬▬) and `<par-loops,seq-gemm>` (▬▬) using subtensors. Tensors are asymmetrically (left plot) and symmetrically shaped (right plot).

ing and parallelization methods can be better understood if performance data of only the eighth case is examined. Figure 2 contains cumulative performance distributions of all the proposed algorithms which are generated `gemm` or `gemm_batch` calls within case 8. As the distribution is empirically computed, the probability y of a point (x,y) on a distribution function corresponds to the number of test cases of a particular algorithm that achieves x or less Tflops. For instance, function `<seq-loops,par-gemm>` with subtensors computes the tensor-matrix product for 50% percent of the test cases with equal to or less than 0.6 Tflops in case of asymmetrically shaped tensor. Consequently, distribution functions with an exponential growth are favorable while logarithmic behavior is less desirable. The test set cardinality for case 8 is 255 for asymmetrically shaped tensors and 91 for symmetrically ones.

In case of asymmetrically shaped tensors, `<par-loops,seq-gemm>` with tensor slices performs best and outperforms `<gemm_batch>`. One unexpected finding is that function `<seq-loops,par-gemm>` with any slicing strategy performs better than `<gemm_batch>` when the tensor order p and contraction mode q satisfy $4 \leq p \leq 7$ and $2 \leq q \leq 4$, respectively. Functions executed with symmetrically shaped tensors reach at most 743 Gflops for the eighth case which is less than half of the attainable peak performance of 1.7 Tflops. This is expected as cases 2 and 3 are not considered. Functions `<par-loops,seq-gemm>` with subtensors and `<gemm_batch>` have almost the same performance distribution outperforming `<seq-loops,par-gemm>` for almost every test case. Function `<par-loops,seq-gemm>` with tensor slices is on average almost as fast as with subtensors. However, if the tensor order is greater than 3 and the tensor dimensions are less than 64, its running time increases by almost a factor of 2.

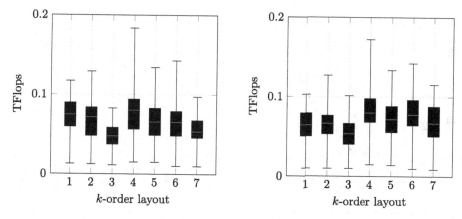

Fig. 3. Box plots visualizing performance statics in double-precision Tflops of `<gemm_batch>` (left) and `<par-loops,seq-gemm>` with subtensors (right). Box plot number k denotes the k-order tensor layout of symmetrically shaped tensors with order 7.

These observations suggest to use `<par-loops,seq-gemm>` with tensor slices for common cases in which the leading and contraction dimensions are larger than 64 elements. Subtensors should only be used if the leading dimension n_{π_1} of $\mathbf{A}_{\pi_1,q}$ and $\mathbf{C}_{\pi_1,q}$ falls below 64. This strategy is different to the one presented in [10] that maximizes the number of modes involved in the matrix multiply. We have also observed no performance improvement if `par-gemm` was used with `par-loops` which is why their distribution functions are not shown in Fig. 2. Moreover, in most cases the `seq-loops` implementations are independent of the tensor shape slower than `par-loops`, even for smaller tensor slices.

Layout-Oblivious Algorithms. Fig. 3 contains two subfigures visualizing performance statics in double-precision Tflops of `<gemm_batch>` (left subfigure) and `<par-loops,seq-gemm>` with subtensors (right subfigure). Each box plot with the number k has been computed from benchmark data with symmetrically shaped order-7 tensors with the k-order tensor layout. The 1-order and 7-order layout, for instance, are the first- and last-order storage formats for the order-7 tensor with $\pi_F = (1, 2, ..., 7)$ and $\pi_L = (7, 6, ..., 1)$. The definition of k-order tensor layouts can be found in Sect. 3.

The low performance of around 70 Gflops can be attributed to the fact that the contraction dimension of subtensors of tensor slices of symmetrically shaped order-7 tensors are 8 while the leading dimension is 8 or at most 48 for subtensors. The relative standard deviation of `<gemm_batch>`'s and `<par-loops,seq-gemm>`'s median values are 12.95% and 17.61%. Their respective interquartile range are similar with a relative standard deviation of 22.25% and 15.23%.

The runtime results with different k-order tensor layouts show that the performance of our proposed algorithms is not designed for a specific tensor layout. Moreover, the performance stays within an acceptable range independent of the tensor layout.

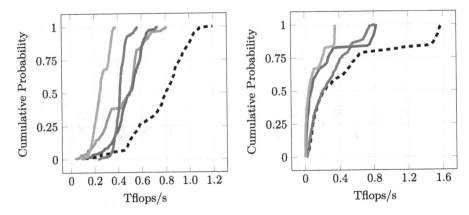

Fig. 4. Cumulative performance distributions of tensor-times-matrix algorithms in double-precision Tflops. Each distribution line belongs to a library: **tlib**[ours] (▪▪▪), **tcl** (▬▬), **tblis** (▬▬), **libtorch** (▬▬), **eigen** (▬▬). Libraries have been tested with asymmetrically-shaped (left plot) and symmetrically-shaped tensors (right plot).

Comparison with Other Approaches. We have compared the best performing algorithm with four libraries that implement the tensor-matrix multiplication.

Library **tcl** implements the TTGT approach with a high-perform tensor-transpose library **hptt** which is discussed in [19]. **tblis** implements the GETT approach that is akin to Blis' algorithm design for the matrix multiplication [12]. The tensor extension of **eigen** (v3.3.7) is used by the Tensorflow framework. Library **libtorch** (v2.3.0) is the C++ distribution of PyTorch. **tlib** denotes our library using algorithm `<par-loops,seq-gemm>` that have been presented in the previous paragraphs.

Figure 2 contains cumulative performance distributions for the complete test sets comparing the performance distribution of our implementation with the previously mentioned libraries. Note that we only have used tensor slices for asymmetrically shaped tensors (left plot) and subtensors for symmetrically shaped tensors (right plot). Our implementation with a median performance of 793.75 Gflops outperforms others' for almost every asymmetrically shaped tensor in the test set. The median performances of tcl, tblis, libtorch and eigen are 503.61, 415.33, 496.22 and 244.69 Gflops reaching on average 74.11%, 61.14%, 76.68% and 39.34% of tlib's throughputs.

In case of symmetrically shaped tensors the performance distributions of all libraries on the right plot in Fig. 2 are much closer. The median performances of tlib, tblis, libtorch and eigen are 228.93, 208.69, 76.46, 46.25 Gflops reaching on average 73.06%, 38.89%, 19.79% of tlib's throughputs[2]. All libraries operate with 801.68 or less Gflops for the cases 2 and 3 which is almost half of tlib's

---

[2] We were unable to run tcl with our test set containing symmetrically shaped tensors. We suspect a very high memory demand to be the reason.

performance with 1579 Gflops. The median performance and the interquartile range of tblis and tlib for the cases 6 and 7 are almost the same. Their respective median Gflops are 255.23 and 263.94 for the sixth case and 121.17 and 144.27 for the seventh case. This explains the similar performance distributions when their performance is less than 400 Gflops. Libtorch and eigen compute the tensor-matrix product, in median, with 17.11 and 9.64 Gfops/s, respectively. Our library tlib has a median performance of 102.11 Gflops and outperforms tblis with 79.35 Gflops for the eighth case.

7 Conclusion and Future Work

We presented efficient layout-oblivious algorithms for the compute-bound tensor-matrix multiplication which is essential for many tensor methods. Our approach is based on the LOG-method and computes the tensor-matrix product in-place without transposing tensors. It applies the flexible approach described in [2] and generalizes the findings on tensor slicing in [10] for linear tensor layouts. The resulting algorithms are able to process dense tensors with arbitrary tensor order, dimensions and with any linear tensor layout all of which can be runtime variable.

Our benchmarks show that dividing the base algorithm into eight different GEMM cases improves the overall performance. We have demonstrated that algorithms with parallel loops over single-threaded GEMM calls with tensor slices and subtensors perform best. Interestingly, they outperform a single batched GEMM with subtensors, on average, by 14% in case of asymmetrically shaped tensors and if tensor slices are used. Both version computes the tensor-matrix product on average faster than other state-of-the-art implementations. We have shown that our algorithms are layout-oblivious and do not need further refinement if the tensor layout is changed. We measured a relative standard deviation of 12.95% and 17.61% with symmetrically-shaped tensors for different k-order tensor layouts.

One can conclude that LOG-based tensor-times-matrix algorithms are on par or can even outperform TTGT-based and GETT-based implementations without loosing their flexibility. Hence, other actively developed libraries such as LibTorch and Eigen might benefit from implementing the proposed algorithms. Our header-only library provides c++ interfaces and a python module which allows frameworks to easily integrate our library.

In the near future, we intend to incorporate our implementations in TensorLy, a widely-used framework for tensor computations [4,8]. Currently, we lack a heuristic for selecting subtensor sizes and choosing the corresponding algorithm. Using the insights provided in [10] could help to further increase the performance. Analysis of different batched GEMM implementations might also reduce the overall runtime. The block interleaved approach, described in [5], is a promising starting point.

Source Code Availability. Project description and source code can be found at https://github.com/bassoy/ttm. The sequential tensor-matrix multiplication of TLIB is part of uBLAS and in the official release of Boost v1.70.0 and later.

References

1. Bader, B.W., Kolda, T.G.: Algorithm 862: MATLAB tensor classes for fast algorithm prototyping. ACM Trans. Math. Softw. **32**(4), 635–653 (2006). https://doi.org/10.1145/1186785.1186794
2. Bassoy, C.: Design of a high-performance tensor-vector multiplication with BLAS. In: Rodrigues, J., et al. (eds.) Computational Science – ICCS 2019: 19th International Conference, Faro, Portugal, June 12–14, 2019, Proceedings, Part I, pp. 32–45. Springer International Publishing, Cham (2019). https://doi.org/10.1007/978-3-030-22734-0_3
3. Bassoy, C., Schatz, V.: Fast higher-order functions for tensor calculus with tensors and subtensors. In: Shi, Y., Fu, H., Tian, Y., Krzhizhanovskaya, V.V., Lees, M.H., Dongarra, J., Sloot, P.M.A. (eds.) Computational Science – ICCS 2018: 18th International Conference, Wuxi, China, June 11–13, 2018, Proceedings, Part I, pp. 639–652. Springer International Publishing, Cham (2018). https://doi.org/10.1007/978-3-319-93698-7_49
4. Cohen, J., Bassoy, C., Mitchell, L.: Ttv in tensorly. Tensor Computations: Applications and Optimization, p. 11 (2022)
5. Dongarra, J., Hammarling, S., Higham, N.J., Relton, S.D., Valero-Lara, P., Zounon, M.: The design and performance of batched blas on modern high-performance computing systems. Proc. Comput. Sci. **108**, 495–504 (2017)
6. Karahan, E., Rojas-López, P.A., Bringas-Vega, M.L., Valdés-Hernández, P.A., Valdes-Sosa, P.A.: Tensor analysis and fusion of multimodal brain images. Proc. IEEE **103**(9), 1531–1559 (2015)
7. Kolda, T.G., Bader, B.W.: Tensor decompositions and applications. SIAM review **51**(3), 455–500 (2009)
8. Kossaifi, J., Panagakis, Y., Anandkumar, A., Pantic, M.: Tensorly: tensor learning in python. J. Mach. Learn. Res. **20**(26), 1–6 (2019)
9. Lee, N., Cichocki, A.: Fundamental tensor operations for large-scale data analysis using tensor network formats. Multidimension. Syst. Signal Process. **29**(3), 921–960 (2018)
10. Li, J., Battaglino, C., Perros, I., Sun, J., Vuduc, R.: An input-adaptive and in-place approach to dense tensor-times-matrix multiply. In: High Performance Computing, Networking, Storage and Analysis, 2015, pp. 1–12. IEEE (2015)
11. Lim, L.H.: Tensors and hypermatrices. In: Hogben, L. (ed.) Handbook of Linear Algebra. Chapman and Hall, 2 edn. (2017)
12. Matthews, D.A.: High-performance tensor contraction without transposition. SIAM J. Sci. Comput. **40**(1), C1–C24 (2018)
13. Napoli, E.D., Fabregat-Traver, D., Quintana-Ortí, G., Bientinesi, P.: Towards an efficient use of the blas library for multilinear tensor contractions. Appl. Math. Comput. **235**, 454–468 (2014)
14. Papalexakis, E.E., Faloutsos, C., Sidiropoulos, N.D.: Tensors for data mining and data fusion: models, applications, and scalable algorithms. ACM Trans. Intell. Syst. Technol. (TIST) **8**(2), 16 (2017)
15. Paszke, A., et al.: Pytorch: an imperative style, high-performance deep learning library. In: Advances in Neural Information Processing Systems, vol. 32 (2019)
16. Pawlowski, F., Uçar, B., Yzelman, A.J.: A multi-dimensional morton-ordered block storage for mode-oblivious tensor computations. J. Comput. Sci. **33**, 34–44 (2019)
17. Shi, Y., Niranjan, U.N., Anandkumar, A., Cecka, C.: Tensor contractions with extended blas kernels on cpu and gpu. In: 2016 IEEE 23rd International Conference on High Performance Computing (HiPC), pp. 193–202 (2016)

18. Solomonik, E., Matthews, D., Hammond, J., Demmel, J.: Cyclops tensor framework: Reducing communication and eliminating load imbalance in massively parallel contractions. In: Parallel & Distributed Processing (IPDPS), 2013 IEEE 27th International Symposium on, pp. 813–824. IEEE (2013)
19. Springer, P., Bientinesi, P.: Design of a high-performance gemm-like tensor-tensor multiplication. ACM Trans. Math. Softw. (TOMS) **44**(3), 28 (2018)

Interpoint Inception Distance: Gaussian-Free Evaluation of Deep Generative Models

Dariusz Jajeśniak[ID], Piotr Kościelniak[ID], Przemysław Klocek[ID],
and Marcin Mazur[✉][ID]

Faculty of Mathematics and Computer Science, Jagiellonian University,
Krakow, Poland
marcin.mazur@uj.edu.pl

Abstract. This paper introduces the Interpoint Inception Distance (IID) as a new approach for evaluating deep generative models. It is based on reducing the measurement of discrepancy between multidimensional feature distributions to one-dimensional interpoint comparisons. Our method provides a general tool for deriving a wide range of evaluation measures. The Cramér Interpoint Inception Distance (CIID) is notable for its theoretical properties, including a Gaussian-free structure of feature distribution and a strongly consistent estimator with unbiased gradients. Our experiments, conducted on both synthetic and large-scale real or generated data, suggest that CIID is a promising competitor to the Fréchet Inception Distance (FID), which is currently the primary metric for evaluating deep generative models.

Keywords: Deep generative model · Evaluation measure · Fréchet Inception Distance (FID) · Cramér Distance

1 Introduction

In recent years, deep generative models (DGMs) have gained tremendous attention. These models are designed and trained to approximate a data distribution via a model distribution. After completing the training, the question arises as to how well this task was accomplished. The research on both of these issues necessitates an appropriate measure that quantifies the difference between the distribution of the training data and the model distribution (a training measure), or the distribution of the test data and the model distribution (an evaluation measure). Figure 1 presents a diagrammatic representation of the general concept of training and evaluating deep generative models.

The purpose of a discrepancy measure in the training process is to construct an objective function that is optimized on the sets of real and fake data. Common choices include the Kulback-Leibler divergence (used in VAEs [17,27]) and the Jensen-Shannon divergence (used in GANs [12]). However, using these measures in a learning process can be challenging due to computational problems

© The Author(s), under exclusive license to Springer Nature Switzerland AG 2024
L. Franco et al. (Eds.): ICCS 2024, LNCS 14832, pp. 272–286, 2024.
https://doi.org/10.1007/978-3-031-63749-0_19

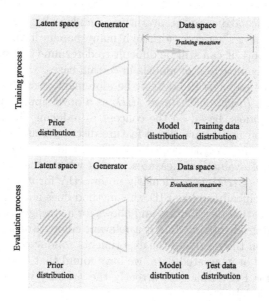

Fig. 1. General concept of training and evaluation of deep generative models. The figures were created using `diagrams.net` software.

such as complexity and vanishing gradient. Other common approaches include using the Optimal Transport (Wasserstein) Distance, as seen in WAEs [34], or a kernel-based distance, as seen in CWAEs [19]. Various authors have made significant efforts to propose novel solutions that outperform state-of-the-art methods. However, finding non-adversarial methods that can compete with GANs is still a challenge. These strategies often require the use of certain techniques, such as hierarchical structure in Nouveau VAE (NVAE) [37], stage training in 2Stage-VAE [7], or Latent Trick in Latent Cramér-Wold (LCW) generator [18].

The evaluation of training results is another issue related to generative modeling that requires an appropriate measure. In recent years, this problem has become even more important as deep generative models have matured enough to be used in downstream tasks. Therefore, better and more nuanced evaluation techniques are necessary [26]. Commonly used measures for evaluating processes include Log-Likelihood (LL) [12], Inception Score (IS) [28], and Fréchet Inception Distance (FID) [14]. Additionally, approaches such as Precision and Recall [21] can provide insight into the model's misspecification. Each of these measures has limitations and weaknesses, as described in Sect. 2. It is important to note that a good evaluation measure should be consistent with human perceptual similarity judgment [5]. However, even for the most popular solutions, including those mentioned above, there are known examples that may not conform to the expected results. This phenomenon may occur despite a well-optimized objective and a good evaluation score [16]. In a critical study provided in [5], the author argues that there is no evaluation method for deep generative models that is sensitive to realistic fake samples, overfitting, mode collapse, transformations, and sample efficiency. Therefore, although many papers have been written on the subject (see [5,6]), it can still be difficult to determine the most appropriate measure for a fair comparison of models in certain cases.

The aim of this paper is to address the aforementioned challenge. We propose the *Interpoint Inception Distance (IID)* as a novel approach for evaluating deep generative models, based on the concept of reducing the measurement of discrepancy between multidimensional feature distributions to one-dimensional interpoint comparisons, as described by [23]. IID provides a general tool for deriving a wide range of evaluation measures, one of which, the *Cramér Interpoint Inception Distance (CIID)*, is particularly noteworthy for its desirable theoretical properties. Specifically, unlike FID, this method does not assume a Gaussian structure for the feature distribution and allows for a strongly consistent estimator with unbiased gradients, making it a relevant competitor to state-of-the-art solutions[1]. Based on the results of the experiments conducted on both synthetic and large-scale real or generated data, we have found that CIID could be a feasible substitute for FID, which is currently the primary metric for evaluating deep generative models [20].

This paper does not attempt to address all potential issues, but rather takes a first step toward improving the evaluation of deep generative models by measuring distributional discrepancy. We believe that the results obtained will have an impact on further studies in related fields, particularly in deepfake detection, which has become increasingly popular due to the development of generative models. Any improvement in generative modeling makes it harder to distinguish between what is real and what is fake. Therefore, it is crucial to evaluate generative models efficiently to reduce potential risks [6].

2   Related Work

Various measures that quantify differences between distributions have been defined in the literature. These include measures based on information theory, optimal transport theory, and kernel theory. In the following paragraphs, we briefly present our selection of the most significant examples with applications in evaluating deep generative models. For a comprehensive analysis and discussion of alternative methods, including adjustments and enhancements to those outlined in this section, refer to [5,6].

Discrepancy Measures Based on the Information Theory. In general, measures based on information theory rely on entropy $\mathrm{H}(\cdot)$ and/or cross-entropy $\mathrm{H}(\cdot,\cdot)$. The Log-Likelihood (or Evidence) [12,33] is calculated using the following formula:

$$\mathrm{LL} = \mathbb{E}_{x\sim p_\mathcal{X}} \log p_G(x) = -\mathrm{H}(p_\mathcal{X}), \tag{1}$$

---

[1] Since our method relies on transferring distributions of real and fake data into the inception (feature) space, we only consider feature-based approaches such as FID. For the same reason, we do not discuss other concepts such as Precision and Recall.

where $p_{\mathcal{X}}$ represents the real data distribution and p_G represents the model distribution induced on data space $\mathcal{X}$ by the generator network G (i.e., the distribution of fake data). As likelihood in higher dimensions is intractable, generated data are often used to approximate $p_G(x)$. This requires the application of suitable estimation techniques, such as the Parzen window approach [33] or the reparametrization trick in VAE [27]. The Log-Likelihood can be considered a universal measure for the training and evaluation of deep generative models [35]. However, due to its low sample efficiency, reliable direct calculation requires large sample sizes. Therefore, for training purposes, it is often substituted with its lower bound, such as the Evidence Lower Bound (ELBO) in VAE, which allows for working with small batches. Furthermore, according to [33], this measure is generally uninformative about the quality of samples. This is because there are known models that produce great samples despite having a poor (low) Log-Likelihood, or vice versa.

The Inception Score [28] is commonly used to evaluate generated images. It can also be useful for training deep generative models, as demonstrated in [29], where a closely related objective for training Category-Aware Generative Adversarial Networks (CatGANs) was proposed. However, to obtain reliable results, it is necessary to evaluate the score on a large number of samples, at least 50k [28]. To calculate the Inception Score, we require the Inception v3 Network [30], which is pre-trained on the ImageNet dataset [8] to capture the desired features of the generated data. The Inception Score is calculated using the following formula:

$$IS = \exp(\mathbb{E}_{x \sim p_G} KL(p_L(\cdot|x) \| p_L) = \exp(H(p_L) - \mathbb{E}_{x \sim p_G} H(\cdot|x)). \qquad (2)$$

Here, $KL(\cdot\|\cdot)$ represents the Kullback-Leibler divergence, while $p_L(\cdot|x)$ represents the label (feature) distribution on the inception (feature) space conditioned on $x \in \mathcal{X}$ (so p_L is respective marginal label distribution). The Inception Score has been found to be reasonably correlated with the quality and diversity of generated images, as well as with human judgment [28]. However, it should be noted that the Inception Score does not take into account real data, which may result in models receiving better (higher) scores simply for producing sharp and diverse images, rather than those that follow the underlying distribution [40]. For a more detailed analysis, see [2].

Discrepancy Measures Based on the Optimal Transport Theory. The Optimal Transport Distance determines the most cost-effective way to transport one probability measure into another. This is expressed by the following formula (see, e.g., [38]):

$$W_c(p, q) = \inf_{\gamma \in \Gamma(p,q)} \int_{\mathbb{R}^k \times \mathbb{R}^k} c(x, y) \, d\gamma(x, y), \qquad (3)$$

where $\Gamma(p, q)$ is the family of joint probability distributions (known as couplings) having p and q as marginals, and $c(\cdot, \cdot)$ is a given transportation cost function. The state-of-the-art deep generative models, Wasserstein GAN [1] and

Wasserstein Autoencoder (WAE) [34], aim to minimize the Wasserstein Distance between $p_{\mathcal{X}}$ and p_G, using either l_1^2 (for WGAN) or l_2^2 (for WAE) as the transportation cost function. However, it is important to note that computing directly from Eq. (3) is difficult or even impossible. Therefore, WGAN adheres to the Kantorovich-Rubinstein duality [38] and minimizes a lower bound for $W_{l_1^2}$. Similarly, WAE utilizes Theorem 1 from [34] to create an objective function that comprises a reconstruction error term on $\mathcal{X}$ and an appropriate regularization term on the latent.

As demonstrated above, utilizing the Optimal Transport Distance directly in data space is not feasible due to its intractability in high dimensions. This limitation also applies to the evaluation of models. In this case, however, the Fréchet Inception Distance (FID) [14] can be used. FID is computed as the Wasserstein Distance $W_{l_2^2}$ between the data and model distributions, which are first transported into feature space by the Inception v3 network pre-trained on the ImageNet dataset, and then approximated by the nearest multidimensional Gaussians. While FID has become a standard for evaluating deep generative models trained on image datasets, it has some significant weaknesses. For example, it may overestimate "strange-looking" samples generated by well-optimized objectives (see, e.g., the figures in [16, Appendix E]). A biased estimator of FID that requires large samples makes it essentially unfeasible in the training process. The imposed Gaussian structure raises reasonable doubts about the reliability of this measure. Therefore, it is justifiable to search for improvements [4].

Discrepancy Measures Based on the Kernel Theory. A kernel-based measure used to compare two probability distributions is the Maximum Mean Discrepancy (MMD) [24]. For a fixed characteristic kernel function k, it is defined as:

$$\mathrm{MMD}_k(p, q) = \mathbb{E}_{x, x' \sim p}\, k(x, x') - 2\mathbb{E}_{x \sim p, y \sim q}\, k(x, y) + \mathbb{E}_{y, y' \sim q}\, k(y, y'). \qquad (4)$$

Because MMD has an unbiased estimator [13], even when used in data space, it has a low sample complexity. This makes it suitable for training models, including generative autoencoders (e.g., MMD-VAE [41]) and GANs (e.g., MMD Net [9]). Additionally, it performs well in a feature space [4,40]. As an example, the Kernel Inception Distance (KID) is provided [4]. KID uses the polynomial characteristic kernel function and the Inception v3 Network, which is pre-trained on the ImageNet dataset. According to the experimental results presented in [4], KID can be considered a computationally efficient evaluation measure that does not require a Gaussian structure of feature distributions, unlike FID. However, both FID and KID do not differentiate between distributions with the same first three moments (refer to Fig. 2 and (citech19tsitsulin2019shape Fig. 1), which indicates room for improvement.

3   Interpoint Inception Distance

This section introduces the Interpoint Inception Distance (IID), a feature-based method for evaluating deep generative models. The approach aims to reduce

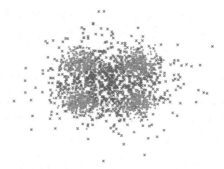

Fig. 2. Example of distributions with identical first three moments (see Sect. 4), resulting in FID and KID scores close to zero [26,36]

the measurement of discrepancy between multidimensional distributions to one-dimensional interpoint comparisons [23]. The statement is general, allowing for the derivation of a broad range of evaluation measures. We present an instance of the Cramér Interpoint Inception Distance (CIID), which has desirable theoretical properties, making it a suitable candidate for an effective evaluation metric.

General Statement. Our approach is based on the following theorem, which is a rewrite of Theorem 1 and Remark 4 in [23].

Theorem 1. *Let X_1, X_2, X_3 and Y_1, Y_2, Y_3 be independent copies of two independent k-dimensional random variables X and Y, respectively. Let h be any real-valued nonnegative function[2] such that $h(x,y) = 0$ if and only if $x = y$. Then the following conditions are equivalent:*

(i) X and Y follow the same (k-dimensional) distribution,
(ii) $h(X_1, X_2)$, $h(Y_1, Y_2)$, and $h(X_3, Y_3)$ follow the same (univariate) distribution,
(iii) $h(X_1, X_2)$, $h(Y_1, Y_2)$, and $h(X_1, Y_1)$ follow the same (univariate) distribution.

Using Theorem 1, we define the Interpoint Distance (ID) with the following formula:

$$\mathrm{ID}(X, Y) = d(h(X_1, X_2), h(Y_1, Y_2)) + d(h(X_1, X_2), h(X_1, Y_1)) \\ + d(h(Y_1, Y_2), h(X_1, Y_1)), \tag{5}$$

where d is an arbitrary one-dimensional statistical distance. Then, applying Eq. (5) to random variables representing features of real and fake data[3] yields a general evaluation measure called the Interpoint Inception Distance (IID). In

---

[2] Although it is not necessary for h to be symmetric, it can be interpreted as a type of semi-metric [39] on $\mathbb{R}^k$.

[3] This means that X and Y represent outputs of the Inception v3 Network, which was pre-trained on the ImageNet dataset.

this case, the function h measures discrepancies between the features of points drawn from the real data distribution or the model distribution. In practice, we use the distance induced by the k-dimensional Euclidean norm $\|\cdot\|$, i.e., $h(x,y) = \|x - y\|$ for $x, y \in \mathbb{R}^k$. In the following paragraph, we discuss a possible choice for d that leads to a special case of IID of particular interest.

Cramér Interpoint Inception Distance (CIID). Our proposal is to use the Cramér Distance [31,32] as d in the IID formula. The p-th Cramér Distance is defined by the following formula:

$$C^p(S,T) = \int_{-\infty}^{\infty} |F_S(t) - F_T(t)|^p \, dt, \tag{6}$$

where F_S and F_T represent the cumulative distribution functions (CDFs) for one-dimensional random variables S and T. This yields the p-th Cramér Interpoint Inception Distance (CIID^p).

The use of the CIID^p evaluation measure involves the application of an appropriate estimator, as described in the following paragraph, along with the corresponding theoretical analysis.

Estimation of CIID. Random variables X and Y^θ (for $\theta \in \Theta$, where Θ is an open subset of the model's parameter space) are taken on the k-dimensional inception space to represent features of real data (following the data distribution p_X) and fake data (following the model distribution p_G^θ), respectively. Sequences of independent copies of X and Y^θ are considered, namely

$$X_{1,n} = (X_{1,1}, \ldots, X_{1,n}), \ X_{2,n} = (X_{2,1}, \ldots, X_{2,n}), \tag{7}$$

and

$$Y_{1,n}^\theta = (Y_{1,1}^\theta, \ldots, Y_{1,n}^\theta), \ Y_{2,n}^\theta = (Y_{2,1}^\theta, \ldots, Y_{2,n}^\theta) \tag{8}$$

(these are interpreted as respective batch samples). The p-th Cramér Interpoint Inception Distance can be estimated using the following formula:

$$\widehat{\text{CIID}_n^p}(X, Y^\theta) = C^p(F_{\|X_{1,n} - X_{2,n}\|}, F_{\|Y_{1,n}^\theta - Y_{2,n}^\theta\|}) + C^p(F_{\|X_{1,n} - X_{2,n}\|}, F_{\|X_{1,n} - Y_{1,n}^\theta\|})$$
$$+ C^p(F_{\|Y_{1,n}^\theta - Y_{2,n}^\theta\|}, F_{\|X_{1,n} - Y_{1,n}^\theta\|}), \tag{9}$$

where

$$F_{\|X_{1,n} - X_{2,n}\|}(t) = \frac{1}{n} \sum_{i=1}^{n} I_{(-\infty,t]}(\|X_{1,i} - X_{2,i}\|), \tag{10}$$

$$F_{\|X_{1,n} - Y_{1,n}^\theta\|}(t) = \frac{1}{n} \sum_{i=1}^{n} I_{(-\infty,t]}(\|X_{1,i} - Y_{1,i}^\theta\|), \tag{11}$$

and

$$F_{\|Y_{1,n}^\theta - Y_{2,n}^\theta\|}(t) = \frac{1}{n} \sum_{i=1}^{n} I_{(-\infty,t]}(\|Y_{1,i}^\theta - Y_{2,i}^\theta\|) \tag{12}$$

are corresponding empirical cumulative distribution functions (ECDFs). (Here, I denotes a set characteristic function.)

Regarding $\widehat{\mathrm{CIID}}_n^p$, our main findings are that it is a strongly consistent estimator and possesses unbiased gradients for $p = 2$, as stated in the following theorem.

Theorem 2. *The Cramér Interpoint Inception Distance estimator defined in Eq. (9) satisfies the following conditions:*

$$\widehat{\mathrm{CIID}}_n^p(X, Y^\theta) \to \mathrm{CIID}^p(X, Y^\theta) \ \text{if } n \to \infty \tag{13}$$

and

$$\mathbb{E}(\nabla_\theta \widehat{\mathrm{CIID}}_n^2(X, Y^\theta)) = \nabla_\theta \mathrm{CIID}^2(X, Y^\theta). \tag{14}$$

The proof of Theorem 2 follows directly from the lemma below.

Lemma 1. *Let* $\mathrm{S}_n = (S_1, \ldots, S_n)$ *and* $\mathrm{T}_n^\theta = (T_1^\theta, \ldots, T_n^\theta)$ *be sequences of independent copies of one-dimensional random variables* S *and* T^θ, *respectively. Then:*

$$C^2(F_{\mathrm{S}_n}, F_{\mathrm{T}_n^\theta}) \to C^2(F_S, F_{T^\theta}) \ \text{if } n \to \infty \tag{15}$$

and

$$\mathbb{E}(\nabla_\theta C^2(F_{\mathrm{S}_n}, F_{\mathrm{T}_n^\theta})) = \nabla_\theta C^2(F_S, F_{T^\theta}). \tag{16}$$

Equation (15) can be derived from the Glivenko-Cantelli Theorem [10] and the Lebesgue Convergence Theorem [11]. To prove Eq. (16), refer to [3].

4   Experiments

This section presents the experimental study that compares our proposed evaluation measure (CIID) with FID. We start by conducting experiments on synthetic data before transitioning to the case of large-scale real or generated data. The source code is available at https://github.com/djajesniak/CIID.

Experiments on Synthetic Data. Let us consider two-dimensional distributions $p \sim N_2(0, I_2)$ and $q_m = q_m^1 \times q_m^2$ for $m \in [0, 1]$, where q_m^1 and q_m^2 are one-dimensional distributions both given by the density function

$$f_m(x) = \frac{1}{2} f_{-m, \sqrt{1-m^2}}(x) + \frac{1}{2} f_{m, \sqrt{1-m^2}}(x) \tag{17}$$

(here $f_{m,s}$ is a density function of the Gaussian $N(m, s)$). Then the first three moments of p and q_m are equal and $p = q_0$. Figure 3 presents estimated FID and CID[4] values (calculated with different sample sizes) between p and $q_{0.95}$ and between p and p. It is clear that FID does not distinguish between p and $q_{0.95}$, while CID does (note that CID[1] is even sensitive to differences resulting

---

[4] The Cramér Interpoint Distance (CID) is derived by using the Cramér Distance in the ID formula, instead of the IID formula as in the case of CIID.

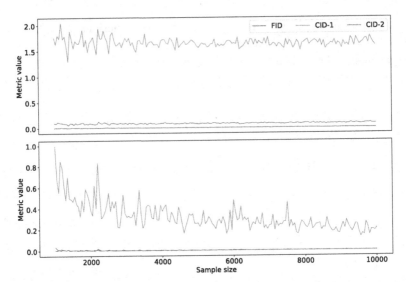

Fig. 3. Estimated FID and CID values (calculated using samples of different sizes) between the two-dimensional distributions p and $q_{0.95}$ (top) or p and p (bottom) that have the same first three moments

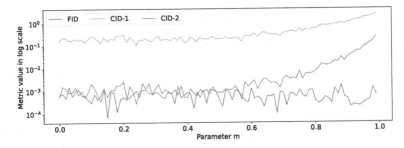

Fig. 4. Estimated FID and CID values on a logarithmic scale (calculated with a sample size of 10000) between the two-dimensional distributions p and q_m with the same first three moments

from sample-based estimation). On the other hand, from Fig. 4 we learn that FID does not indicate the difference between p and q_m for any $m \in [0, 1]$, while CID does (but CID^2 starts to discriminate at $m \approx 0.6$).

Sensitivity to Implemented Disturbances. We performed experiments inspired by those initially performed in [14] and then continued in [4]. We applied several different disturbances (i.e., "salt and pepper", Gaussian noise, black rectangles, Gaussian blur, and elastic transform) to 8k-sized data samples from the CelebA [22] and ImageNet [8] datasets, to examine resilience to the noise of CIID compared to FID. Figures 5, 6, 7, 8, and 9 present our experimental results. Each score was scaled to $[0, 1]$ to be plotted on one vertical axis. Generally, CIID behaved comparably to FID, but it seemed to better maintain small disturbances, as shown in Fig. 9, where the FID score for null disturbance is positive, which may be attributed to the curse of dimensionality and the strong bias of its estimator.

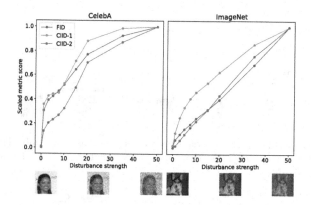

Fig. 5. Salt and pepper. The x-axis represents the percentage of pixels changed to black or white. On the y-axis is the current value of the metric divided by its maximum value

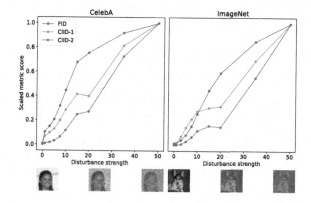

Fig. 6. Gaussian noise. The x-axis represents one-third of the standard deviation of the noise. On the y-axis is the current value of the metric divided by its maximum value

Impact of Suboptimal Weights. We conducted an empirical investigation employing a DCGAN [25] architecture trained on the CelebA dataset, to evaluate the effectiveness of CIID compared to FID throughout the network's training phase. We scrutinized the evaluation measures performance under suboptimal model weights. We run a training procedure 5 times computing evaluation scores on 8k-sized samples every 1000 batches and then taking the average. The mean values are presented in Fig. 10 (for generated samples refer to Fig. 11). During training, a similar behavior of FID and CIID was observed. Notably, CIID

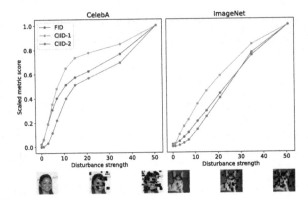

Fig. 7. Black rectangles. The x-axis represents the number of small black rectangles randomly added to the image. On the y-axis is the current value of the metric divided by its maximum value

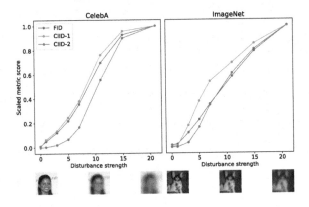

Fig. 8. Gaussian blur. The x-axis represents the strength of the disturbance. Kernel size is equal to $(3x, 3x)$ and sigma parameter to (x, x). On the y-axis is the current value of the metric divided by its maximum value

demonstrated a faster descent, reaching lower relative values[5] in comparison to FID.

Variability. During the experiments, it was observed that CIID has lower variability than FID. To test it, we computed the CIID and FID ten times between samples from the CelebA dataset and sets of entirely black images (represented by tensors containing only zeros). The coefficient of variation for each evaluation measure was then computed. The results showed a coefficient of variation of 0.00135 for FID, 0.00066 for CIID^1, and 0.00055 for CIID^2. This demonstrates that our proposed metrics have roughly half the variability of FID.

---

[5] Compared to their values at epoch 0, which are maximal because they are represented as metric values between the original data samples and pure Gaussian noise.

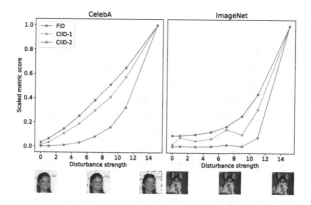

Fig. 9. Elastic transform. The x-axis represents the strength of the disturbance. The alpha parameter is equal to $10x$. On the y-axis is the current value of the metric divided by its maximum value

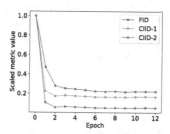

Fig. 10. Scaled values of the metrics during DCGAN training (averaged over 5 runs). The x-axis represents subsequent epochs and the y-axis denotes the current metric value divided by its maximum value

Fig. 11. Examples of images generated after 1 epoch (left) and after 12 epochs (right) by DCGAN trained on the CelebA dataset

5   Conclusions

In this paper, the Interpoint Inception Distance (IID) is introduced as a new method for the evaluation of deep generative models. It is shown that one of its instances, the Cramér Interpoint Inception Distance (CIID), exhibits remarkable theoretical properties, such as a non-Gaussian feature distribution structure and an estimator that yields unbiased gradients and is strongly consistent. Experiments on synthetic and large-scale real or generated data suggest that CIID is a promising competitor to FID, distinguishing well between distributions with the same first three moments, having lower variability, and appearing more objective to small differences.

Limitations and Future Directions. Thus far, our solution has been validated on a limited number of experimental setups that only involve image data. However, we believe that it could also prove useful in the context of text or signal data. Moreover, we have not yet explored the use of CIID to account for *sample novelty*. This is a recent concept introduced in [15], which incorporates vulnerability to overfitting. On the other hand, the existence of an unbiased gradient estimator permits the justification of potential applications of our proposed measure in training processes. The aforementioned factors will serve to guide future directions of our research.

Acknowledgment. The work of D. Jajeśniak and M. Mazur was supported by the National Centre of Science (Poland) Grant No. 2021/41/B/ST6/01370. Some experiments were performed on servers purchased with funds from the flagship project entitled "Artificial Intelligence Computing Center Core Facility" from the DigiWorld Priority Research Area within the Excellence Initiative – Research University program at Jagiellonian University in Krakow. *D. Jajeśniak and M. Mazur acknowledge co-first authorship.*

Disclosure of Interests. The authors have no competing interests to declare that are relevant to the content of this article.

References

1. Arjovsky, M., Chintala, S., Bottou, L.: Wasserstein generative adversarial networks. In: International Conference on Machine Learning, pp. 214–223. PMLR (2017)
2. Barratt, S., Sharma, R.: A note on the inception score. arXiv preprint arXiv:1801.01973 (2018)
3. Bellemare, M.G., et al.: The Cramer distance as a solution to biased Wasserstein gradients. arXiv preprint arXiv:1705.10743 (2017)
4. Bińkowski, M., Sutherland, D.J., Arbel, M., Gretton, A.: Demystifying MMD GANs. In: International Conference on Learning Representations (2018)
5. Borji, A.: Pros and cons of GAN evaluation measures. Comput. Vis. Image Underst. **179**, 41–65 (2019)
6. Borji, A.: Pros and cons of GAN evaluation measures: new developments. Comput. Vis. Image Underst. **215**, 103329 (2022)

7. Dai, B., Wipf, D.: Diagnosing and enhancing VAE models. In: International Conference on Learning Representations (2018)
8. Deng, J., Dong, W., Socher, R., Li, L.J., Li, K., Fei-Fei, L.: ImageNet: a large-scale hierarchical image database. In: 2009 IEEE Conference on Computer Vision and Pattern Recognition, pp. 248–255. IEEE (2009)
9. Dziugaite, G.K., Roy, D.M., Ghahramani, Z.: Training generative neural networks via maximum mean discrepancy optimization. In: Proceedings of the Thirty-First Conference on Uncertainty in Artificial Intelligence, pp. 258–267 (2015)
10. Ferguson, T.S.: A Course in Large Sample Theory. Routledge (2017)
11. Gariepy, L.E.R., Evans, L.: Measure theory and fine properties of functions, revised edition. In: Studies in Advanced Mathematics. CRC Press, Boca Raton (2015)
12. Goodfellow, I., et al.: Generative adversarial nets. In: Advances in Neural Information Processing Systems, vol. 27 (2014)
13. Gretton, A., Borgwardt, K.M., Rasch, M.J., Schölkopf, B., Smola, A.: A kernel two-sample test. J. Mach. Learn. Res. 13(1), 723–773 (2012)
14. Heusel, M., Ramsauer, H., Unterthiner, T., Nessler, B., Hochreiter, S.: GANs trained by a two time-scale update rule converge to a local Nash equilibrium. Adv. Neural Inf. Process. Syst. 30 (2017)
15. Jiralerspong, M., Bose, J., Gemp, I., Qin, C., Bachrach, Y., Gidel, G.: Feature likelihood score: Evaluating the generalization of generative models using samples. Adv. Neural Inf. Process. Syst. 36 (2024)
16. Jung, S., Keuper, M.: Internalized biases in Fréchet inception distance. In: NeurIPS 2021 Workshop on Distribution Shifts: Connecting Methods and Applications (2021)
17. Kingma, D.P., Welling, M.: Auto-encoding variational Bayes. arXiv preprint arXiv:1312.6114 (2013)
18. Knop, S., Mazur, M., Spurek, P., Tabor, J., Podolak, I.: Generative models with kernel distance in data space. Neurocomputing 487, 119–129 (2022)
19. Knop, S., Spurek, P., Tabor, J., Podolak, I., Mazur, M., Jastrzebski, S.: Cramer-Wold auto-encoder. J. Mach. Learn. Res. 21(1), 6594–6621 (2020)
20. Kynkäänniemi, T., Karras, T., Aittala, M., Aila, T., Lehtinen, J.: The role of ImageNet classes in Fréchet inception distance. arXiv preprint arXiv:2203.06026 (2022)
21. Kynkäänniemi, T., Karras, T., Laine, S., Lehtinen, J., Aila, T.: Improved precision and recall metric for assessing generative models. Adv. Neural Inf. Process. Syst. 32 (2019)
22. Liu, Z., Luo, P., Wang, X., Tang, X.: Deep learning face attributes in the wild. In: Proceedings of the IEEE International Conference on Computer Vision, pp. 3730–3738 (2015)
23. Maa, J.F., Pearl, D.K., Bartoszyński, R.: Reducing multidimensional two-sample data to one-dimensional interpoint comparisons. Ann. Stat. 24(3), 1069–1074 (1996)
24. Muandet, K., Fukumizu, K., Sriperumbudur, B., Schölkopf, B., et al.: Kernel mean embedding of distributions: a review and beyond. Found. Trends® Mach. Learn. 10(1-2), 1–141 (2017)
25. Radford, A., Metz, L., Chintala, S.: Unsupervised representation learning with deep convolutional generative adversarial networks. arXiv preprint arXiv:1511.06434 (2015)

26. Ravuri, S., Rey, M., Mohamed, S., Deisenroth, M.P.: Understanding deep generative models with generalized empirical likelihoods. In: Proceedings of the IEEE/CVF Conference on Computer Vision and Pattern Recognition, pp. 24395–24405 (2023)

27. Rezende, D.J., Mohamed, S., Wierstra, D.: Stochastic backpropagation and approximate inference in deep generative models. In: Proceedings of the 31st International Conference on Machine Learning. Proceedings of Machine Learning Research, vol. 32, pp. 1278–1286 (2014)

28. Salimans, T., Goodfellow, I., Zaremba, W., Cheung, V., Radford, A., Chen, X.: Improved techniques for training GANs. Adv. Neural Inf. Process. Syst. **29** (2016)

29. Springenberg, J.T.: Unsupervised and semi-supervised learning with categorical generative adversarial networks. arXiv preprint arXiv:1511.06390 (2015)

30. Szegedy, C., Vanhoucke, V., Ioffe, S., Shlens, J., Wojna, Z.: Rethinking the inception architecture for computer vision. In: Proceedings of the IEEE Conference on Computer Vision and Pattern Recognition, pp. 2818–2826 (2016)

31. Székely, G.J.: E-statistics: the energy of statistical samples. Bowling Green State University, Department of Mathematics and Statistics Technical Report **3**(05), 1–18 (2003)

32. Székely, G.J., Rizzo, M.L.: Energy statistics: a class of statistics based on distances. J. Statist. Plan. Inference **143**(8), 1249–1272 (2013)

33. Theis, L., Oord, A.v.d., Bethge, M.: A note on the evaluation of generative models. arXiv preprint arXiv:1511.01844 (2015)

34. Tolstikhin, I., Bousquet, O., Gelly, S., Schoelkopf, B.: Wasserstein auto-encoders. In: International Conference on Learning Representations (2018)

35. Tolstikhin, I.O., Gelly, S., Bousquet, O., Simon-Gabriel, C.J., Schölkopf, B.: AdaGAN: boosting generative models. Adv. Neural Inf. Process. Syst. **30** (2017)

36. Tsitsulin, A., et al.: The shape of data: intrinsic distance for data distributions. In: International Conference on Learning Representations (2019)

37. Vahdat, A., Kautz, J.: NVAE: a deep hierarchical variational autoencoder. Adv. Neural. Inf. Process. Syst. **33**, 19667–19679 (2020)

38. Villani, C.: Optimal Transport, Old and New, vol. 338. Springer, Heidelberg (2009)

39. Wilson, W.A.: On semi-metric spaces. Am. J. Math. **53**(2), 361–373 (1931)

40. Xu, Q., et al.: An empirical study on evaluation metrics of generative adversarial networks. arXiv preprint arXiv:1806.07755 (2018)

41. Zhao, S., Song, J., Ermon, S.: InfoVAE: balancing learning and inference in variational autoencoders. In: Proceedings of the AAAI Conference on Artificial Intelligence, vol. 33, pp. 5885–5892 (2019)

Elliptic-Curve Factorization and Witnesses

Jacek Pomykała[ID] and Olgierd Żołnierczyk[✉][ID]

Faculty of Cybernetics, Military University of Technology, Warsaw, Poland
{jacek.pomykala,olgierd.zolnierczyk}@wat.edu.pl

Abstract. We define the EC (Elliptic Curve)-based factorization witnesses and prove related results within both conditional and unconditional approaches. We present experimental computations that support the conjecture of behavior of related admissible elliptic curves in relation to the deterministic complexity of suitable factoring algorithms based on the parameters of the witnesses. This paper features three main results devoted to the factorization of RSA numbers $N = pq$, where $q > p$. The first result of computational complexity of elliptic curve factorization is improved by the factor D^σ, comparing to previously known result $O\left(D^{2+o(1)}\right)$, where D is smoothness bound, assuming additional knowledge of the admissible elliptic curve. The second result demonstrates the feasibility of achieving factorization in deterministic, polynomial time, based on knowledge obtained at a specific step in the elliptic curve method (ECM), a feat previously considered impossible. The third result establishes deterministic time for conditional factorization using the elliptic version of Fermat method. It has the magnitude order $(\log N)^{O(1)}\left(1 + \left(\frac{|a_p|+|a_q|}{D}\right)^2\right)$, provided $\frac{q}{p} \ll 1$. Here a_p, a_q are the Frobenius traces of the corresponding curves $(E(\mathbb{F}_p), E(\mathbb{F}_q))$, and D indicates the approximation of the quotient p/q by the quotient a_p/a_q, assuming that the order of the group of points over a pseudo elliptic curve $E(\mathbb{Z}_N)$ is known.

Keywords: EC factorization · B-smooth numbers · Factor bases

1 Introduction

The Fermat-Euclide (compositness) witness $a \in \mathbb{Z}_N^*$ in relation to deterministic Pollard's $p-1$ algorithm was introduced and investigated by Źrałek in [17]. It satisfies the condition

$$\nu_l(\mathrm{ord}_p a) \neq \nu_l(\mathrm{ord}_q a),$$

for some prime $l \mid \mathrm{ord}_N a$. The more detailed definition and treatment for the sake of oracle factoring methods was developed in [7,14] with the aid of Dirichlet's characters reinforced by the large and shifted sieve, respectively.

© The Author(s), under exclusive license to Springer Nature Switzerland AG 2024
L. Franco et al. (Eds.): ICCS 2024, LNCS 14832, pp. 287–301, 2024.
https://doi.org/10.1007/978-3-031-63749-0_20

The first challenge in general is to focus here on small values $a \leq A = A(N)$, reducing the problem to small values of Dirichlet character's nonresidues or more generally small generating sets of $\mathbb{Z}_N^*$, as applied in the investigation of the least witness for N in [1].

The second one concerns small values of $l \leq B = B(N)$, related to well known Pollard's $p-1$ (see [13]) and Williams $p+1$ (see [16]) method of factoring. Both A and B-aspects are important in the deterministic approach to factoring integers. The particular interest concerns the semiprime numbers $N = pq\,(p < q)$ where p and q are large unknown primes applied in RSA cryptosystem [15].

In this paper we investigate factoring of integers with the aid of elliptic curves as proposed in [12], called unconditional and one based on the knowledge of the order of elliptic curve E over $\mathbb{Z}_N$ called conditional (cf. [6]). Both approaches are related to each other because in the deterministic approach of searching the related pairs (E, Q), where $E = E(\mathbb{Z}_N)$, $Q \in E(\mathbb{Z}_N)$ and $B = B(N)$ is suitably chosen parameter. In particular the classical Fermat factoring method has its "elliptic" version which can be adopted to "quantum annealing" method applied in [18]. Such approach is not the aim of the present work and is postponed to be investigated in another paper. Here we define the related notions of decomposition witnesses and prove the results concerning the factoring of N in time depending on the set of parameters X of the witnesses.

The notion of elliptic decomposition (factorization) witnesses (called in short witnesses) that we discuss below is new and relates to elliptic curve $E(\mathbb{Q})$ of Weierstrass equation with integer coefficients of the form $E := E_{\bar{b}} : y^2 = x^3 + b_1 x + b_2$, where $\bar{b} = (b_1, b_2) \in \mathbb{Z}^2$. In what follows we assume that N is coprime to 6, similarly as in Sect. 2.1 [12] we define

$$\Delta_{\bar{b}} := 6(4b_1^3 + 27b_2^2) \tag{1}$$

and in order to have the elliptic curves $E(\mathbb{F}_p)$ and $E(\mathbb{F}_q)$ we assume that $\gcd(\Delta_{\bar{b}}, N) = 1$. Here and in the sequel we will use the abbreviation (Ell, N) for the unconditional approach and (Ell, E_N) for the conditional approach.

In the second approach we apply the elliptic curves and Fermat's factorization method, reducing decomposition $N = pq$ to the knowledge about E_N and the approximation of $\frac{p}{q}$ by $\frac{E_p}{E_q}$ or $\frac{p+1-E_p}{q+1-E_q} := \frac{a_p(E)}{a_q(E)}$, respectively.

The deterministic time needed to decompose $N = pq$ depends on the witness $(u, v) = (b_1, b_2) \in \mathbb{Z}_N^2$ and a set of parameters $X = X_{u,v}$ of type $\{D\}, \{D, s, \alpha\}$ or $\{B, K\}$, where $1/D$ stands for the precision of approximation of p/q by the related quotient of Frobenius traces, $d \neq s \mid d^2, d \mid \gcd(E_p, E_q)$, α defines the precision of approximation of d/s by $1 + \frac{1}{D}$, while K measures the distance of E_r from the relevant B-factor of E_r.

In turn, the first approach is based on the largest B-smooth factor of E_r denoted by $s_B(E_r)$, and the lower bound for the reduced point $Q_{r'} \in E(\mathbb{F}_{r'})$, where $\{r, r'\} = \{p, q\}$. The related parameter $\beta \in [0, 1]$ indicates the lower bound for the related exponent of B-smooth factor of E_r, where $E = E_{\bar{b}}$ is the elliptic curve over $\mathbb{Z}_N$. On the other hand $\gamma \in [0, 1]$ points out the lower bound for $\mathrm{ord}Q_{r'}$ and occurs as an element of the set X of witness parameters, which is important in the investigation of the special case of witnesses called the nonseparating witnesses. On the other hand the separating witnesses are applied with the additional parameter $\sigma \in [1, 2]$ of E_r, which allows to improve the deterministic time of decomposition $N = pq$ on the factor D^σ, compared to the conventional approach, when E is admissible curve.

The most characteristic example concerns the case $(\beta, \gamma) = (1, 1)$ and was applied in the seminal paper [12]. In the recent paper [4] the analysis concerned the more general case $(1, \gamma)$. Here we focus on the case when $\beta \in [0, 1], \gamma \in [0, 1]$ are arbitrary and prove the deterministic time of factoring N (depending on admissibility parameters B, β of E_r and set of parameters $X = X_{u,v}$ of the witness $(u, v) \in \mathbb{Z}_N^2$).

More accurately we analyze (B, β, D, Δ)−admissible numbers E_r which are (B, β)− smooth, D−smooth (where $D \geq B$) and such that the largest squarefree factor of $E_r/s_B(E_r)$ is $\leq \Delta$.

If $\Delta = \Delta(\beta, \sigma) = \min\left(D^{2-\sigma}, (r + 1 - \sqrt{r})^{1-\beta}\right)$, where $\sigma \in [0, 1]$ then we call such number (B, β, D, σ)−admissible. Clearly the bigger β is, the less space there is for the contribution of primes $q \in (B, D]$ (counting with multiplicity), while the number of prime divisors $q > B$ is (essentially) restricted by the parameter σ. On the other hand the bigger value of σ is, the better estimate of the deterministic time of factoring N, depending on the set X and parameters B, β satisfying the inequality $D^{2-\sigma} \leq (r + 1 - \sqrt{r})^{1-\beta}$. The witness parameters X play the significant role in deterministic factorization of N provided E_r is the admissible number.

We have made some numerical support to the extension of conjecture assumed in [12] for $B = L(\alpha_1, r)$, $D = L(\alpha_2, r)$, where $L(\alpha, r) := \exp\left(\alpha\sqrt{\log r \log \log r}\right)$ and suitable values of parameters β and σ below.

Summarizing, in this work we consider the witnesses $(u, v) \in \mathbb{Z}_N^2$ called $(\mathcal{A}, X)$−elliptic witness for N, where $X = X_{u,v}$ is the set of parameters applied in the algorithm $\mathcal{A}$ if

- The algorithm has N, (u, v) and X as input and decomposition $N = pq$ depending on E_r or $a_r(E)$ for $r \in \{p, q\}$, as output
- The factorization $N = pq$ may be conditional meaning that $E_N = E_p E_q$ is known
- The complexity $t_\mathcal{A}$ of $\mathcal{A}$ depends on X and admissibility parameters of E_r or $a_r(E)$ for $r \in \{p, q\}$
- We consider two types of witnesses, namely $(u, v) = \bar{b}$ or $(u, v) = (x, y) := Q$, where $Q \in E_{\bar{b}}(\mathbb{Z}_N)$.

2 Separating and Nonseparating Witnesses in (Ell, N) Factorization

2.1 Notation and Basic Facts Concerning the Arithmetic in $E(\mathbb{Z}_N)$

In this section we recall basic facts on elliptic curves over $\mathbb{Z}_N$, where $N = \prod_{i=1}^s p_i$ is coprime to 6 (see [11, 12]). The projective plane $\mathbb{P}^2(\mathbb{Z}_N)$ is defined to be the set of equivalence classes of primitive triples in $\mathbb{Z}_N^3$ (i.e., triples (x_1, x_2, x_3) with $\gcd(x_1, x_2, x_3, N) = 1$) with respect to the equivalence $(x_1, x_2, x_3) \sim (y_1, y_2, y_3)$ if $(x_1, x_2, x_3) = u(y_1, y_2, y_3)$ for a unit $u \in \mathbb{Z}_N^*$. An elliptic curve over $\mathbb{Z}_N$ is given by the short Weierstrass equation $E : y^2 z = x^3 + axz^2 + bz^3$, where $a, b \in \mathbb{Z}_N$ and the discriminant $-16(4a^3 + 27b^2) \in \mathbb{Z}_N^*$. The point $O = (0 : 1 : 0)$ called the zero point belongs to $E(\mathbb{Z}_N)$. Let $V(E(\mathbb{Z}_N)) = \{(x, y) \in E(\mathbb{Z}_N)\} \cup \{O\}$ be the set of finite points in $E(\mathbb{Z}_N)$ with the zero point O. For each point $(x : y : z) \in E(\mathbb{Z}_N) \setminus V(E(\mathbb{Z}_N))$ the $\gcd(z, N)$ is a nontrivial divisor of n.

Let $E(\mathbb{F}_{p_i})$ be the group of $\mathbb{F}_{p_i}$-rational points on the reduction E mod p_i for primes $p_i \mid N$. For the set $E(\mathbb{Z}_N)$ of points in $\mathbb{P}^2(\mathbb{Z}_N)$ satisfying the equation of E there exists by the CRT the bijection

$$\varphi : E(\mathbb{Z}_N) \to \prod_{i=1}^{s} E(\mathbb{F}_{p_i}) \tag{2}$$

induced by the reductions mod p_i The points $(x : y : z) \in E(\mathbb{Z}_N)$ with $z \in \mathbb{Z}_N^*$ can be written $(x/z : y/z : 1)$ and are called finite points. The set $E(\mathbb{Z}_N)$ is a group with the addition for which φ is a group isomorphism, which in general can be defined using the so-called complete set of addition laws on E (see [11]).

To add two finite points $P, Q \in E(\mathbb{Z}_N)$ we can also use the same formulas as for elliptic curves over fields in the following case: for $\varphi(P) = (P_1, \ldots, P_s)$ and $\varphi(Q) = (Q_1, \ldots, Q_s) \in \prod_i E(\mathbb{F}_{p_i})$ either $Q_i \neq \pm P_i$ for each i or $Q_i = P_i$ and $Q_i \neq -P_i$ for each i. Then

$$\begin{cases} x_{P+Q} = \lambda^2 - x_P - x_Q \\ y_{P+Q} = \lambda(x_P - x_{P+Q}) - y_P, \end{cases} \tag{3}$$

where

$$\lambda = \begin{cases} \frac{y_Q - y_P}{x_Q - x_P} & \text{if } Q_i \neq \pm P_i \text{ for each } i \\ \frac{3x_P^2 + a}{2y_P} & \text{if } Q_i = P_i \text{ and } Q_i \neq -P_i \text{ for each } i. \end{cases}$$

Let $P, Q \in E(\mathbb{Z}_N)$ and if $R = P + Q$ is finite then the formulas (3) give the coordinates of the resulted point R. Otherwise either we find the nontrivial divisor of N or prove that all local orders $\mathrm{ord}R_i$ are equal each other for $i = 1, 2, \ldots, s$ (see e.g. [5,12] for details).

In what follows we assume that $N = pq$ has two distinct prime divisors (both > 3) and $B = B(N)$ is fixed. We apply the above formulas to compute the point $mQ \in E(E_N)$. The computation of finite point mQ takes $O(\log m)$ adding operations in $E(\mathbb{Z}_N)$. For $B-$smooth number $m = m_B$ represented as $m = p_k^{e_k} \ldots 3^{e_3} 2^{e_2}$, where $e_i = e_i(m)$ is the highest exponent in which p_i does not exceed $\min(p, q) + 2\sqrt{\min(p, q)} + 1$ the computation of $m_B Q$ takes

$$\ll \log N \sum_{i \leq k} \log(p_i) = O(B \log N) \tag{4}$$

adding operations in $E(\mathbb{Z}_N)$.

2.2   Admissible Numbers and Witnesses Definitions

Here and in the sequel we will use the symbol $o(1)$ as p and q tend to infinity. We let

Definition 1. *The number $m \in \mathbb{N}$ is called $B-$ smooth if all prime factors of m are $\leq B$. Moreover it is called $(B, \beta)-$smooth if the largest $B-$smooth divisor of m, denoted by $s_B(m)$ is $\geq m^\beta$. If additionally m is $D-$smooth then we call it $(B, \beta, D)-$smooth number.*

Let m^* stand for the largest squarefree divisor of m.

Definition 2. *The number $m \in \mathbb{N}$ is called $(r, B, \beta, D, \sigma)-$admissible if it is $(B, \beta, D)-$smooth and the following conditions hold*

$$\beta \leq 1 - \frac{(2 - \sigma) \log D}{\log I_r^-}, \tag{5}$$

$$\left(\frac{m}{s_B(m)} \right)^* \leq D^{2-\sigma}. \tag{6}$$

Directly from 6 it follows that

$$\omega \left(\frac{m}{s_B(m)} \right) \leq \kappa := (2 - \sigma) \frac{\log D}{\log B}, \tag{7}$$

In what follows we assume that $\kappa \geq 2$, which implies that $B^2 \leq D^{2-\sigma}$. We call the related $(r, B, \beta, D, \sigma)-$admissible numbers shortly $(\beta, \sigma)-$admissible, if the remaining parameters are clear from the context.

Let $N(N = pq)$ be fixed and $E = E_{\bar{b}}$ be an elliptic curve over $\mathbb{Z}_N$. Then E is called $(B, \beta, D, \sigma)-$admissible if E_r is $(r, B, \beta, D, \sigma)-$admissible for some $r \in \{p, q\}$.

The value σ indicates an improvement in the exponent of the standard complexity bound $D^{2+o(1)}$, when searching the factors p and q of N provided E_r is $(r, B, \beta, D, \sigma)-$admissible number.

We prove the results concerning the factorization of N in deterministic time $t = t_A$ depending on the set of parameters X of the witness. In unconditional case we distinct 2 types of witnesses corresponding to $D-$smooth number E_r:

- Separating witnesses $Q = (x, y) \in \mathbb{Z}_N^2$ such that $\mathrm{ord}Q_r \neq \mathrm{ord}Q_{r'}$, where $\{r, r'\} = \{p, q\}$.
- Nonseparating witnesses $Q = (x, y) \in \mathbb{Z}_N^2$ such that $\mathrm{ord}Q_r = \mathrm{ord}Q_{r'}$, where $\{r, r'\} = \{p, q\}$.

Let N and $E = E_{\bar{b}}$ over $\mathbb{Z}_N$ be given. Below we consider the pairs (E, Q), where $Q \in E(\mathbb{Z}_N)$, such that E is admissible curve and Q is suitable witness for N. The separating witnesses below are applied only for the value $\gamma = 1$ (hence the related restriction can be suppressed), but in order to keep the consistency of presentation we will maintain the following general definition.

In what follows we let $\vartheta > 1$.

Definition 3. *Let* $N = pq$, $\vartheta > 1$, $p < q < \vartheta p$, $0 \leq \gamma \leq 1$, E *be* $(B, \beta, D, \sigma)-admissible$ *and* $\{r, r'\} = \{p, q\}$. *The pair* $Q = (x, y)$ *is called* $(E, \gamma)-strong$ *separating witness for* N *if* $Q \in E(\mathbb{Z}_N)$ *and we have*

$$P^+(\mathrm{ord}Q_{r'}) \text{ does not divide } E_r \tag{8}$$

$$\mathrm{ord}Q_{r'} \geq 4\vartheta^{1/3} \min(r, r')/N^{\gamma/2}. \tag{9}$$

The first condition above was applied in [12], for the factorization of N in expected subexponential time, while the second in [4] in the context of oracle factoring (deterministic approach).

Definition 4. *Let* $N = pq$, $p < q$, $0 \leq \gamma \leq 1$, E *be* $(B, \beta, D, \sigma)-admissible$, *and* $\{r, r'\} = \{p, q\}$. *The pair* $Q = (x, y)$ *is called* $(E, \gamma)-weak$ *separating witness for* N *if* $Q = (x, y) \in E(\mathbb{Z}_N)$ *and we have*

$$\mathrm{ord}Q_{r'} \neq \mathrm{ord}Q_r \tag{10}$$

$$\mathrm{ord}Q_{r'} \geq 4\vartheta^{1/3} \min(r, r')/N^{\gamma/2}. \tag{11}$$

Given the admissible curve E and the suitable separating witness we discover the factorization $N = pq$ in deterministic time $O\left((B^2 + D^{2-\sigma})^{1+o(1)}\right)$, thus improving t_A(for $(\beta, \sigma)-$admissible values of E_r) on the power D^σ.

In the following definition we restrict ourselves to the range $\gamma \in [0, 1/4]$ in view of the application to Theorem 2).

Definition 5. *Let* $N = pq$, $p < q < \vartheta p$, $0 \leq \gamma < 1/4$ *and* E *be* $(B, \beta, D, \sigma)-admissible$ *and* $\{r, r'\} = \{p, q\}$. *The pair* $Q = (x, y) \in E(\mathbb{Z}_N)$ *is called* $(E, \gamma)-nonseparating$ *witness for* N *if*

$$\mathrm{ord}Q_r = \mathrm{ord}Q_{r'} \geq 4\vartheta^{1/3} \min(r, r')/N^{\frac{\gamma}{2}} \tag{12}$$

and either

$$1 \leq \min(a_r(E), a_{r'}(E)) \leq \min(r, r')^{1/2-\gamma} \tag{13}$$

or

$$\gamma \leq 2(2 - \sigma)\frac{\log D}{\log N}. \tag{14}$$

In Sect. 4 we state the main results regarding factorization $N = pq$ with the aid of nonseparating witness in deterministic time $O(D^{2-\sigma+o(1)})$.

3   Decomposition Witnesses in (Ell, E_N) factorization

The classical Fermat's factoring method allows to efficiently factor the positive integer N provided $N = n_1 n_2$, where the absolute value of $n_1 - n_2$ is of order of magnitude $N^{1/4}$, since then we are able to find the related divisor close to $\sqrt{ab} = \sqrt{N}$. In the context of factoring N, the linear forms of type $ap + bq$ with integral coefficients a, b were considered in [9], and in [10] the factorization of N in deterministic time $O(N^{1/3+o(1)})$ was proved.

In this work we consider the special linear forms of type $F^{\pm} := F^{\pm}(a, b) = ap \pm bq$, where the coefficients a, b are related to the given elliptic curves $E(\mathbb{F}_q)$ and $E(\mathbb{F}_p)$ respectively. Namely we analyse the cases when a, b are either the Frobenius traces $a_q(E)$ and $a_p(E)$ or the values of E_q and E_p respectively. We search for the good approximation for F^+ by the geometric mean $\sqrt{abpq} = \sqrt{NE_N}$ in terms of $|F^-|$ also in the case when p and q are not of the similar order of magnitude. This leads to the definition of the relevant witnesses parameters $X \in \{D, \{D, s, \alpha\}, \{B, K\}\}$, provided E_N is known.

In [6] the authors applied the Coppersmith factoring method [2] to prove that $N = pq$ can be factored in random polynomial time provided the factorization of E_N is known. The authors also proved that factoring can be derandomized provided the number of prime divisors of E_N is not too large. Here we follow another (conditional) approach which can be expressed in terms of the witnesses and their parameters X. Certainly they are also connected to non-separating witnesses since $\mathrm{ord}Q_r = \mathrm{ord}Q_{r'} \geq 4\vartheta^{1/3}\min(r, r')/N^{\gamma/2}$ implies that $d = \gcd(E_r, E_{r'}) \geq \gcd(\mathrm{ord}Q_r, \mathrm{ord}Q_{r'}) \geq 4\vartheta^{1/3}\min(r, r')/N^{\gamma/2}$ and moreover E_r has large B−smooth factor. But here we focus rather on the precision of approximating p/q by the related quotients of $a_q(E_q)$ and $a_p(E_p)$ respectively.

The somewhat more accurate Definition 7 below, refers to finding a "good" approximation of $(s/d)^2$ by $1 + 1/D$, which allows the time of decomposition $N = pq$ to be expressed in terms of two stages of appropriate approximations in Theorem 3. Finally the last definition below refers to the direct application of Coppersmith result (see Lemma 3).

Definition 6. *Let N and $M = M(N) \leq N^{1/2}$ be given. Let $D = D(N)$ and $q \in [Mp, 2Mp]$, where $pq = N$ and assume that the condition (1) holds true. The pair $\bar{b} \in \mathbb{Z}_N^2$ is called $D-$ factoring witness if*

$$\left| \frac{p}{q} - \frac{a_p(E_{\bar{b}})}{a_q(E_{\bar{b}})} \right| \leq \frac{1}{D} \tag{15}$$

for some $r \in \{p, q\}$ (if $a_p(E) = a_q(E) = 0$ then we set $0/0 := 0$).

Definition 7. *Let N, E_N and $d \mid \gcd(E_p, E_q) > 1$ be given. The pair $\bar{b} \in \mathbb{Z}_N^2$ is called $(D, s, \alpha)-$ factoring witness if the condition (1) holds true and $s \neq d$, $s \mid d^2$ satisfies the condition*

$$\left(\frac{d}{s} \right)^2 = 1 + \frac{1}{D} + \left(\frac{\alpha\theta}{D^2} \right), \tag{16}$$

for some $|\theta| \leq 1$.

Definition 8. *Let N be given, $M = 1$ and $B = B(N), K = K(N)$. The pair $\bar{b} \in \mathbb{Z}_N^2$ is called $(B, K)-$ factoring witness if the condition (1) holds true and $s_B(E_r) > r/K$ for some $r \in \{p, q\}$, where $s_B(m)$ denotes the largest B−smooth factor of m.*

4   Main Results for (Ell, N) Factorization

Below we state the results for separating ($\gamma = 1$) and nonseparating ($\gamma \in [0, 1/4)$) witnesses separately.

4.1   Separating Witnesses

We proceed the main result (Theorem 1 below) by some definitions and auxiliary results. The separation witnesses for admissible elliptic curve E give the benefits in saving the factor D^σ in computational cost in comparison to the standard approach. The following lemma follows directly from Proposition 2.5 of [4].

Lemma 1. *Let $Q \in E_{\bar{b}}(\mathbb{Z}_N)$ be a finite point on the elliptic curve $E_{\bar{b}}$ over $\mathbb{Z}_N$, where $\gcd(\Delta_{\bar{b}}, N) = 1$. Assume that the reduction point $Q_r \in E(\mathbb{F}_r)$ has a $B-$smooth order $\mathrm{ord}Q_r$ for some $r \in \{p, q\}$. Then either one can discover the factorization $N = pq$ or compute $\mathrm{ord}Q_r$ and conclude that $s_B(\mathrm{ord}Q_r) = \mathrm{ord}Q_r = \mathrm{ord}Q_{r'}$ in deterministic time $O(B^{2+o(1)})$.*

Let $m^* = \prod_i q_i$, where $q_1 > q_2 > \ldots > q_k$ are distinct prime numbers. We say that the sequence $(q_1, ..., q_k)$ belongs to the tuple $\bar{l} = (l_1, ..., l_k)$ if $q_i \in [2^{l_i}, 2^{l_i+1})$ for $i = 1, 2, ..., k$.

Definition 9. *The tuple $\bar{l} = (l_1, l_2, ..., l_k)$ is called $\Delta-$admissible if $l_1 \geq l_2 \geq ... \geq l_k$ and*

$$\sum_{i \leq k} l_i \leq \log \Delta / \log 2. \tag{17}$$

The following lemma allows to reduce counting the sequences $(q_1, ..., q_k)$ with the coordinates depending only on the values of l_i, $i = 1, ..., k$, ($k \leq \lfloor \kappa \rfloor$) of $\Delta-$admissible tuples $\bar{l}$.

Lemma 2. *The number of sequences $(q_1, ..., q_k)$ belonging to a fixed $\Delta-$admissible tuple $\bar{l}$ is $\leq \Delta$. Moreover the number of $\Delta-$ admissible tuples $\bar{l} = (l_1, ..., l_k)$, to which some sequence $(q_1, ..., q_k)$ may belong is equal to $O\left((\log \Delta)^k\right)$, where the constant implied by the symbol O does not depend on Δ.*

Proof. Each $q_1 q_1 \cdots q_k \leq \Delta$ belongs to exactly one $\Delta-$admissible tuple $\bar{l}$. Since $q_i \in [2^{l_i}, 2^{l_i+1})$ the number of relevant sequences $(q_1, ..., q_k)$ belonging to a fixed tuple $\bar{l}$ is $\leq 2^{l_1} \cdots 2^{l_k} \leq \Delta$, by 17. Moreover the number of $\Delta-$admissible tuples $\bar{l}$ is bounded by the product of choices for l_i, where $0 \leq l_i \leq \log q_i / \log 2$, for $i = 1, 2, ..., k$, that is by $\leq \prod_{i \leq k} (\log q_i / \log 2 + 1) \leq (\log \Delta / \log 2 + 1)^k = O\left((\log \Delta)^k\right)$, where the constant implied by the symbol O does not depend on Δ. This completes the proof of the Lemma 2.

We are now in a position to state and prove the first main result of this section.

Theorem 1. *Let N ($N = pq, p < q$) be given, $\{r, r'\} = \{p, q\}$ and E be $(r, B, \beta, D, \sigma)-$ admissible curve and $k \leq \kappa$.*

Let $\gamma = 1$ and the point $Q = (x, y) \in E(\mathbb{Z}_N)$ is $(E, \gamma)-$strong or $(E, \gamma)-$weak separating witness for N. Then one can find p and q in deterministic time $O\left((B^2 + D^{2-\sigma})^{1+o(1)}\right)$.

Proof. Let E be elliptic curve over $\mathbb{Z}_N$ which is $(B, \beta, D, \sigma)-$admissible. Assume that $\gamma = 1$ and the point $Q = (x, y) \in E(\mathbb{Z}_N)$ is $(E, \gamma)-$strong or $(E, \gamma)-$weak separating witness for N. We claim that one can find p and q in deterministic time $O\left((B^2 + D^{2-\sigma})^{1+o(1)}\right)$.

Assume that $\text{ord}Q_r \neq \text{ord}Q_{r'}$ and that $s_D/s_B = E_r/s_B(E_r) = \prod_{i \leq k} q_i^{\nu_i}$. If $s_D/s_B = 1$ ($k = 0$) then E_r is $B-$smooth and the result follows from Lemma 1. Otherwise let $\Delta = D^{2-\sigma}$ and assume that for some $k \leq \kappa$ the sequence$(q_1, ..., q_k)$ belongs to $\Delta-$admissible tuple $\bar{l} = (l_1, ..., l_k)$.

Let $q^\nu = q_1^\nu$ vary over all primes in the interval $[2^{l_1}, 2^{l_1+1})$ in exponent $0 \leq \nu \leq \log \Delta / \log q$. We compute the multiples $(q^\nu m_B)Q := q^\nu R_0 \in E(\mathbb{Z}_N)$, where $R_0 = m_B(Q)$ and m_B is the product of all primes $p \leq B$ in maximal powers that are less than I_r^+.

If for some power q^ν the point $q^\nu R_0 \in E(\mathbb{Z}_N)$ is not finite then in view of Lemma 1 we will discover the decomposition $N = pq$ in $\ll D \log(q^\nu) \leq D^{1+o(1)}$ adding operations in $E(\mathbb{Z}_N)$. Hence if $\omega(E_r/s_B(E_r)) \leq 1$ we decompose N in deterministic time $\ll (B^2 + D)^{1+o(1)}$.

Otherwise we conclude that $\text{ord}Q_r$ ($\text{ord}Q_r \mid E_r$) has at least two prime divisors $q_1 > q_2$, that is $\omega(E_r/s_B(E_r)) \geq 2$. Now rising $q_2 \in [2^{l_2}, 2^{l_2+1})$ to maximal possible powers ν_2 we follow the computation of $q_2^{\nu_2} R_1$ similarly as above, where $R_1 = q_1^{\nu_1} m_B(Q)$ is a finite point in $E(\mathbb{Z}_N)$.

Since $\bar{l}$ is $\Delta-$admissible the number of possible choices for the pairs (q_1, q_2) with the related exponents (ν_1, ν_2) is in view of Lemma 2 bounded by $\Delta^{1+o(1)} = D^{2-\sigma+o(1)}$. We continue this procedure for all $k \leq \kappa$ and by the assumption that $\text{ord}Q_{r'} \neq \text{ord}Q_r \mid E_r$ we infer that this procedure terminates after at most $\lfloor \kappa \rfloor$ steps, giving the deterministic total time $O\left((B^2 + D^{2-\sigma})^{1+o(1)}\right)$, as required. ∎

4.2   Nonseparating Witnesses

From now on we let $\vartheta > 1$ and $0 \leq \gamma < 1/4$. In the proof of Theorem 2 we apply the the following

Lemma 3. *(Coppersmith)   (see [2]) If we know N and the high order $(1/4)$ $(\log_2 N) + O_\vartheta(1)$ bits of q, where $p < q < \vartheta p$, then in polynomial time in $\log N$ we can discover p and q.*

Corollary 1. *Assume that $p < q < \vartheta p$ and E_r is known. Then one can discover the decomposition $N = pq$ in deterministic polynomial time.*

Proof. Since $E_r \in [I_r^-, I_r^+]$ where $I_r^\pm = r + 1 \pm \sqrt{r}$ we deduce that E_r distincts from r at most on $\log_2 r + O(1) = \frac{1}{4} \log_2 N + O(1)$ least significant bits. Hence by Lemma 3 we get the conclusion. ∎

The nonseparating witnesses allow to handle also the case when $0 \leq \gamma < 1/4$. Moreover if $\gamma = O(\log \log N)/\log N$ then the direct application of Coppersmith result gives the polynomial time factorization $N = pq$, provided we have $Q \in E(\mathbb{Z}_N)$ satisfying $\mathrm{ord}Q_r \geq \min(r,r')/N^{\gamma/2}$ for some $r \in \{p,q\}$, whenever E_r is D–smooth with $D = (\log N)^{O(1)}$.

Theorem 2. *Let N ($N = pq$, $p < q < \vartheta p$) be given, $\gamma < 1/4$, $\{r,r'\} = \{p,q\}$ and E be $(B, \beta, D, \sigma)-$ admissible curve and $c(\vartheta) \geq 4\vartheta^{5/4}$.*

Assume that the point $Q = (x,y) \in E(\mathbb{Z}_N)$ is $(E,\gamma)-$nonseparating witness for N and satisfy the conditions (12) and (13). Then one can find p and q in deterministic polynomial time.

On the other hand if (12) and (14) hold true, then one can compute the decomposition of $N = pq$ in deterministic time $O(D^{2-\sigma+o(1)})$, provided

$$\gamma \leq \min\left(2(2-\sigma)\frac{\log D}{\log N}, \frac{1}{4}\right).$$

Moreover if E_r is $D-$smooth where $D = (\log N)^{O(1)}$ and we have the point $Q \in E(\mathbb{Z}_N)$ such that $\mathrm{ord}Q_r \geq c(\vartheta)\min(r,r')/N^{\gamma/2}$ for some $r \in \{p,q\}$ and $\gamma = O(\log \log N/\log N)$, then the decomposition $N = pq$ can be computed in polynomial time.

Proof. In order to prove the Theorem 2 we first show that if we are given $Q \in E(\mathbb{Z}_N)$ that is (E,γ) nonseparating witness, ($0 \leq \gamma < 1/4$) that is the conditions 12 and 13 hold true, then we can decompose $N = pq$ ($p < q$) in deterministic polynomial time. Next we show that if the conditions 12 and 14 hold true then we can decompose $N = pq$ in deterministic time $O(D^{2-\sigma+o(1)})$.

To prove the first assertion we apply the assumption that

$$1 \leq \min(a_p(E), a_q(E)) \leq 2p^{1/2-\gamma}$$

and the condition

$$m := \mathrm{ord}Q_p = \mathrm{ord}Q_q \geq c(\vartheta)\frac{\min(q,p)}{N^{\gamma/2}} = c(\vartheta)p/N^{\gamma/2},$$

where $c(\vartheta) \geq 4\vartheta^{5/4}$. Moreover by the assumption $\gamma < 1/4$ we have that $m > N^{3/8} > N^{1/3}$ for sufficiently large $N \geq N_0 = N_0(\vartheta, \gamma)$. Therefore we can represent the number N in base m,

$$N = c_0 + c_1 m + c_2 m^2.$$

Since $E_p = p + 1 - a_p(E) = mr_p$ and $E_q = q + 1 - a_q(E) = mr_q$, letting $t_p = a_p(E) - 1$ and $t_q = a_q(E) - 1$ we have

$$N = pq = (mr_p + a_p(E) - 1)(mr_q + a_q(E) - 1) = (mr_p + t_p)(mr_q + t_q)$$

$$= r_p r_q m^2 + (r_p t_q + r_q t_p)m + t_p t_q$$

where $t_p, t_q \geq 0$. Thus the coefficients $r_p r_q$, $r_p + r_q$, and $t_p t_q$ are uniquely defined by c_i, $i = 0, 1, 2$ provided all they are in the interval $[0, m)$.

Since all they are nonnegative it remains to check that they are $< m$. We have that $r_p r_q = (E_p E_q)/m^2 \leq N/m^2 < m$, since $m > N^{1/3}$ for sufficiently large $N \geq N_0(\vartheta)$ by the above. Moreover we have $r_p t_q + r_q t_p \leq 2\max(\sqrt{p}, \sqrt{q})(E_p + E_q)/m = 4\sqrt{q}(q/m) \leq 4q^{3/2}/m \leq 4\vartheta^{3/2}(N^{1/2})^{3/2}/m < m$, since $N^{3/4} < m^2/4\vartheta^{3/2}$ if $\gamma < 1/4$ and N is sufficiently large.

Finally we have $t_p t_q \leq 2(p^{(1/2)-\gamma})(2q^{1/2}) \leq 4N^{1/2}/p^\gamma \leq 4\,N^{1/2}/(q/\vartheta)^\gamma) \leq 4\vartheta^{1/4} N^{1/2-\gamma/2} < m$, since $m \geq 4\vartheta^{5/4} p/N^{\gamma/2}$

To complete the argument it remains to remark that $c_1 = r_p + r_q = c_0 u^{-1} + c_2 u$, where $u = \frac{t_p}{r_p} \in \mathbb{Q}$ is assumed to be in the reduced form. Now we have that $N = pq = (mr_p + t_p)(mr_q + t_q)$ and both factors are ≥ 2. Hence we discover p by computing $\gcd(N, mr_p + t_p)$. This implies the required factorization $N = pq$ in the case when the condition 13 holds.

On the other hand the condition 14 implies that E_r for any $r \mid N$, differ from $m = \operatorname{ord} Q_r$ at most on the factor $\ll N^{\gamma/2}$ which is $\leq D^{2-\sigma}$ by the assumption.

Therefore we are in a position to apply Lemma 3. Since for any $r \in \{p, q\}$, $E_r \in I_r$ differs from r on at most $\log_2 r^{1/2} \leq \log N^{1/4}$ least significant bits, the application of Lemma 3 to detect the factor $r \mid N$ in deterministic polylog(N) time, which implies that the total deterministic time in this case is $O(D^{2-\sigma+o(1)})$, as required. If $D = \log N)^{O(1)}$, the conclusion follows immediately from Corollary 1. This completes the proof of Theorem 2.

5   Separating Witnesses for Subexponential B and D

Here we investigate the distribution of random elliptic curves $E = E_{\bar{b}}$ that are (B, β, D, σ)−admissible, where $B = L(\alpha_1, r)$ and $D = L(\alpha_2, r)$, where $\alpha_1 < \alpha_0 < \alpha_2$ and $\alpha_0 = 1/\sqrt{2}$. By [12] the expected fraction of triples $(x, y, b_1) \in \mathbb{Z}_N^3$ such that $E_r \in I_r$ is $B-$ smooth number for some $r \in \{p, q\}$ is equal to $1/L_B$, where

$$L_B = L\left(\frac{1}{2\alpha_1} + o(1), r\right).$$

Then the expected time of finding the separating witness is $L(\alpha_1 + o(1), r)$ giving the optimal choice of α_1 satisfying the equality $\frac{1}{2\alpha_1} = \alpha_0 = 1/\sqrt{2}$. The expected time of factoring N is then equal to

$$L(2\alpha_0 + o(1), r) = \exp\left(\sqrt{(2 + o(1))\log r \log \log r}\right).$$

The analogous question for (B, β, D, σ)−admissible numbers E_r is more delicate, since we have to enter the additional parameters $\beta, D = L(\alpha_2, r)$ and σ satisfying the following inequality

$$\omega(E_r/s_B(E_r)) \leq \kappa = (2 - \sigma)\frac{\log D}{\log B}, \qquad (18)$$

giving the deterministic time $\ll (B^2 + D^{2-\sigma})^{1+o(1)} \ll D^{2-\sigma+o(1)}$.

If $\omega(E_r/s_B(E_r)) \leq 1$ then the decomposition $N = pq$ is by Lemma 2 discovered in deterministic time $O\left((B^2 + D)^{1+o(1)}\right)$. The contribution of all (B, β, D, σ)−admissible numbers $E_r \in I_r$ comes from the numbers of type $s_B(E_r) \prod_{i \leq \lfloor \kappa \rfloor} q_i^{\nu_i}$, where $\prod_{i \leq \lfloor \kappa \rfloor} q_i^{\nu_i} \leq D^{2-\sigma}$ and $\lfloor \kappa \rfloor$ denotes the integer part of κ.

Bearing in mind [8] (Theorem 5.2) we state the analogous conjecture as in [12] called β−conjecture for (B, β, D, σ)−admissible numbers. Namely consider the admissible triples $T = (x, y, b_1) \in \mathbb{Z}_N^3$ such that $E_r \in I_r$.

Namely let

$$f_\beta(B, D, \sigma) = \frac{\#\{T \in \mathbb{Z}_N^3 : E_r \in I_r : E_r \text{ is } (r, B, \beta, D, \sigma) - \text{admissible}\}}{\#\{T \in \mathbb{Z}_N^3 : E_r \in I_r\}} \quad (19)$$

$\beta-Conjecture:$

Let $0 < \beta < 1$. Let $B = L(\alpha_1, r) \leq L(\alpha_0, r) \leq D = L(\alpha_2, r)$. Selecting randomly the triples $(x, y, b_1) \in \mathbb{Z}_N^3$ with $E_r \in I_r$ we conjecture that

$$f_\beta(B, D, \sigma) = 1/L\left(\frac{1 - \theta(1 - \beta)}{2\alpha_1} + o(1), r\right), \quad (20)$$

where $\theta = \theta(\alpha_1, \alpha_2, \sigma) > 0$.

5.1   Computational Support

β-Conjecture. Assume that $\alpha_1 = \alpha_0 - \delta$ and $\alpha_2 = \alpha_0 + \delta$. Let us illustrate the β-Conjecture through computational support by examining $f_\beta(B, D, \sigma)$ - the density levels of (B, β, D, σ)-admissible numbers. Experiments returning the value of f_β are conducted as procedures $\texttt{Exp}(\lambda, c, \delta, \beta, \sigma)$, where λ denotes the bit length of the numbers p, q, and c determines the accuracy of the search, that is,

$$\#\{\mathbb{Z}_N^3 \xrightarrow{\$} (x, y, b_1)\} \geq L(c, r).$$

The $\texttt{Exp}$ procedure returns $f_\beta(B, D, \sigma)$ - the ratio of the counts of orders from these two categories, according to the Formula (19).

The ranges of the remaining parameters define relationships from the definition 2 of admissible number. More precisely, these ranges are determined from the inequality (5): and from the inequality (7):

$$\omega\left(\frac{E_r}{s_B(E_r)}\right) \leq (2 - \sigma)\frac{\log D}{\log B} = (2 - \sigma)\frac{\alpha_2}{\alpha_1},$$

because inequality (6) can be replaced by the last one. The value of $\omega\left(\frac{E_r}{s_B(E_r)}\right)$ we fixed at 3, to control ranges of dependent parameters. Let the selection of parameters, fully consistent with these dependencies, be as follows:

$$\lambda = 30, c = 1.1, \sigma = 0.45, \delta = 0.35,$$
$$\beta \in \{0.24, 0.27, 0.30, 0.33, 0.36, 0.39, 0.42, 0.45, 0.48\} = \bar{\beta}.$$

This entails repeating the experiment $\text{Exp}(30, 1.1, 0.35, \beta_i, 0.45)$ for all $\beta_i \in \bar{\beta}$.

To illustrate the computational support for the (β, σ)−conjecture, in this case, we want to show that there exists a constant value $\theta(\alpha_1, \alpha_2, \sigma)$ for which the density function $f_\beta(B, D, \sigma)$ behaves in accordance with the predictions presented in equality (20), with accuracy to the error indicated there, asymptotically equal to $o(1)$.

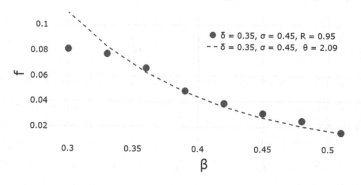

Fig. 1. The graph shows experimentally obtained densities $f_\beta(B, D, \sigma)$, with fixed $\sigma = 0.45, \delta = 0.35$ for the bit length p, q equal to 30 bits. For $\theta \approx 2.09$, the consistency coefficient R of the sample approximated by the function of θ was equal to 0.95

In the Fig. 1, we observe that there is no dependency of the parameter θ on the parameter β, which is consistent with the β−conjecture. The key conclusion from the numerical example is that the behavior of the function $f_\beta(B, D, \sigma)$ is well approximated by the constant value θ, resulting from the fact that θ does not depend on other parameters than $\alpha_1, \alpha_2, \sigma$. This means that the form of the function, presented in equation (20), is consistent with the experimentally obtained results. Furthermore, the optimal value of $\theta \approx 2.09$, meaning that the examined densities of (B, β, D, σ)-admissible orders E_r exceed the B-smooth densities from these ranges.

We conclude that the experimental searches conducted align with the proposed hypothesis.

6   Main Result for (Ell, E_N) Factorization

We recall that $a_p = a_p(E), a_q = a_q(E)$ are the related Frobenius traces modulo p and q respectively. By by $\omega(m), (\Omega(m))$ we denote the number of distinct (all) prime divisors of m, respectively.

In what follows we use the convention that $\text{polylog}(N)$ means the time bounded by some power (which can be explicitly given) of $\log N$, as N tends to infinity.

The proof of Theorem 3 below is based on two lemmas. The first, is commonly known starting point in the Fermat factoring method and will be applied in the proof of (i) of Theorem 3.

Lemma 4. *Let $X \leq Y$ be real positive numbers and XY be given. Then $X + Y$ can be approximated from below by $2\sqrt{XY}$ with precision $\geq \frac{(X-Y)^2}{4\sqrt{XY}}$.*

The second is directly applied in the proof of (ii) of Theorem 3.

Lemma 5. *Let $N = pq$, where $p < q < \vartheta p$, $1 < a_q := q + 1 - E_q$, $1 < d \mid \gcd(E_p, E_q)$. Assume that N, E_N and $d \neq s \mid d^2$ are known and for some $c = c(\vartheta) > 0$ the following condition holds*

$$\left| \frac{p}{q} - \frac{E_p}{E_q}\left(\frac{d}{s}\right)^2 \right| < \frac{c}{D^2} \tag{21}$$

Then p and q can be computed in deterministic time $(polylogN)(D^{-4}NM^2)$.

Now, with these two lemmas, we are in a position to state the main result of this section. Proofs of Lemma 5 and Theorem 3 and are available here: in the full-version paper).

Theorem 3. *Let N and E_N be given, where $N = pq$, $q \in [Mp, 2Mp]$. Then*

(i) *Assume that $D = D(N)$ and $\bar{b} \in \mathbb{Z}_N^2$ is $D-$factoring witness. Then we can factor N in deterministic time*

$$t = t(N, M, D, a_p, a_q) = (polylogN)\left(M\left(1 + \left(\frac{|a_p| + |a_q|}{D}\right)^2\right)\right). \tag{22}$$

Unconditionally we have $t = (polylogN)M$, provided $D > (MN)^{1/4}$.

(ii) *Let $N = pq$, $p < q < \vartheta p$, $d \mid \gcd(E_p, E_q)$ and $d \neq s \mid d^2$ be given. Assume that $\bar{b} \in \mathbb{Z}_N^2$ is $(D, s, \alpha)-$ factoring witness. Let $\beta = \beta(\alpha)$ be a positive constant such that*

$$\frac{p}{q} - \frac{E_p}{E_q} = \frac{p}{q}\left(\frac{1}{D} + \frac{\beta\theta}{D^2}\right) \tag{23}$$

for some $|\theta| \leq 1$. Then we can detect p, q in deterministic time $O\left(polylogN \frac{NM^2}{D^4}\right)$, where the constant implied by the symbol O depends on α, β and ϑ. Hence $t = polylogN$, provided $D > N^{1/4}M^{1/2}$ (which is stronger than the bound for t in (i) if $M = M(N)$ and $D = D(N)$ are suitably large).

(iii) *$N = pq$, $p < q < \vartheta p$. Assume that $d \mid \gcd(E_p, E_q)$ be such that $N^{1/2}/d < (\log N)^{O(1)}$. Then one can factor N in deterministic polynomial time (depending on ϑ).*

References

1. Burthe, R.J., Jr.: The average least witness is 2. Acta Arith. **80**, 327–341 (1997)
2. Coppersmith, D.: Finding a small root of a bivariate integer equation; Factoring with high bits known. In: Maurer, U. (ed.) EUROCRYPT 1996. LNCS, vol. 1070, pp. 178–189. Springer, Heidelberg (1996). https://doi.org/10.1007/3-540-68339-9_16
3. Dąbrowski, A., Pomykała, J., Shparlinski, I.: On oracle factoring of integers. J. Complex. (76) (2023)
4. Dryło, R., Pomykała, J.: Smooth factors of integers and elliptic curve based factoring with an oracle. In: Grześkowiak, M., Pieprzyk, J., Pomykała, J. (eds.) Number Theoretic methods in Cryptology, vol. 126, pp. 73–88. Banach Center Publications (2023)
5. Dryło, R., Pomykała, J.: Integer factoring problem and elliptic curves over the ring Z_n. Colloq. Math. **159**, 259–284 (2020)
6. Dieulefait, L.V., Jimenez Urroz, J.: Factorization and malleability of RSA moduli, and counting points on elliptic curves modulo N. Mathematics **8**(12), 2126 (2020). https://doi.org/10.3390/math8122126
7. Hittmeir, M., Pomykała, J.: Deterministic integer factorization with oracles for Euler's totient function. Fund. Inform. **172**(1), 39–51 (2020)
8. Konyagin, S., Pomerance, C.: On primes recognizable in deterministic polynomial time. In: Graham, R.L. (ed.) The mathematics of Paul Erdös. Vol. I. Springer, Berlin (1997). [Algorithms Comb. **13**, 176–198]
9. Lawrence, F.W.: Factorisation of numbers. Messenger Math. **24**, 100–109 (1895)
10. Lehman, R.S.: Factoring large integers. Math. Comp. **28**(126), 637–646 (1974)
11. Lenstra, H.W.: Elliptic curves and number-theoretic algorithms. In: Proceedings of the International Congress of Mathematics, pp. 99–120. American Mathematical Society, Berkeley (1986)
12. Lenstra, H.W.: Factoring integers with elliptic curves. Ann. of Math. **126**, 649–673 (1987)
13. Pollard, J.M.: Theorems on factorization and primality testing. Math. Proc. Cambridge Philos. Soc. **76**(3), 521–528 (1974)
14. Pomykała, J., Radziejewski, M.: Integer factoring and compositeness witnesses. J. Math. Cryptol. **14**, 346–358 (2020)
15. Rivest, R., Shamir, A., Adleman, L.: A method for obtaining digital signatures and public-key cryptosystems. Commun. ACM **21**(2), 120–126 (1978)
16. Williams, H.C.: A p+1 method of factoring. Math. Comput. **39**(159), 225–234 (1982)
17. Żrałek, B.: A deterministic version of Pollard's p-1 algorithm. Math. Comp. **79**(269), 513–533 (2010)
18. Żołnierczyk, O., Wroński, M. (2023). Searching B-smooth numbers using quantum annealing: applications to factorization and discrete logarithm problem. In: Mikyška, J., de Mulatier, C., Paszynski, M., Krzhizhanovskaya, V.V., Dongarra, J.J., Sloot, P.M. (eds.) Computational Science - ICCS 2023. LNCS, vol. 10477. Springer, Cham (2023)

Data-Driven 3D Shape Completion
with Product Units

Ziyuan Li, Uwe Jaekel, and Babette Dellen$^{(\boxtimes)}$

Department of Mathematics and Technology, University of Applied Sciences Koblenz,
53424 Remagen, Germany
{zli,jaekel,dellen}@hs-koblenz.de

Abstract. Three-dimensional point clouds play a fundamental role in
a wide array of fields, spanning from computer vision to robotics and
autonomous navigation. Modeling the 3D shape of objects from these
point clouds is important for various applications, including 3D shape
completion and object recognition. This paper presents a complex-valued
product-unit network for data-driven 3D shape completion. Using product units, sparse superpositions of complex power laws, including sparse
polynomial functions, are fitted to incomplete 3D point clouds and used
for extrapolating the data in the 3D space. In computer-vision applications, this task occurs frequently, e.g., when only partial views are
available or occlusions hinder the acquisition of the full point cloud.
We conduct a comparative analysis with a standard neural network to
emphasize the superior extrapolation capabilities of product-unit networks within the 3D space. Furthermore, we present a real-world task
that serves as a tangible demonstration of the proposed method's utility
in the context of completing incomplete point cloud data acquired with
a 3D scanner. This research contributes new insights into the field of
neural network applications for 3D point cloud processing, revealing the
broad potential of product-unit networks in this domain.

Keywords: Product units · 3D shape completion · Data-driven
methods

1 Introduction

Shape completion is the process of inter- or extrapolating missing data that
describes object shape. Incomplete shapes and point clouds are often encountered in computer-vision applications during data acquisition [2]. Often, only
partial views are available or a part of the object is hidden due to self-occlusion
or other objects placed in the viewing path. The missing data hinders both object
recognition and point-cloud matching required for pose estimation. Hence, finding approaches to recover the complete object from partial data is a topic of
high importance in current computer vision research, since it is needed for many
applications, e.g., robotic grasping [1,3,15,22,23,28].

Approaches for 3D shape completion can be distinguished in some simplification as either data driven or learning based. Learning-based approaches learn a

© The Author(s), under exclusive license to Springer Nature Switzerland AG 2024
L. Franco et al. (Eds.): ICCS 2024, LNCS 14832, pp. 302–315, 2024.
https://doi.org/10.1007/978-3-031-63749-0_21

set of model shape classes from training data and match them directly to observations to perform shape completion [2,27] or pose estimation [28]. Depending on the scenario, this may limit the method to some extent to the shape classes that have been included in the training data.

Different from that, data-driven approaches fit parametric models to partial data to derive a representation of the surface *ad hoc*. Polynomial surface models are a common choice for this task [4,19,20], but they also require some prior knowledge about the object shape to select a suitable model type [3,10,16,17, 26,29]. When the model type is unknown beforehand, many possibilities have to be considered to find a suitable sparse polynomial representation, which can render this task intractable.

Similar problems arise in machine-learning approaches when transforming the data into a higher-dimensional space [9,24] or including nonlinear units representing higher-order polynomials in the network [8] to create nonlinear models (in a data-driven approach). Here, either the basis of this nonlinear space or a nonlinear kernel must be provided beforehand. Moreover, calculations in higher dimensions can be very complex and difficult [25]. The inclusion of nonlinear units in the network also allows better modeling of nonlinear relationships, but, again, the number of combinations of polynomial terms increases exponentially with the order of the polynomial terms, also leading to considerable computational complexity [7].

To address this problem, we propose complex-valued product-unit networks to generate shape models *ad hoc* from the available 3D point-cloud data. These shape models are sparse superpositions of complex power laws, including sparse polynomial functions. This is not an overly restrictive assumption on the model in many cases, since much more complex functions can often be well approximated by Taylor or fractional Taylor series expansions, i.e., linear combinations of power laws. Since the required leading terms of the sparse polynomial can be learned from the data directly, the problem of the complexity of choice is avoided.

Product-unit networks have already been shown before to have advantages towards standard neural networks in extrapolation tasks [12]. Complex-valued product units have been used previously to model the nonlinear properties of nuclear masses [11]. Standard neural network typically make piecewise, quasilinear approximations of the functions or patterns they are meant to learn [12] which restricts the model's ability to extrapolate nonlinear relationships into regions of the feature space that the training data does not cover. In this paper, we advance the product-unit network presented in [11,12] by adapting it to the problem of 3D shape completion and working in the complex-valued domain. We show that the extrapolation capabilities are of the complex-valued product-unit network are superior to the ones of a comparable standard neural network.

2   Methods

Briefly, our methodology involves the transformation of existing 3D point cloud data into the spherical coordinate system. Subsequently, we employ a complex-valued product-unit network for the purpose of predicting missing components and modeling shape as a 2D-function denoted by $r(\theta, \phi)$, where θ and ϕ represent the polar and azimuth angles, respectively, r denotes the radial distance. Here the network utilizes θ and ϕ as inputs, with r serving as output based on the training data.

2.1 Mathematical Model

In contrast to a neuron in a standard neural network, the product unit operates by calculating the product of the powers of its input values [11–13,21]. Mathematically, its output is expressed as follows:

$$y = \prod_{i=1}^{n} x_i^{\omega_i} \quad , \tag{1}$$

where x and y denote the input and output, respectively; ω represents the weight associated with each input, and n signifies the total number of input variables involved. Mathematically, the product operation can be substituted with a summation operation for computational efficiency [11–13,21], leveraging the faster computational speed of addition over multiplication in digital systems. This substitution involves transforming the input values into their logarithmic equivalents, which are subsequently processed through a summation operation. The resulting summation output is then passed through an exponential function, enabling the neural network to produce comparable results. This transformation can be expressed mathematically as:

$$y = e^{\sum_{i=1}^{n} \omega_i \log x_i} \quad . \tag{2}$$

If the input x_i is negative, the logarithm $\log_e x_i$ becomes complex, represented as $\log_e |x_i| + i\pi$ [13]. To support scenarios where the network's operations benefit from complex-valued parameters, we extend the weights and biases to the complex-valued domain, characterized by the expression $\omega = a + bi$. Here, both a and b represent real numbers, with i introducing a complex component to the weight parameter. In alignment with the complex-valued product unit, the weights and biases associated with other neurons, the summation units, within the network are configured in the corresponding complex form.

It is noteworthy that the extension of a neural network into the complex-valued space results in a doubling of both the weights and bias parameters compared to their original quantities, while the number of neural connections remains unchanged. However, when modelling real-valued functions, the imaginary parts of the weights often converge towards zero, effectively reducing the number of parameters by a factor again by a reality condition.

Instead of replacing all neurons in a standard neural network with complex-valued product units, our proposed network integrates product units to replace the standard summation units in a specific layer. This architecture harnesses the distinct strengths inherent in each type of computational unit. Conventional summation units within neural networks are adept at linear and simple nonlinear transformations of input data through activation functions. They specialize in learning fundamental patterns and features within the data. Complex-valued product units can capture more intricate nonlinear relationships within data, owing to their capability to compute products of powered inputs. This unique feature allows them to discern and model intricate dependencies and relationships among input variables. In the context of 3D shape completion, the architecture can be understood as a hierarchical approach to feature extraction. The standard summation units handle linear data transformations, while the complex-valued units build upon these to extract higher-level, more nuanced representations related to the geometry and structure of the 3D shapes. The complex nature of weights and biases in these complex-valued product units allows for a richer representation of data. By incorporating complex numbers, the product units are capable of processing negative inputs without necessitating the introduction of a threshold at the preceding layer.

2.2 Data Sets and Data Acquisition

Firstly, we generated four different 3D point clouds to represent canonical geometric shapes, namely a cone, cylinder, ellipsoid, and cuboid. Each of these point clouds comprised precisely 30,000 data points, expressed in a Cartesian coordinate system defined by the coordinates x, y, and z. The representation of shapes in the spherical coordinate exhibits varying levels of complexity. For instance, this coordinate system is quite suitable to represent a cone. Conversely, accurately modeling a cuboid becomes exceedingly intricate due to the non-linear mapping of points within the spherical coordinate. To address this variability, we have categorized these shapes into three distinct levels of complexity. Shapes suitable for representation in the spherical coordinate system, such as the cone and cylinder, allocated 45% of their respective point clouds for training and interpolation, while the remaining 55% was earmarked for extrapolation. This distribution is visually depicted in Fig. 1(a) and (b). The ellipsoid, positioned at an intermediate complexity level, had an allocation of 70% and 30%, as shown in Fig. 1(c). In the case of the most challenging shape, the cuboid, 95% were used for training and interpolation, leaving 5% for extrapolation, as presented in Fig. 1(d).

Moreover, for the practical application of our methodology, we generated 3D models representing two irregularly shaped real-world objects: a bottle and a computer mouse, as illustrated in Fig. 2(a) and (c). This process involved utilizing a 3D scanner, specifically the Artec Space Spider [5], and its corresponding processing software. The object underwent an initial scanning phase using the 3D scanner, and subsequently, the acquired scan data was employed to construct a 3D model within the Artec Studio software. Following this, selected portions of the model were intentionally removed to simulate missing parts, and a 3D point cloud was generated through the utilization of CloudCompare software. The resulting point clouds are presented in Fig. 2(b) and (d), each comprising 50,000 points.

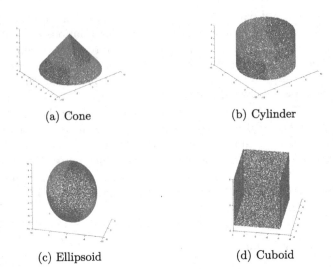

(a) Cone

(b) Cylinder

(c) Ellipsoid

(d) Cuboid

Fig. 1. The point clouds representing the four distinct shapes have been partitioned into two subsets: training data and extrapolation data. The blue segment represents the data allocated for training and interpolation purposes, and the orange segment represents the data used for extrapolation and the subsequent validation of extrapolation results. (Color figure online)

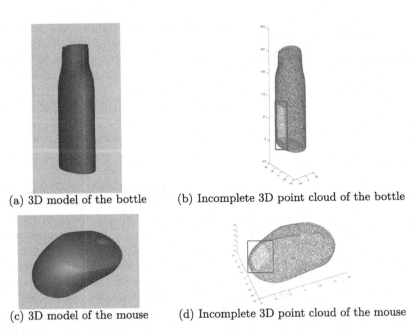

(a) 3D model of the bottle

(b) Incomplete 3D point cloud of the bottle

(c) 3D model of the mouse

(d) Incomplete 3D point cloud of the mouse

Fig. 2. 3D models and incomplete 3D point clouds of the objects, where the missing parts are marked by red boxes. (Color figure online)

2.3   Data Processing

In practical applications, the acquired 3D point cloud may be situated at arbitrary positions within the field of view. To address this variability, we initially calculate the average of all point coordinates, designating this computed value as the central point of the point cloud. Subsequently, a novel Cartesian coordinate system is established with this central point serving as the origin. Conceptually, this procedure can be interpreted as a realignment of the point cloud to coincide with the origin of the pre-existing Cartesian system. Following this spatial adjustment, the point cloud data undergoes a transformation from the Cartesian system to the spherical coordinate system, which is characterized by the variables θ, ϕ, and r, and is computed by the following equations [18]:

$$\theta = \arccos \frac{z}{\sqrt{x^2 + y^2 + z^2}} = \arccos \frac{z}{r} = \text{arcot} \frac{z}{x^2 + y^2} \quad , \tag{3}$$

$$\phi = \text{atan2}\,(y, x) = \begin{cases} \arctan\left(\frac{y}{x}\right) & \text{if } x > 0 \\ \frac{\pi}{2}\,\text{sgn}\,y & \text{if } x = 0 \\ \arctan\left(\frac{y}{x}\right) + \pi & \text{if } x < 0 \wedge y \geq 0 \\ \arctan\left(\frac{y}{x}\right) - \pi & \text{if } x < 0 \wedge y < 0 \end{cases} \quad , \tag{4}$$

$$r = \sqrt{x^2 + y^2 + z^2} \quad . \tag{5}$$

In this way, the association between r and (θ, ϕ) exhibits a one-to-one correspondence in the majority of cases, rendering it amenable to regression networks. Consequently, θ and ϕ serve as inputs for a regression network, with r being the output to be predicted. This underscores the necessity of transforming data from Cartesian coordinates to spherical coordinates. Retaining the data in Cartesian coordinates can lead to ambiguities, as the triplet (x, y, z) does not ensure a one-to-one mapping. For example, using x and y as neural-network inputs and z as the output can result in multiple points sharing identical x and y values but differing z values, thereby complicating the training of the network.

2.4   Network Training

The employed complex-valued product-unit network is characterized by a streamlined three-layer architecture. The initial layer is composed of 10 complex-valued summation units, followed by a subsequent layer housing 120 complex-valued product units, the output layer features a singular complex-valued summation unit, notably absent of any activation function, as visually depicted in Fig. 3(a). In contrast, the reference network is a standard three-layer neural network operating with ReLU activation functions, as illustrated in Fig. 3(b). ReLU is a widely adopted activation function in neural networks, defined by the mathematical expression $f = \max(0, x)$, where x represents the input to the unit [6,14]. As previously indicated, a complex-valued neural network exhibits twice the weights and bias parameter count compared to the standard neural network with the identical architecture, but the number of connections between neurons is the same. This increases the total number of refresh parameters. So in order to make a fair comparison, we not only employed a standard neural network mirroring the exact architecture of the complex-valued product-unit network, both featuring an equivalent number of neurons per layer, but we also introduced an even lager standard neural network with double the network size in terms of number of neurons. Specifically, the network comprises 20 standard summation units in the first layer, 240 standard summation units in the second layer, and the output layer still contains only 1 standard summation unit. The larger standard neural network not only has a comparable number of weights and bias parameters to the number of complex-valued product units, but also has more connections between neurons. Their total number of refresh parameters is much greater than that of the complex-valued product-unit network.

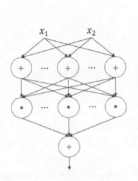

(a) Complex-valued product-unit network

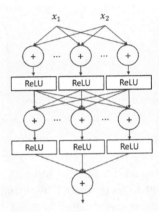

(b) Standard neural network

Fig. 3. Neural networks. Neural units labelled in blue represent complex-valued neural units. (Color figure online)

All neural networks are configured with mean square loss functions. However, the complex-valued product-unit network extends this loss function to accommodate the complex-valued domain, giving

$$L_{\text{CMSE}} = \frac{1}{N} \sum_{i=1}^{N} (y_i - \hat{y}_i)(y_i - \hat{y}_i)^* \quad , \tag{6}$$

where y_i represents the ground-truth and $\hat{y}_i$ represents the predicted values. This formulation considers the complex conjugate $(\cdot)^*$. In contrast, the standard neural networks employ the mean square error (L_{MSE}) given by

$$L_{\text{MSE}} = \frac{1}{N} \sum_{i=1}^{N} (y_i - \hat{y}_i)^2 \quad . \tag{7}$$

In addition, the three models underwent similar training procedures, employing the Adam optimizer for 2500 epochs. A linear learning rate scheduler was applied throughout these epochs, gradually reducing the learning rate from its initial value to 0.01 times that rate. However, to ensure optimal performance from the networks, we did not standardize the learning rates. Given that the product-unit network is more sensitive to weight changes compared to the standard neural network [12], it was assigned a smaller initial learning rate of 0.01. For the standard neural networks, when training with the cone and cylinder point clouds, a larger initial learning rate of 0.1 was used. However, during training with the ellipsoid and cuboid point clouds, we observed that the standard neural network performed more efficiently with an initial learning rate of 0.01 as opposed to 0.1. Consequently, we retained the 0.01 initial learning rate for these scenarios. In the real task, the initial learning rates were also all 0.01.

2.5 Identify the Missing Parts of the Point Cloud in the Real-World Task

In the real-world task, we cannot easily obtain input data for the missing sections of the point cloud. These data must be computationally derived. To solve this problem, we devised a density-based algorithm specifically tailored to identify these missing portions within 3D point clouds.

Our approach is grounded in the spherical coordinate system. Initially, we generate numerous small-scale grids covering plausible ranges of θ and ϕ values. Each grid represents a distinct section on the surface of the 3D model, effectively segmenting the point cloud into discrete regions. These regions are associated with specific intervals of θ and ϕ. The subsequent step involves quantifying the number of points residing within each demarcated region, enabling the calculation of point density for that specific area. Subsequently, we introduce a predefined threshold, expressed as a percentage. If the point density within a given region falls below this threshold, relative to the average density of its neighboring regions, the region is identified as missing.

Employing this algorithm enables the efficient identification of areas within the 3D model lacking sufficient data points. This process ensures that the reconstructed model is more accurate, enhancing this method overall robustness in practical applications.

3 Results

3.1 Synthetic Objects

The neural networks were initially trained independently using the incomplete point cloud data. Subsequently, inputs from the complete point cloud were fed into the trained networks for prediction. Given that the outputs of the complex-valued product-unit network are also complex valued, it necessitates computing the absolute value of these complex numbers. The subsequent result is a 3D point cloud as predicted by the trained model. It is important to emphasize that this 3D point cloud is entirely derived from the model's predictions. The predicted outcomes generated from input data that are part of the training set are termed "interpolation", while those from unseen data are referred to as "extrapolation".

Each point cloud underwent five test runs on each network. Loss values were computed for interpolation, extrapolation, and the overall total loss in each test, while training time was recorded. The average values from these five tests are comprehensively outlined in Table 1. Furthermore, for each shape, Fig. 4 illustrates the best outcomes, chosen from the five obtained through the complex-valued product-unit network and the ten acquired via the standard neural networks, respectively.

Table 1. Results of predictions for the synthetic objects. "avg." signifies average, "interp." and "extrap." denote interpolation and extrapolation, respectively. "CPUN" is an acronym for the complex-valued product-unit network, while "NN" refers to the standard neural network with an architecture identical to that of CPUN. Furthermore, "NN-L" denotes the larger standard neural network with twice as many neurons per layer except for the output layer as the NN. The best results are marked in bold.

Object	Cone			Cylinder		
Network	CPUN	NN	NN-L	CPUN	NN	NN-L
Avg. interp. loss $[10^{-04}]$	**0.065**	1.516	10.874	**0.099**	5.988	32.601
Avg. extrap. loss $[10^{-04}]$	**2.939**	5.565	17.909	**2.061**	16.190	42.456
Avg. total loss $[10^{-04}]$	**1.638**	3.732	14.724	**1.178**	11.601	38.024
Avg. training time $[s]$	285.2	**270.75**	272.81	277.48	**261.78**	281.69
Object	Ellipsoid			Cuboid		
Network	CPUN	NN	NN-L	CPUN	NN	NN-L
Avg. interp. loss $[10^{-04}]$	**0.297**	4.615	1.012	**17.538**	74.498	25.050
Avg. extrap. loss $[10^{-04}]$	**75.618**	947.080	639.980	**63.950**	272.860	178.540
Avg. total loss $[10^{-04}]$	**26.052**	326.844	219.500	**19.452**	82.660	31.370
Avg. training time $[s]$	417.14	**368.05**	401.38	594.53	**535.46**	583.17

As summarized in Table 1, for more straightforward geometries, such as the cone and cylinder, the complex-valued product-unit network has significantly lower average interpolation and extrapolation loss values in contrast to the two standard neural networks. When it comes to intricate geometries like the ellipsoid and cuboid, all networks have higher loss values. However, the complex-valued product-unit network remains superior, outperforming the standard neural networks across every performance metric, especially in terms of extrapolation capabilities. The observed variance in losses across different geometries aligns with expectations, considering the inherent challenges tied to representing complex structures such as the cuboid within a spherical coordinate system.

Additionally, a marginal difference in training duration was observed. Specifically, for networks with a comparable number of parameters, such as the complex-valued product-unit network and the standard neural network with

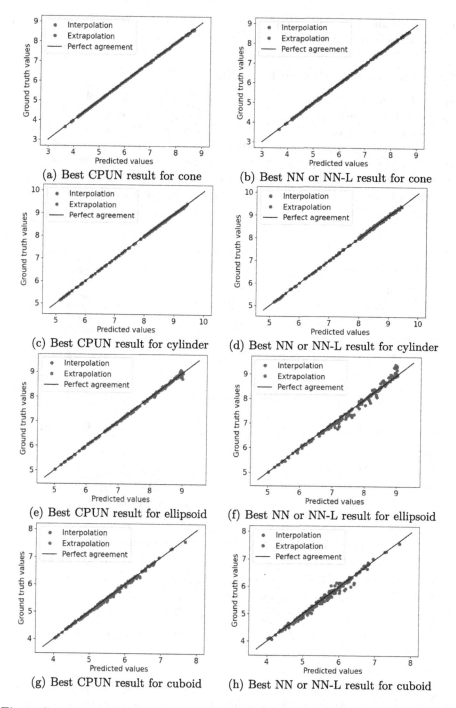

(a) Best CPUN result for cone   (b) Best NN or NN-L result for cone

(c) Best CPUN result for cylinder   (d) Best NN or NN-L result for cylinder

(e) Best CPUN result for ellipsoid   (f) Best NN or NN-L result for ellipsoid

(g) Best CPUN result for cuboid   (h) Best NN or NN-L result for cuboid

Fig. 4. Comparison of predicted results with ground-truth. Each plot incorporates 200 data points, which are evenly randomly sourced from interpolated and extrapolated results. If the predicted value is exactly the same as the ground-truth value, the point should be on the black line.

312   Z. Li et al.

a larger architecture, the complex-valued product-unit network necessitated slightly longer training times in the majority of scenarios.

3.2 Real-World Task with Real Objects

The real-world task diverges from theoretical task for synthetic objects in their approach. For the real-world task, we retained the original incomplete point cloud data, focusing solely on predicting its missing segments. That is, we engaged only in extrapolation, excluding interpolation. Ultimately, the extrapolated results were merged with the original data to reconstruct a complete point cloud.

We trained the complex-valued product-unit network and the standard neural network with a larger architecture described above using two incomplete point

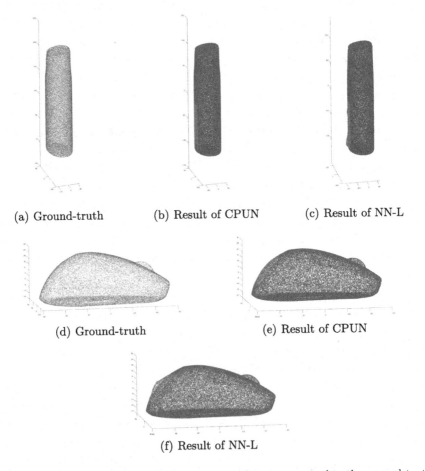

(a) Ground-truth   (b) Result of CPUN   (c) Result of NN-L

(d) Ground-truth   (e) Result of CPUN

(f) Result of NN-L

Fig. 5. The outcomes predicted by the two networks are compared to the ground-truth. The blue segment denotes the data acquired from scanning and subsequent processing (raw data), while the orange segment illustrates the extrapolated results derived from both networks. (Color figure online)

clouds. The absent segments of these point clouds were pinpointed utilizing the density-based method outlined above. Following this, the trained networks were employed to predict the designated missing segments of the point clouds. The predictions were subsequently contrasted with the 3D point cloud derived from the complete model, as depicted in Fig. 5.

The Real-world task offers a more realistic representation of the uncertainties that the real world presents, as compared to the theoretical task for synthetic objects counterparts. In our observations, when dealing with irregular objects such as a bottle or computer mouse, the complex-valued product-unit network delivers more accurate results than the standard neural network. As depicted in Fig. 5, the point cloud segments predicted by the standard neural network displays considerable distortions. This is in agreement with the results in 1D and 2D investigated in [12]. In contrast, the predictions made by the complex-valued product-unit network align closely with the point cloud derived from the complete model, showing minimal deviations.

4   Conclusion and Discussion

Based on our findings, it is evident that, having identical architectures, the complex-valued product-unit network exhibits superior predictive capabilities compared to the standard neural network. This disparity is particularly pronounced in terms of extrapolation ability. Even when the standard neural network is augmented with a doubling of neurons, its performance in both interpolation and extrapolation still falls short of that achieved by the complex-valued product-unit network.

The discernible discrepancy in performance becomes even more pronounced in the real-world task, particularly when dealing with real-world objects characterized by irregular shapes and structures. Standard neural networks consistently exhibit more severe distortions in the results they predict, in stark contrast to the highly accurate structures predicted by the complex-valued product-unit network in the face of such complex scenarios.

For this issue, we also analysed the reasons behind it. We believe that the effectiveness of neural network architectures is intricately tied to their inherent design and their capacity to meet specific challenges. While the standard neural network stands as a versatile tool with broad applications, its performance, especially in domains like 3D point cloud analysis, may be surpassed by more specialized networks. The improved performance of the complex-valued product-unit network compared to the standard neural network observed in our experiments can presumably be attributed to its distinctive architecture. The initial layer, composed of standard summation units, allows linear transformation of the data to be learned. As the data progresses through subsequent layers containing complex-valued product units, the model can capture more intricate and contextually rich information, including nonlinear coupling between inputs, potentially crucial for understanding and completing complex 3D shapes.

However, it is worth noting the slightly prolonged training times for the complex-valued product-unit network. This delay indicates the computational demands of handling complex-valued operations. But, given the marked performance improvement, this trade-off will be justifiable for many applications, particularly those where accuracy is paramount.

In the course of our research, it became evident that the performance of the neural networks in 3D point-cloud completion tasks is susceptible to rotations of the object's point cloud. Specifically, when an object's point cloud is centered around the y-axis and then rotated to the x-axis by a certain angle, the predictive accuracy deteriorates. This decline in accuracy can be attributed to the fact that the mathematical representation of the point cloud undergoes a significant transformation after the rotation. In this regard, we believe that combining the method for identifying the 6D pose of an object with our current method is a good direction for our future research. Furthermore, we plan to explore the extension of our application to the challenge of complementing the incomplete contour of a 3D object in a camera view, particularly in situations where occlusion issues may arise.

References

1. Achlioptas, P., Diamanti, O., Mitliagkas, I., Guibas, L.: Learning representations and generative models for 3d point clouds. In: International Conference on Machine Learning, pp. 40–49. PMLR (2018)
2. Achlioptas, P., Diamanti, O., Mitliagkas, I., Guibas, L.J.: Learning representations and generative models for 3d point clouds. arXiv preprint arXiv:1707.02392 (2017)
3. Alenya, G., Dellen, B., Foix, S., Torras, C.: Robotized plant probing: leaf segmentation utilizing time-of-flight data. IEEE Robot. Automat. Magaz. **20**(3), 50–59 (2019)
4. Allen, B., Curless, B., Popović, Z.: The space of human body shapes: reconstruction and parameterization from range scans. ACM Trans. Graph. **22**(3), 587–594 (2003)
5. Artec 3D. Artec Space Spider (2024). https://www.artec3d.com/portable-3d-scanners/artec-spider. Accessed 16 Jan 2024
6. Behnke, S.: Hierarchical Neural Networks for Image Interpretation, vol. 2766. Springer, Heidelberg (2003)
7. Bishop, C.M., et al.: Neural Networks for Pattern Recognition. Oxford University Press (1995)
8. Clark, J.W., Gernoth, K.A., Dittmar, S., Ristig, M.: Higher-order probabilistic perceptrons as bayesian inference engines. Phys. Rev. E **59**(5), 6161 (1999)
9. Cortes, C., Vapnik, V.: Support-vector networks. Mach. Learn. **20**, 273–297 (1995)
10. Dai, A., Ruizhongtai Qi, C., Nießner, M.: Shape completion using 3d-encoder-predictor cnns and shape synthesis. In: Proceedings of the IEEE Conference on Computer Vision and Pattern Recognition, pp. 5868–5877 (2017)
11. Dellen, B., Jaekel, U., Freitas, P.S., Clark, J.W.: Predicting nuclear masses with product-unit networks. Phys. Lett. B **852**, 138608 (2024)
12. Dellen, B., Jaekel, U., Wolnitza, M.: Function and pattern extrapolation with product-unit networks. In: Rodrigues, J.M.F., et al. (eds.) ICCS 2019. LNCS, vol. 11537, pp. 174–188. Springer, Cham (2019). https://doi.org/10.1007/978-3-030-22741-8_13
13. Durbin, R., Rumelhart, D.E.: Product units: a computationally powerful and biologically plausible extension to backpropagation networks. Neural Comput. **1**(1), 133–142 (1989)

14. Hahnloser, R.H., Sarpeshkar, R., Mahowald, M.A., Douglas, R.J., Seung, H.S.: Digital selection and analogue amplification coexist in a cortex-inspired silicon circuit. Nature **405**(6789), 947–951 (2000)
15. Husain, F., Colome, A., Dellen, B., Alenya, G., Torras, C.: Realtime tracking and grasping of a moving object from range video. In: IEEE International Conference on Robotics and Automation (ICRA), pp. 2617–2622 (2014)
16. Husain, F., Dellen, B., Torras, C.: Consistent depth video segmentation using adaptive surface models. IEEE Trans. Cybernet. **45**, 266–278 (2014)
17. Husain, F., Dellen, B., Torras, C.: Robust surface tracking in range image sequences. In: Digital Signal Processing, pp. 37–44 (2014)
18. Itō, K.: Encyclopedic Dictionary of Mathematics, vol. 1. MIT Press (1993)
19. Kazhdan, M., Bolitho, M., Hoppe, H.: Poisson surface reconstruction. In: Proceedings of the Fourth Eurographics Symposium on Geometry Processing, vol. 7 (2006)
20. Kazhdan, M., Hoppe, H.: Screened Poisson surface reconstruction. ACM Trans. Graph. **32**(3), 1–13 (2013)
21. Leerink, L., Giles, C., Horne, B., Jabri, M.: Learning with product units. Adv. Neural Inf. Process. Syst. **7** (1994)
22. Mahler, J., et al.: Dex-net 2.0: deep learning to plan robust grasps with synthetic point clouds and analytic grasp metrics. arXiv preprint arXiv:1703.09312 (2017)
23. Schmidt, P., Vahrenkamp, N., Wächter, M., Asfour, T.: Grasping of unknown objects using deep convolutional neural networks based on depth images. In: 2018 IEEE International Conference on Robotics and Automation (ICRA), pp. 6831–6838. IEEE (2018)
24. Schölkopf, B., Smola, A.J., Bach, F., et al.: Learning with Kernels: Support Vector Machines, Regularization, Optimization, and Beyond. MIT Press (2002)
25. Shawe-Taylor, J., Cristianini, N., et al.: Kernel Methods for Pattern Analysis. Cambridge University Press (2004)
26. Smith, E., et al.: Active 3d shape reconstruction from vision and touch. Adv. Neural. Inf. Process. Syst. **34**, 16064–16078 (2021)
27. Stutz, D., Geiger, A.: Learning 3d shape completion from laser scan data with weak supervision. In: IEEE Conference on Computer Vision and Pattern Recognition (CVPR). IEEE Computer Society (2018)
28. Wolnitza, M., Kaya, O., Kulvicius, T., Wörgötter, F., Dellen, B.: 6d pose estimation and 3d object reconstruction from 2d shape for robotic grasping of objects. In: 2022 Sixth IEEE International Conference on Robotic Computing (IRC), pp. 67–71 (2022). https://doi.org/10.1109/IRC55401.2022.00018
29. Zhang, Y., Liu, Z., Li, X., Zang, Y.: Data-driven point cloud objects completion. Sensors **19**(7), 1514 (2019)

Optimizing BIT1, a Particle-in-Cell Monte Carlo Code, with OpenMP/OpenACC and GPU Acceleration

Jeremy J. Williams[1(✉)], Felix Liu[1], David Tskhakaya[2], Stefan Costea[3], Ales Podolnik[2], and Stefano Markidis[1]

[1] KTH Royal Institute of Technology (KTH), Stockholm, Sweden
jjwil@kth.se
[2] Institute of Plasma Physics of the CAS (IPP CAS), Prague, Czech Republic
[3] LeCAD, University of Ljubljana (UL), Ljubljana, Slovenia

Abstract. On the path toward developing the first fusion energy devices, plasma simulations have become indispensable tools for supporting the design and development of fusion machines. Among these critical simulation tools, BIT1 is an advanced Particle-in-Cell code with Monte Carlo collisions, specifically designed for modeling plasma-material interaction and, in particular, analyzing the power load distribution on tokamak divertors. The current implementation of BIT1 relies exclusively on MPI for parallel communication and lacks support for GPUs. In this work, we address these limitations by designing and implementing a hybrid, shared-memory version of BIT1 capable of utilizing GPUs. For shared-memory parallelization, we rely on OpenMP and OpenACC, using a task-based approach to mitigate load-imbalance issues in the particle mover. On an HPE Cray EX computing node, we observe an initial performance improvement of approximately 42%, with scalable performance showing an enhancement of about 38% when using 8 MPI ranks. Still relying on OpenMP and OpenACC, we introduce the first version of BIT1 capable of using GPUs. We investigate two different data movement strategies: unified memory and explicit data movement. Overall, we report BIT1 data transfer findings during each PIC cycle. Among BIT1 GPU implementations, we demonstrate performance improvement through concurrent GPU utilization, especially when MPI ranks are assigned to dedicated GPUs. Finally, we analyze the performance of the first BIT1 GPU porting with the NVIDIA Nsight tools to further our understanding of BIT1's computational efficiency for large-scale plasma simulations, capable of exploiting current supercomputer infrastructures.

Keywords: OpenMP · Task-Based Parallelism · OpenACC · Hybrid Programming · GPU Offloading · Large-Scale PIC Simulations

1 Introduction

Plasma simulations are vital for understanding complex interactions between plasma and wall materials, which present significant modeling challenges, including the need of resolving different simulation time and spatial scales or modeling accurately atomic and collision processes.

L. Franco et al. (Eds.): ICCS 2024, LNCS 14832, pp. 316–330, 2024.
https://doi.org/10.1007/978-3-031-63749-0_22

Particularly, these challenges are notable when modeling plasma-loaded divertors in fusion devices, such as the ITER tokamak, a major nuclear fusion project. In summary, the divertor manages the heat and particle fluxes that occur during the operation of the tokamak. In fact, during a fusion reaction, high energy neutrons are produced, and these can cause damage to the first wall of the tokamak. Additionally, impurities from the plasma, need to be efficiently removed to maintain optimal conditions for the fusion process. The divertor accomplishes these tasks by diverting the flow of plasma to a specific region of the tokamak. This region, known as the divertor region, is usually located at the bottom of the toroidal chamber (see Fig. 1).

Among the tools used to address these challenges, BIT1 is a specialized plasma simulation tool, focusing on describing accurately atomic processes and collisions in plasmas during plasma-wall interactions. In particular, BIT1 is widely used to analyze how power is distributed on divertors in these devices. BIT1 plays a critical role as a massively parallel PIC code for studying complex plasma systems and their interactions with various materials.

Initially introduced by D. Tskhakaya and collaborators [8,9], BIT1 has distinctive capabilities. It models plasmas confined between two conducting walls and includes collision modeling to capture complex plasma dynamics. What makes BIT1 unique is its capability of modeling accu-

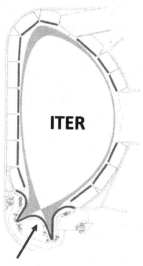

Fig. 1. BIT1 simulates plasma behavior in the tokamak divertor region (blue arrow), such as in the ITER fusion device. (Color figure online)

rately processes occurring at the interface of plasma and a wall, such as sputtering from the wall, emissions, and collisions. While it has shown that BIT1 is scalable for thousands MPI processes [13], it has two major limitations. The first one is that BIT1 relies only on MPI for parallel communication, even for on-node communication, where a shared-memory computing approach is more convenient and can decrease the memory usage and allow for task-based approaches. The second limitation is the lack of support for running on GPU-accelerated supercomputers. Given the fact that most of the top supercomputers in the world, such as Frontier, Aurora, Eagle and LUMI, the lack of support for GPUs is a major limitation that hinders the usage of BIT1 in the largest supercomputers available. This work addresses these BIT1 limitations by designing and implementing hybrid MPI and OpenMP/OpenACC version that can exploit shared-memory and GPUs.

Recently, findings presented in [13] have led to a detailed investigation of the BIT1 code performance, pointing out performance bottlenecks, and identifying a roadmap for BIT1 performance optimization. The work highlighted areas for optimizing BIT1 to enhance the performance, suggesting a focus on the particle mover function, particularly the particle pusher, as an initial step, given the challenges posed by arranging particles into cells and MPI ranks. This work's primary focus is on the enhancement of BIT1's performance with a goal to optimize the particle mover function, one of the most computationally intensive parts of the code. This optimization utilizes OpenMP and OpenACC for multicore CPUs and GPUs, enabling researchers to harness the full potential of modern hardware for plasma simulations, ultimately improving our understanding of plasma dynamics and advancing research in fusion and plasma science. The contributions of this work are the following:

- We design and implement hybrid MPI+OpenMP and MPI+OpenACC versions of the BIT1 code to improve on-node performance. This implementation uses a task-based approach to address potential issues with load-imbalance.
- We develop the first GPU porting of the BIT1 code to NVIDIA GPUs with OpenMP and OpenACC in the particle mover stage of the BIT1 code.
- We critically analyze and discuss the performance of the newly ported BIT1 code, showing major performance improvements, and identify the next performance optimization steps.

2   Background

Particle-in-Cell (PIC) methods are among the most crucial tools for plasma simulations. They find applications in a diverse range of plasma environments, from space and astrophysical plasma to laboratory settings, industrial processes, and fusion devices. BIT1, in particular, is a PIC code specifically tailored for plasma-material simulations. What sets BIT1 apart is its handling of collisions and interactions with material boundaries, including models for phenomena like sputtering at material interfaces.

BIT1 is a 1D3V PIC code, allowing simulations in one dimension while considering particles with three-dimensional velocities. The foundation of BIT1's computational approach is the PIC method, widely adopted in plasma physics. This method involves tracking the trajectories of millions of particles within a field consistent with density and current distributions, while abiding to Maxwell's and Poisson equations. Figure 2 provides a visual representation of BIT1's explicit PIC method. To initiate the PIC simulations, BIT1 configures the computational grid and sets up particle positions and velocities for various species, including electrons and ions. Subsequently, a computational cycle iteratively updates the electric field, particle positions, and velocities, accurately representing the dynamic interactions within the plasma. BIT1 employs advanced Monte Carlo techniques to simulate collisions and ionization processes. It is important to note that one of the most computationally intensive stages in BIT1 is the

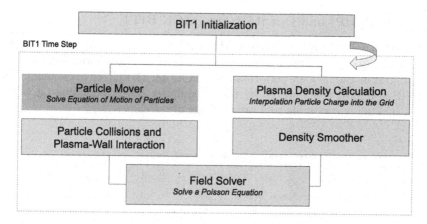

Fig. 2. A diagram representing the algorithm used in BIT1. After the initialization the PIC algorithm cycle is repeated at each time step [13]. In orange, we highlight the particle mover step that we parallelize with OpenMP and OpenACC.

particle mover, which is responsible for calculating the trajectories of millions of particles [9].

In the current implementation, BIT1 employs domain decomposition for parallelization, utilizing MPI for efficient parallel communication. MPI point-to-point communication is essential for managing information exchange at domain boundaries, crucial for tasks like the smoother, Poisson solver, and handling particles exiting the computational domain. However, the existing BIT1 implementation relies solely on MPI, lacking support for hybrid parallel computing capabilities, such as MPI+OpenMP, MPI+OpenACC, GPU offloading or acceleration.

One of the main and distinctive features of the BIT1 code is the data layout particle information, such as positions and velocities, are stored in memory. BIT1 associates the particles with the cells they are located in, e.g. each grid cell has an associated list with particle information. As soon a particle move from a cell to a neighboring one, then a particle information is removed from one list and added to another one. In plasma simulations, there may be regions of space where plasma particles concentrate, resulting in situations where certain cells have a large number of particles while others have only a few. This results in work imbalance in the particle mover.

3   Methodology and Experimental Setup

In this work, our focus is on investigating the porting of the BIT1 particle mover to leverage OpenMP and OpenACC for shared memory programming and GPU acceleration.

3.1   Hybrid MPI and OpenMP/OpenACC BIT1

OpenMP Tasks Particle Mover Parallelization. OpenMP is one of the most widely used programming model designed to facilitate shared-memory parallel programming in high-performance computing (HPC) environments. The OpenMP standard is supported by major compiler suites, including GCC and LLVM, making it accessible to a broad range of developers.

```
1  #pragma omp parallel shared(chsp, sn2d, dinj, nstep, np, x, yp, yx, vy)
2      private(isp, i, j) firstprivate(nsp, nc)
3  {
4      #pragma omp single
5      {
6          for (isp = 0; isp < nsp; isp++) {
7              ...
8              #pragma omp taskloop grainsize(500) nogroup
9              for (j = 0; j < nc; j++) {
10                 #pragma omp simd
11                 for (i = 0; i < np[isp][j]; i++)
12                     x[isp][j][i] += nstep[isp] * vx[isp][j][i];
13             }
14             ...
15             #pragma omp taskloop grainsize(500) nogroup
16             for (j = 0; j < nc; j++) {
17                 #pragma omp simd
18                 for (i = 0; i < np[isp][j]; i++)
19                     x[isp][j][i] += nstep[isp] * vx[isp][j][i];
20             }
21         }
22     }
23 }
```

Listing 1.1. Simplified C code snippet illustrating the OpenMP parallelization for CPU in the particle mover.

Listing 1.1 showcases our OpenMP port of the core computational function for the particle mover. In the code, x[][][] and vx[][][] represent particle position and velocity in one dimension. nsp is the number of plasma species, and nc is the number of cells in the one-dimensional grid. np[][] denotes the number of particles per species per cell.

Our parallelization approach uses the OpenMP taskloop construct. The outermost loop, with a small number of iterations (species present in the simulation), is deemed unsuitable for traditional parallelization methods. The taskloop construct dynamically distributes loop iterations among available threads, ensuring effective load balancing and optimizing the parallelization strategy for enhanced performance on multicore CPUs.

When examining the code, the #pragma omp parallel pragma initiates a parallel region with shared variables (chsp, dinj, sn2d, nstep, np, x, yp, vx, vy), while isp and i are private to each thread. The firstprivate clause ensures private and initialized values for nsp and nc for each parallel thread.

Within the parallel region, the single construct ensures that the subsequent block of code is executed by a single thread, crucial for the initialization section. The taskloop grainsize(500) nogroup pragma parallelizes the subsequent loop, dividing iterations into tasks, while optimizing task granularity for efficient parallel execution based on empirical testing and adjustments to achieve optimal performance. The nogroup clause allows for dynamic scheduling.

Finally, the `simd` pragma within the innermost loop exploits SIMD parallelism, improving vectorization and the efficiency of particle movement calculations.

OpenACC Multicore Particle Mover Parallelization. OpenACC, akin to OpenMP, is a directive-based programming model primarily designed for GPU accelerators. However, the directive-based approach for GPU offloading can also be advantageous for CPUs, offering a straightforward method for porting codes to CPUs with minimal code changes.

```
1  #pragma acc parallel loop present(chsp[:lenA], sn2d[:lenA], dinj[:lenA], nstep[:lenA],
2        np[:lenA][:lenB], x[:lenA][:lenB][:lenC], yp[:lenA][:lenB][:lenC],
3        vx[:lenA][:lenB][:lenC], vy[:lenA][:lenB][:lenC])
4  {
5      for (isp = 0; isp < nsp; isp++) {
6          ...
7          #pragma acc loop gang vector
8          for (j = 0; j < nc; j++) {
9              #pragma acc loop vector
10                 for (i = 0; i < np[isp][j]; i++)
11                     x[isp][j][i] += nstep[isp] * vx[isp][j][i];
12             }
13         ...
14         #pragma acc loop gang vector
15         for (j = 0; j < nc; j++) {
16             #pragma acc loop vector
17                 for (i = 0; i < np[isp][j]; i++)
18                     x[isp][j][i] += nstep[isp] * vx[isp][j][i];
19             }
20     }
21 }
```

Listing 1.2. Simplified C code snippet illustrating the OpenACC Multicore CPU parallelization in the particle mover.

Listing 1.2 shows our optimized OpenACC parallelization for the particle mover on multicore CPUs. The `#pragma acc parallel loop` directive initiates concurrent execution, specifying essential data arrays. Utilizing `gang` and `vector` directives enhances parallel processing in a nested loop structure (`#pragma acc loop gang vector`) for particle and grid index iterations. Particle positions are updated based on `nstep` and `vx`.

In the else clause, a similar nested loop (`#pragma acc loop vector`) optimally executes particle indices (`i`). Further efficiency is achieved with a conditional statement triggering an additional nested loop to update yp and x.

This version strategically uses OpenACC directives (`#pragma acc loop gang vector` and `#pragma acc loop vector`) to optimize multicore CPUs for the particle mover function, enhancing parallel performance in nested loops.

3.2   Accelerating BIT1 with OpenMP and OpenACC

Accelerating BIT1 with OpenMP Target. Listing 1.3 illustrates our use of the OpenMP target construct to parallelize BIT1's particle mover function for GPU offloading. The code strategically employs OpenMP target directives to optimize particle movement computations by offloading them to GPUs.

```
1  #pragma omp target enter data map(to: chsp[:lenA], sn2d[:lenA], dinj[:lenA], nstep[:
       lenA],
2     np[:lenA][:lenB], x[:lenA][:lenB][:lenC], yp[:lenA][:lenB][:lenC],
3     vx[:lenA][:lenB][:lenC], vy[:lenA][:lenB][:lenC])
4  {
5     for (isp = 0; isp < nsp; isp++) {
6        ...
7        #pragma omp target teams distribute parallel for thread_limit(256) num_teams
       (391)
8           for (j = 0; j < nc; j++) {
9              #pragma omp simd
10                for (i = 0; i < np[isp][j]; i++)
11                   x[isp][j][i] += nstep[isp] * vx[isp][j][i];
12           }
13        ...
14        #pragma omp target teams distribute parallel for thread_limit(256) num_teams
       (391)
15           for (j = 0; j < nc; j++) {
16              #pragma omp simd
17                for (i = 0; i < np[isp][j]; i++)
18                   x[isp][j][i] += nstep[isp] * vx[isp][j][i];
19           }
20        #pragma omp target exit data map(from: x[:lenA][:lenB][:lenC]...)
21     }
22  }
```

Listing 1.3. Simplified C code snippet illustrating the OpenMP (OMP target) parallelization with data clauses and array "shape" for GPU porting in the particle mover.

The pragma #pragma omp target enter data initiates the data transfer from the host to the GPU, encompassing crucial arrays such as chsp, sn2d, dinj, and the arrays for particle position and velocity (x, yp, vx, vy).

Within the unstructured data mapping region, the #pragma omp target teams distribute parallel for directive initiates worksharing across multiple levels of parallelism using combined constructs, thereby enabling the parallel execution of the nested loops for particle movement calculations on the GPU. To fine-tune parallelism, directives thread_limit(256) and num_teams(391) are used to set thread and team limits based on our experimental setup's system specifications and workload requirements.

In the nested loops, the #pragma omp simd directive provides a hint to the compiler for potential vectorizations of the inner loops, optimizing SIMD parallelism. The calculations, updating particle positions based on velocities and time steps, are concurrently distributed across GPU threads.

Finally, the #pragma omp target exit data directive ensures seamless transfer of modified data, specifically particle positions, from the GPU back to the host for efficient GPU offloading and parallelization of particle mover computations.

Accelerating BIT1 with OpenACC Parallel. OpenACC, designed for GPU acceleration, simplifies the task of offloading functions to GPUs, offering an accessible solution without complex GPU programming. Supported by platforms like NVIDIA and GCC, OpenACC empowers developers to harness GPU parallel processing efficiently.

In Listing 1.4, we showcase our OpenACC parallelization of the particle mover function for GPU acceleration. Data movement between CPU and GPU is facilitated using #pragma acc enter data and #pragma acc exit data direc-

```
 1  #pragma acc enter data copyin(chsp[:lenA], sn2d[:lenA], dinj[:lenA], nstep[:lenA],
 2      np[:lenA][:lenB], x[:lenA][:lenB][:lenC], yp[:lenA][:lenB][:lenC],
 3      vx[:lenA][:lenB][:lenC], vy[:lenA][:lenB][:lenC])
 4  {
 5      for (isp = 0; isp < nsp; isp++) {
 6          ...
 7          #pragma acc parallel loop gang worker vector vector_length(128)
 8              present(np[:lenA][:lenB], nstep[:lenA], x[:lenA][:lenB][:lenC],
 9              vx[:lenA][:lenB][:lenC]) firstprivate(nc,isp,nsp) private(i)
10              for (j = 0; j < nc; j++) {
11                  #pragma acc loop
12                      for (i = 0; i < np[isp][j]; i++)
13                          x[isp][j][i] += nstep[isp] * vx[isp][j][i];
14              }
15          ...
16          #pragma acc parallel loop gang worker vector vector_length(128)
17              present(np[:lenA][:lenB], nstep[:lenA], x[:lenA][:lenB][:lenC],
18              vx[:lenA][:lenB][:lenC]) firstprivate(nc,isp,nsp) private(i)
19              for (j = 0; j < nc; j++) {
20                  #pragma acc loop
21                      for (i = 0; i < np[isp][j]; i++)
22                          x[isp][j][i] += nstep[isp] * vx[isp][j][i];
23              }
24      #pragma acc exit data copyout(x[:lenA][:lenB][:lenC]...)
25      }
26  }
```

Listing 1.4. Simplified C code snippet illustrating the OpenACC (ACC parallel) parallelization with data clauses and array "shape" for GPU porting in the particle mover.

tives with `copyin` and `copyout` clauses, managing the transfer of relevant arrays (`chsp`, `sn2d`, `dinj`, `nstep`, `np`, `x`, `yp`, `vx`, `vy`).

In the GPU parallel unstructured data region, the `#pragma acc parallel loop` directive is employed to parallelize the outer loop over j for components (nc). The `present` clause ensures availability of specified arrays, while the `first private` and `private` clauses handle variables nc, isp, nsp, and i appropriately.

The loop parallelization strategy uses `gang`, `worker`, and `vector` directives. The `gang` directive divides loop iterations into gangs, potentially assigned to different cores. Within each gang, `worker` directive enables concurrent execution, and `vector` directive subdivides each worker, specifying simultaneous processing with `vector_length(128)` determining vector size.

The innermost loop over i is parallelized with `#pragma acc loop` directive, optimizing GPU capacity for parallel computations on inner loops while minimizing data transfer overhead.

The `#pragma acc exit data copyout` directive ensures copied back modified data to the CPU after GPU computations are complete.

3.3 Experimental Setup

In this work, we use the following two systems:

- **Dardel**, an HPE Cray EX supercomputer, features a robust **CPU** partition with 1,270 compute nodes. Each node is equipped with two AMD EPYC™ Zen2 2.25 GHz 64-core processors, 256GB DRAM, and interconnected using an HPE Slingshot network with Dragonfly topology providing 200 GB/s bandwidth. The Lustre file system has a 12 PB capacity, and the operating system is SUSE Linux Enterprise Server 15 SP3. We load GNU compiler suite option "PrgEnv-gnu" for compiler "gcc v11.2.0" and MPI library, "cray-mpich v8.1.17".

- **NJ**, an HPC system, has an AMD EPYC 7302P 16-Core Processor with 32 CPU cores. It operates on the x86_64 platform, 2 threads per core, and 16 cores per socket, running at a 3.0 GHz base clock. **NJ** also hosts two NVIDIA A100 GPUs with 40 GB HBM2e memory, 6912 Shading Units, 432 Tensor Cores, and 108 SM Count. GPUs have 192 KB L1 Cache per SM, 40 MB L2 Cache, and offer double-precision matrix (FP64) performance of approximately 9.746 TFLOPs. We load CUDA Driver v11.0, NVIDIA HPC SDK v23.7 with compiler version "gcc v12.2.0," and MPI "openmpi v4.1.5".

In this work, we focus on optimizing the particle mover function in BIT1, with a particular emphasis on closely monitoring and analyzing BIT1 performance. As a test case, we simulate neutral particle ionization resulting from interactions with electrons in upcoming magnetic confinement fusion devices like ITER and DEMO. The scenario involves an unbounded unmagnetized plasma consisting of electrons, D^+ ions and D neutrals. Due to ionization, neutral concentration decreases with time according to $\partial n/\partial t = n n_e R$, where n, n_e and R are neutral particles, plasma densities and ionization rate coefficient, respectively. We use a one-dimensional geometry with 100K cells, three plasma species (e electrons, D^+ ions and D neutrals), and 10M particles per cell per species. The total number of particles in the system is 30M. Unless differently specified, we simulate up to 1K time steps. An important point of this test is that it does not use the Field solver and smoother phases (shown in the diagram of Fig. 2).

4 Performance Results

4.1 Hybrid MPI and OpenMP/OpenACC BIT1

Focusing on intra-node testing, an in-depth investigation into how BIT1 performs in terms of "execution time" has been conducted, to explore the advantages of utilizing hybrid approaches on the two HPC systems. The aim was to determine if using both MPI and OpenMP/OpenACC, instead of just MPI, would make a significant difference. Figure 3 shows executions of hybrid BIT1 total execution vs. optimized mover function using 2 and 16 ranks per node for 1000 times steps on *NJ*. For both 2 ranks and 16 ranks, our hybrid MPI+OpenMP version of BIT1 shows a reduction in both total simulation and mover function time. This suggests that parallelizing BIT1 with OpenMP threads improves performance by enabling multiple threads to work on the problem concurrently. Similar to OpenMP, our hybrid MPI+OpenACC version for multicore CPUs also results in a reduction in both total simulation and mover function time. This demonstrates that BIT1 benefits when used with multicore CPUs, due to better utilization of CPU cores through parallelization.

Investigating the scalability of hybrid BIT1 on CPUs and Fig. 3, it is easy to see that with an increase in MPI ranks from 2 to 16, there is a significant improvement in performance for both total simulation and optimized mover

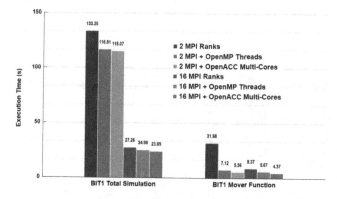

Fig. 3. Hybrid BIT1 total simulation and optimized mover function using 2 and 16 ranks per node on *Dardel* for 1000 times steps.

function execution times. This shows that our hybrid BIT1 scales well. However, to confirm these findings, our second system, *Dardel*, was used for further investigation with a focus on the optimized mover function using OpenMP (since OpenACC was not used).

On *Dardel*, as seen in Fig. 4, the execution time significantly decreases as the number of MPI ranks and OpenMP threads increases, showcasing the potential for efficient parallel executions. For instance, with 128 MPI ranks versus 8 MPI ranks with 16 OpenMP threads, the execution time reduces to 10.55 s, a notable improvement from the original 12.72 s.

4.2 Accelerating BIT1 with GPUs

The challenge of enhancing the performance of the particle mover function in BIT1 by tapping into the computational power of GPUs has been systematically deliberated, revealing valuable insights. In addressing this challenge, our focus shifts to improving the particle mover function's `execution time` by offloading it to the GPU using OpenACC and OpenMP. In doing this, we are one step closer for BIT1 being ready for Exascale platforms. To achieve this target, two main strategies provided by OpenACC and OpenMP were investigated: the explicit approach, where data regions for GPU offloading are clearly defined using directives, and the unified memory method, which simplifies memory management by sharing a common space between the CPU and GPU.

BIT1 OpenACC GPU Explicit. Initial work began by using OpenACC's explicit approach on the GPUs on *NJ* for up to 10 time steps for better visualization of profiling results, particularly focusing on the particle mover function. The profiling results, using NVIDIA Nsight Systems, reveal crucial insights. The CUDA kernel statistics showed that the primary kernel responsible for the particle mover function consumes 99.7% of the GPU execution time, emphasizing its significance in the overall computation. In Fig. 5, the memory operation statistics shed light on the substantial time spent on `memcpy` operations. Specifically, 80%

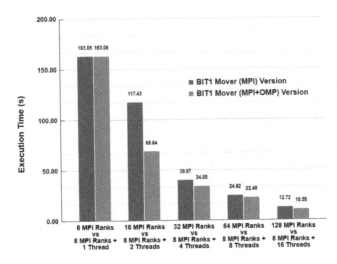

Fig. 4. Optimized mover function - scaling up to 128 MPI ranks on *Dardel* for 1000 times steps.

of the GPU time is allocated to copying data from host to device, emphasizing the importance of efficient data transfer strategies.

These findings emphasize the importance of optimizing data movement and kernel execution in the particle mover function. Strategies such as overlapping computation and communication, along with exploring ways to minimize data transfer size, can be instrumental. Additionally, considering the high memory bandwidth of the NVIDIA A100 GPUs on *NJ*, enhancing the efficiency of memory operations becomes pivotal for achieving optimal performance.

BIT1 OpenACC GPU Unified Memory. Next, the particle mover function has been initially investigated on *NJ* using OpenACC Unified Memory. The pro-

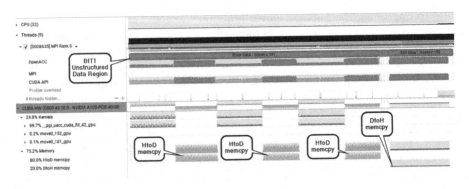

Fig. 5. NVIDIA Nsight Systems OpenACC Explicit View - GPU porting of BIT1 mover function on *NJ* with 1 time step.

filing results revealed critical insights into the dynamics of memory management and kernel execution.

The CUDA kernel statistics showed the primary mover kernel $move0_{152\_gpu}$ dominated GPU execution time, accounting for 66.7% with a total execution time of 1.623457 s. Similarly the second kernel $move0_{181\_gpu}$ contributed 33.3% with an average execution time of 0.8106292 s, highlighting the significance of these kernels in the overall computation.

For hybrid BIT1 data movement and the implementation of a unified memory strategy, the runtime system manages the seamless transfer of data between the host and the device. One key advantage, aside from programmer convenience, is the opportunity for the runtime to automatically detect instances of overlapping computation and communication. Figure 6 displays NVIDIA Nsight Systems' view of such overlapping, revealing that the unified memory version exhibited faster overall runtime than explicit copies. This indicates automatic overlapping of computation and communication. However, we observe that BIT1 OpenACC Unified Memory performance is still hindered by substantial data movement between the host and device.

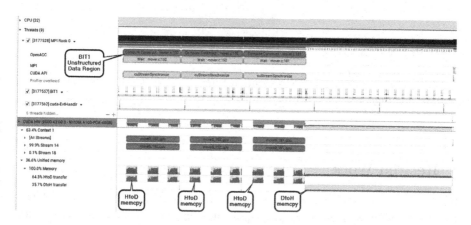

Fig. 6. NVIDIA Nsight Systems OpenACC Unified Memory View - GPU porting of BIT1 mover function on *NJ* with 1 time step.

BIT1 OpenMP GPU Offloading. Due to the observed performance gain with hybrid BIT1 on multicore CPU, an investigation was conducted using the OpenMP target construct for further BIT1 GPU offloading. Figure 7 and 8 provides performance evaluation insights into GPU utilization compared to the CPU-only baseline with 2 MPI and 16 MPI Ranks. Implementations using OpenMP or OpenACC on GPUs exhibit increased execution times, indicating that parallelization strategies may introduce overhead, potentially outweighing the benefits of parallel processing. Among BIT1 GPU implementations, OpenMP Target with 2 GPUs stands out for delivering a substantial reduction in execution time, demonstrating the performance improvement through concurrent GPU utilization, especially when MPI ranks are assigned to dedicated GPUs.

Yet, with promising results, a critical challenge emerges: data transfer constraints during each PIC cycle. Profiling results from Fig. 6 expose the impact of copying substantial data from the CPU to the GPU at each time step, leading to notable performance bottlenecks. Addressing this challenge involves avoiding large data transfers from the CPU to the GPU at each iteration. The exploration of CUDA streams and particle batch processing with OpenMP Target across Multi-GPUs per node emerges as a promising avenue to streamline data transfer and processing, as also observed in [3].

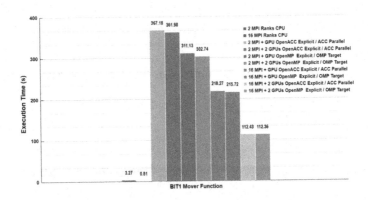

Fig. 7. Optimized mover function execution time(s) on *NJ* for 100 time steps using OpenMP and OpenACC Explicit on NVIDIA GPUs.

5   Related Work

BIT1, an advanced PIC code, is designed to simulate plasma-material interaction [9] and is utilized in various applications, including fusion devices such as tokamaks. BIT1 builds upon the XPDP1 code [11], initially developed by Verboncoeur's team at Berkeley, and incorporates optimized data layout to efficiently handle collisions [8]. Recently, Williams, Jeremy J., et al. profiled the performance of BIT1, highlighting the particle mover as one of the most computationally intensive parts of the code [13]. Several works have been devoted to hybrid parallelization using MPI and OpenMP for PIC codes, including Smilei [4], iPIC3D [6], and Warp-X [10], to mention a few. In contrast to earlier studies, we employ task-based shared-memory parallelization techniques [1], which are commonly used in linear solvers [5], to specifically address load-imbalance issues in BIT1's particle mover [13]. Additionally, BIT1 was ported to NVIDIA GPUs using OpenACC [2]. This approach builds upon previous endeavors, including the successful OpenACC porting of the GTC-P PIC code [12] and iPIC3D [7] to NVIDIA GPUs.

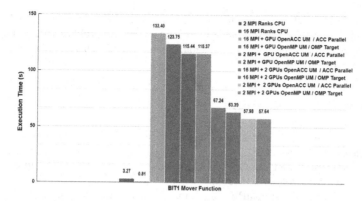

Fig. 8. Optimized mover function execution time(s) on *NJ* for 100 time steps using OpenMP and OpenACC Unified Memory on NVIDIA GPUs.

6 Discussion and Conclusion

Our primary goal was to enhance the particle mover function in BIT1, focusing on node-level efficiency and GPU offloading. Hybrid MPI and OpenMP/OpenACC approaches significantly improved on-node performance, showcasing the potential to leverage multicore CPUs and GPUs efficiently.

Results on multicore CPUs demonstrated the effectiveness of the hybrid approaches. The scalability of the MPI+OpenMP version indicates its potential for large-scale plasma simulations, crucial for efficient use of current supercomputer infrastructure.

GPU porting using OpenMP and OpenACC unveiled challenges and opportunities in tapping into GPU resources for the first time. Emphasizing a balanced hybrid approach for optimal GPU performance, our findings suggest that implementing OpenMP or OpenACC on GPUs may increase execution times, potentially outweighing parallel processing benefits. Notably, the OpenMP Target with 2 GPUs demonstrated a significant reduction in execution time among GPU results, highlighting potential performance improvement through concurrent GPU utilization, especially when dedicated GPUs are assigned to MPI ranks.

Future research can enhance BIT1's capabilities by fine-tuning GPU optimization and integrating advanced algorithms. Exploring CUDA approaches and batch processing shows promise in optimizing particle movement, while collaborative efforts with experimental data can bolster simulation reliability.

Hybrid MPI, OpenMP, and OpenACC approaches hold promise for comprehensive parallelization. Exploring synergies between these paradigms ensures BIT1's adaptability to various computing environments.

Acknowledgments. Funded by the European Union. This work has received funding from the European High Performance Computing Joint Undertaking (JU) and Sweden, Finland, Germany, Greece, France, Slovenia, Spain, and Czech Republic under grant agreement No 101093261.

References

1. Ayguadé, E., et al.: The design of OpenMP tasks. IEEE Trans. Parallel Distrib. Syst. **20**(3), 404–418 (2008)
2. Chandrasekaran, S., et al.: OpenACC for Programmers: Concepts and Strategies. Addison-Wesley Professional (2017)
3. Chien, S.W., et al.: sputniPIC: an implicit particle-in-cell code for multi-GPU systems. In: 2020 IEEE 32nd International Symposium on Computer Architecture and High Performance Computing (SBAC-PAD), pp. 149–156. IEEE (2020)
4. Derouillat, J., et al.: SMILEI: a collaborative, open-source, multi-purpose particle-in-cell code for plasma simulation. Comput. Phys. Commun. **222**, 351–373 (2018)
5. Liu, F., et al.: Parallel Cholesky factorization for banded matrices using OpenMP tasks. In: Euro-Par 2023: Parallel Processing. LNCS, vol. 14100, pp. 725–739. Springer, Cham (2023). https://doi.org/10.1007/978-3-031-39698-4_49
6. Markidis, S., et al.: The EPiGRAM project: preparing parallel programming models for Exascale. In: Taufer, M., Mohr, B., Kunkel, J.M. (eds.) ISC High Performance 2016. LNCS, vol. 9945, pp. 56–68. Springer, Cham (2016). https://doi.org/10.1007/978-3-319-46079-6_5
7. Peng, I.B., et al.: Acceleration of a Particle-in-Cell code for space plasma simulations with OpenACC. In: EGU General Assembly Conference Abstracts, p. 1276 (2015)
8. Tskhakaya, D., et al.: Optimization of pic codes by improved memory management. J. Comput. Phys. **225**(1), 829–839 (2007)
9. Tskhakaya, D., et al.: PIC/MC code bit1 for plasma simulations on HPC. In: 2010 18th Euromicro Conference on Parallel, Distributed and Network-Based Processing, pp. 476–481. IEEE (2010)
10. Vay, J.L., et al.: Warp-X: a new Exascale computing platform for beam-plasma simulations. Nucl. Instrum. Methods Phys. Res., Sect. A **909**, 476–479 (2018)
11. Verboncoeur, J.P., et al.: Simultaneous potential and circuit solution for 1D bounded plasma particle simulation codes. J. Comput. Phys. **104**(2), 321–328 (1993)
12. Wei, Y., et al.: Performance and portability studies with OpenACC accelerated version of GTC-P. In: 2016 17th International Conference on Parallel and Distributed Computing, Applications and Technologies (PDCAT), pp. 13–18. IEEE (2016)
13. Williams, J.J., et al.: Leveraging HPC profiling & tracing tools to understand the performance of particle-in-cell monte Carlo simulations. Euro-Par 2023: Parallel Processing Workshops, arXiv preprint arXiv:2306.16512 (2023)

Brain-Inspired Physics-Informed Neural Networks: Bare-Minimum Neural Architectures for PDE Solvers

Stefano Markidis$^{(\boxtimes)}$ (ID)

KTH Royal Institute of Technology, Stockholm, Sweden
markidis@kth.se

Abstract. Physics-Informed Neural Networks (PINNs) have emerged as a powerful tool for solving partial differential equations (PDEs) in various scientific and engineering domains. However, traditional PINN architectures typically rely on large, fully connected multilayer perceptrons (MLPs), lacking the sparsity and modularity inherent in many traditional numerical solvers. An unsolved and critical question for PINN is: What is the minimum PINN complexity regarding nodes, layers, and connections needed to provide acceptable performance? To address this question, this study investigates a novel approach by merging established PINN methodologies with brain-inspired neural network techniques. We use Brain-Inspired Modular Training (BIMT), leveraging concepts such as locality, sparsity, and modularity inspired by the organization of the brain. With brain-inspired PINN, we demonstrate the evolution of PINN architectures from large, fully connected structures to bare-minimum, compact MLP architectures, often consisting of a few neural units! Moreover, using brain-inspired PINN, we showcase the spectral bias phenomenon occurring on the PINN architectures: bare-minimum architectures solving problems with high-frequency components require more neural units than PINN solving low-frequency problems. Finally, we derive basic PINN building blocks through BIMT training on simple problems akin to convolutional and attention modules in deep neural networks, enabling the construction of modular PINN architectures. Our experiments show that brain-inspired PINN training leads to PINN architectures that minimize the computing and memory resources yet provide accurate results.

Keywords: Brain-Inspired PINN · Bare-Minimum PINN Architectures · Spectral Bias Phenomenon · Modular PINN

1 Introduction

Scientific Machine Learning (SciML) is a discipline harnessing machine learning methods, such as neural networks (NNs) [2] and operators [18], to solve scientific computing problems, including scientific simulations, linear and non-linear solvers, inverse problems, and equation discovery [8]. One of the most active

© The Author(s), under exclusive license to Springer Nature Switzerland AG 2024
L. Franco et al. (Eds.): ICCS 2024, LNCS 14832, pp. 331–345, 2024.
https://doi.org/10.1007/978-3-031-63749-0_23

research areas in SciML is the development of Partial Differential Equations (PDE) solvers that are the backbone of scientific simulations. The SciML PDE solvers are part of the so-called SciML approaches with *learning bias* as the PDE is embedded into the loss function, and the solution is determined by the NN training or learning. This is opposed to SciML approaches with *inductive bias* where the given knowledge about the modeled system, e.g., symmetries and conservation laws, influences the NN architecture design.

The core concept of SciML PDE solvers revolves around encoding the governing PDE equation into the NN loss function, facilitating numerical differentiation on NN graphs through automatic differentiation, and optimizing the loss function using techniques like the first-order Adam [9] or second-order BFGS [14] optimizers. Prominent SciML PDE solvers include Physics-Informed Neural Networks (PINN) [25], deep Galerkin, and Ritz methods [26,30]. PINNs have rapidly evolved and found wide-ranging applications from computational fluid dynamics to material science and chemistry and are the focus of this work.

These PDE solvers, termed physics-informed, encode physics conservation laws into the loss function to guide the learning process toward its minimization. Conceptually, PINN serves as an extension and non-linear version of Finite Element Methods (FEM) [31], with non-linear activation functions acting as piecewise basis functions and the loss function representing the residual or equation error at each solver iteration. Its analogy and equivalence with kernel regression and other traditional methods have facilitated investigations into PINN's numerical properties, such as consistency and convergence aspects [21,28], including the PINN spectral bias (higher convergence rate for low-frequency solution components). An increased understanding of the PINN fundamental numerical properties allows us to develop numerical methods further and integrate new SciML techniques into traditional approaches [19].

Despite the PINN efficacy, there is a gap in understanding the impact of PINN architecture and the development of minimal architectures in terms of neural units, layers, and connectivity while ensuring accurate results. Most PINN studies analyze fully connected multi-layer perceptron (MLP) architectures with minimal sparsity and lack of structural modularity. In contrast, traditional numerical solvers, like finite-difference linear solvers, exhibit high structure and sparsity, which is evident in scenarios such as sparse matrices arising from Poisson equation discretization. The absence of macroscale structure in PINN poses challenges for interpretability and explainability, obfuscating the manifestation of intrinsic PDE nature within the network architecture.

In the search for PINN solvers and their building blocks that are resource-efficient and compact, we explore the application of brain-inspired neural network techniques [4,13]. These techniques draw loose inspiration from models of brain computations, mapping neurons and synapses to neural networks and weights/biases. A key distinction between traditional and brain-inspired NN architectures lies in the learning rule, typically local instead of global back-propagation, and plasticity, reflecting the adaptivity and dynamicity of connection strengths and connectivity during training. Traditional NN architectures lack the concept of locality, whereas brain-inspired NN architectures prioritize it, leading to modularity, with specific parts of the network specialized in distinct operations. Moreover, brain-inspired NNs tend to exhibit high sparsity.

Brain-Inspired Modular Training (BIMT) is one of the most successful brain-inspired neural network techniques [15,16]. Its fundamental idea introduces neural network locality by associating a spatial coordinate with each neural unit, enabling the rearrangement of neuron positions to enhance locality and modularity. This approach trains the network for increased locality by adding a loss function that penalizes non-local connections.

An important aspect concerns the number of resources that should be used to solve a given PDE with PINN. Typically, a few layers and tens of neural units, all fully connected, are used for solving simple PDE problems. Still, no insights exist on the bare minimum PINN architecture in terms of computing units capable of producing the PDE solutions. This study addresses the fundamental research questions: *What is the minimal or reduced number of PINN neural units required to produce PDE solutions using PINN?* The answer might surprise many: simple PDE requires only a few neural units to encode the PDE solution. We call this PINN *bare-minimum architectures*, as they consist only of a few neural units and are only affected by a small loss of accuracy when compared to large fully-connected MLP. Through BIMT, we can derive bare-minimum architectures for solving given PDEs. We demonstrate that the spectral bias phenomenon [8] (also called F-principle [29]) manifests in the bare-minimum architectures: solving high-frequency problems requires more neural units than PINN solving problems with low-frequency components. Finally, we combine bare-minimum PINN architectures into modular and compact PINN architectures. These basic bare-minimum modules resemble other common deep-learning building blocks like convolutional kernels/filters and attention modules. We show that modular PINN provides promising results in terms of performance.

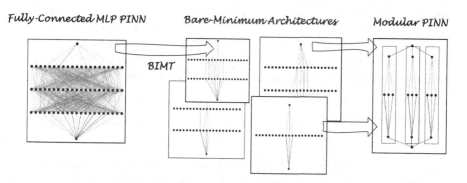

Fig. 1. A graphical illustration of the contributions of this work: *i)* Brain-Inspired Modular Training (BIMT) [15] allows us to obtain bare-minimum PINN architectures *ii)* we use bare-minimum architectures as basic modules to be combined to build compact modular PINN architectures.

The primary objective of this study is illustrated in Fig. 1: first, we derive PINN bare-minimum architectures from fully connected MLP; second, we use bare-minimum architectures to build modular PINN. Overall, we demonstrate that BIMT leads to dynamically changing architectures with highly sparse and modular features akin to traditional numerical approaches, albeit with a slight compromise in accuracy. Furthermore, employing brain-inspired PINNs, we derive primitive building blocks that can be reused to construct larger PINNs, referred to as modular PINNs.

The contributions of this work can be summarized as follows:

- We demonstrate that BIMT leads to bare-minimum and compact architectures for PINN PDE solvers. For instance, we illustrate that PINN for solving simple differential equations, such as the logistic equation, requires only one neural unit in the hidden layer.
- We identify several neural network architecture primitives capable of serving as modules in larger neural networks. By examining different basic primitives for solving the Poisson equation with source terms of increasing higher frequencies, we observe a manifestation of the PINN spectral bias on the number of PINN connections needed to represent accurately a solution, indicating that higher-frequency source terms necessitate denser PINN architectures.
- We develop and implement a modular PINN architecture based on the identified building blocks from training brain-inspired neural networks. Our results demonstrate that modular PINN architectures built on PINN primitive modules exhibit lower test errors than fully connected MLP PINNs utilizing the same number of neural units.

The remainder of this paper is organized as follows. In Sect. 2, we provide background information on PINNs and introduce the concept of brain-inspired neural networks. Section 3 details the methodology employed in this study, including the implementation of PINN based on BIMT and the experimental setup. Section 4 presents the results of our investigation, including the evolution of PINN architectures, comparisons with analytical solutions, and the derivation of modular PINN architectures. Section 5 discusses previous work in related areas of PINNs and BIMT. Finally, Sect. 6 concludes the paper with a summary of key findings and suggestions for future research directions.

2   Background

PINNs are neural networks that take a point t_i in the equation domain (referred to as the collocation point) as input and provide the approximated solution $\tilde{x}(t_i)$. The solution is encoded into the neural network with l layers. At any given time, the PINN network acts as a surrogate solver, providing the approximated solution $\tilde{x}$ by running:

$$\tilde{x}(t_i) = a \circ Z_l \circ \ldots \circ a \circ Z_2 \circ a \circ Z_1(t_i), \tag{1}$$

where ∘ denotes the composition operation, a represents the non-linear activation function, and the affine-linear maps Z_i are expressed as:

$$Z_i(t_l) = W_i t_l + b_i. \tag{2}$$

Here, W_i and b_i are the weights and biases of layer i, respectively.

The training process aims to determine the weights and biases through iterative steps. It begins with a forward pass, where Eq. 1 is applied to several collocation points, typically chosen at random positions or according to a specific distribution. At each iteration, the forward pass yields an approximate solution at the collocation point, $\tilde{x}(t_i)$. Subsequently, the error or loss function is measured to adjust the weights and biases, a process known as *back-propagation*. An optimizer then modifies the weights and biases of different layers to minimize the loss function.

The basic training process of PINNs is unsupervised, as it does not necessarily require labeled data, such as solutions obtained from other simulation techniques. This is made possible by encoding the differential equation and boundary conditions into the loss function, utilizing the residual to guide the training process. This approach is akin to Krylov subspace solvers, where the residual is minimized iteratively.

For instance, consider solving a Poisson equation $d^2x(t)/dt^2 = \sin(t)$. The residual at a certain point t_i can be calculated as:

$$r_i = \left.\frac{d^2\tilde{x}}{dt^2}\right|_i - \sin(t_i). \tag{3}$$

The second-order derivative at point t_i is computed using automatic differentiation, which allows for calculating derivatives on the neural network, exploiting the chain rule. Unlike finite difference differentiation, automatic differentiation enables derivative calculation at any given collocation point without requiring a grid or associated spacing.

Without boundary conditions, PINNs converge to one of the infinite solutions. We impose two boundary conditions to obtain a unique solution for our second-order PDE. For example, with boundary conditions $x(0) = 0$ and $x(2\pi) = 0$, two additional residuals are introduced at the boundary collocation points:

$$r_{BC_0} = \tilde{x}(0), \quad r_{BC_1} = \tilde{x}(2\pi). \tag{4}$$

These residuals are incorporated as Mean Squared Error (MSE) into the PINN loss function, which includes the following terms:

$$\mathcal{L}_{PINN} = \frac{1}{N}\sum_i^N r_i^2 + \frac{1}{NBC0}\sum_i^{NBC0} r_{i,BC0}^2 + \frac{1}{N_{BC1}}\sum_i^{NBC1} r_{i,BC1}^2, \tag{5}$$

where N_{BC0} and N_{BC1} represent the number of samples taken at points 0 and 2π, respectively. This approach serves as a soft constraint on solving a multi-learning task, minimizing the residual both within the system and at the boundary. In scenarios involving experimental or simulation data, an additional component can be added to the loss function to consider the error between the neural network and observational data. However, in this study, we adopt a fully unsupervised approach without utilizing experimental data.

3   Brain-Inspired Physics-Informed Neural Networks

This work employs the BIMT approach to determine the PINN architecture, as detailed in Ref. [15]. For clarity, we summarize the key features of BIMT:

- **L1 Penalty (Lasso Regularization)**: BIMT utilizes L1 penalty or Lasso regularization during network training to prevent overfitting and enhance generalization. This regularization induces sparsity in the weight matrix, wherein some weights may become exactly zero during training. The Lasso regularization employed in BIMT aims to increase the sparsity and modularity of the neural network architecture. A hyperparameter λ governs the strength of the L1 penalty and can be adjusted during the simulation.
- **Introduction of Geometry and Distance**: BIMT incorporates the notion of geometry and distance into the neural network architecture by associating a coordinate with each neural unit across different layers. In this work, we adopt a two-dimensional Euclidean space, with the x-direction spanning along the neural units within the same layer (input, hidden, and output layers) and the y-direction spanning across layers. The distance $d_{i,j}$ between two neural units is leveraged to scale the weights and biases when computing the L1 penalty. By integrating this technique, we can optimize for increased locality or minimize the distance between neural units by incorporating a loss function component that accounts for total distance. A hyperparameter A, related to the network size, regulates the importance of locality. For $A = 0$, the L1 penalization does not consider locality.
- **Neural Unit Swapping**: BIMT permits swapping different neural units within the same layer if it enhances locality, i.e., decreases the distance between neural units. BIMT introduces the concept of neural unit importance, computed as the sum of input and output weights to determine which neural units to swap and select critical swaps. This importance metric is utilized to swap the most significant neural units within the layer if it improves locality. However, this operation incurs computational overhead and is typically not performed at every iteration.

Consequently, BIMT yields neural network architectures that dynamically evolve during training, characterized by high sparsity due to Lasso penalization and local structure due to the additional penalty on distances between neural units. These features align with principles of brain-inspired computing, albeit resulting in slightly reduced performance (in terms of training and test accuracy) compared to fully connected MLP counterparts [15].

3.1   Implementation

The implementation of brain-inspired PINNs extends the Python and PyTorch [23] BIMT implementation [15] to solve differential equations. We leverage the PyTorch `autograd` [22] for automatic differentiation.

Unless otherwise specified, our implementation initiates with 21 neural units per layer, starts with a fully connected network comprising one or two hidden layers, and executes 100,000 epochs. We employ 1,000 collocation points within the domain and 50 boundary points. The weights are initialized using Xavier initialization [10], and biases are set to a constant value of 0.01. The learning rate is fixed at 0.002.

Among various activation functions tested, the `sinLU` activation function [24] $(a(x) = x \sin(x)\sigma(x)$, where $\sigma(x)$ is the sigmoid function) yields the most compact brain-inspired architecture. We utilize the AdamW optimizer [17], which decouples weight decay and gradient update, as it provides optimal performance, sparsity, and modularity. Notably, unlike fully connected MLP PINNs, we observed that second-order optimizers such as L-BFGS do not enhance accuracy in brain-inspired PINNs; they quickly converge to local minima without achieving higher precision.

Following the approach outlined in the seminal BIMT paper [15], we divide the training into three phases with varying L1 penalization regimes: λ starts at 0.001 with no bias penalization, increases to 0.01 to enhance locality at one-fourth of the total training, and finally, at three-fourths of the total training, switches to bias penalization while reducing λ back to 0.001. We set A to 2, and neural unit swaps occur every 200 epochs. Prior to the final error evaluation, we prune the PINN weights to eliminate weights and biases below 10^{-3} in absolute value.

For demonstration purposes, this work focuses on solving the one-dimensional Poisson equation with a harmonic source term:

$$\frac{d^2 x(t)}{dt^2} = \sin(t) + 4\sin(2t) + 9\sin(3t) + 16\sin(4t), \tag{6}$$

with boundary conditions $x(0) = 0$ and $x(2\pi) = 0$ in the simulation domain $[0, 2\pi]$. The Poisson equation is omnipresent in scientific computing: it is used for electromagnetics to solve electrostatic problems and in incompressible flow in computational fluid dynamics.

Regarding Eq. 6, This particular choice of harmonic source term allows us to evaluate different and higher spectral components. It has been observed that PINNs exhibit a spectral bias, converging rapidly to low-frequency parts of the solution while requiring more time to resolve high-frequency components accurately. By introducing various components with differing spectral characteristics, we can assess the performance of the brain-inspired architecture. We utilize 100 test collocation points to test and compare the results against the analytical solution. MSE and Euclidean error metrics are used throughout training to evaluate the performance of different PINN models.

The code utilized in this study is openly accessible on GitHub[1].

4   Results

As the first step, we analyze the evolution of the brain-inspired PINN architecture during the training and assess its performance. In Fig. 2, we show the results of the brain-inspired PINN network training applied to solve the 1D Poisson equation $d^2x(t)/dt^2 = \sin(t) + 4\sin(2t) + 9\sin(3t) + 16\sin(4t)$ with $x(0) = 0$ and $x(2\pi) = 0$ for 400,000 epochs. The different inserts show the evolution of brain-inspired architecture. The red and blue edges connect neural units with positive and negative weights, respectively. We have evolved from a fully connected MLP PINN to a sparse and modular computer architecture, only utilizing a small part of the total capacity of the original network. In the background plot, the train loss and test error (calculated against the analytical solution) are represented in blue and orange colors.

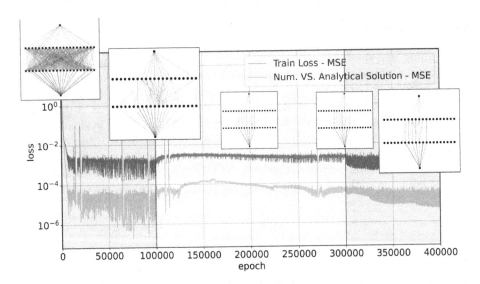

Fig. 2. Evolution of the brain-inspired PINN network architecture during the training for the solution of $d^2x(t)/dt^2 = \sin(t) + 4\sin(2t) + 9\sin(3t) + 16\sin(4t)$ with $x(0) = 0$ and $x(2\pi) = 0$. The PINN architecture evolves from being fully connected to being highly sparse and modular. The red and blue lines represent connections associated with positive and negative weights. The training occurs in three phases where the strength of the L1 penalty (related to the importance of locality) changes. (Color figure online)

By analyzing the losses, we can identify the three phases of the training as we change the value of λ in the three phases. These three phases are represented

---

[1] https://github.com/smarkidis/BrainInspiredPINN/.

as different background colors. Until the epoch of 100,000, the L1 penalty is relatively low, and the loss function decreases quickly; after that, we increase the importance of locality and note significant changes in the PINN architecture. At epoch 300,000, we decrease λ and turn on bias penalization.

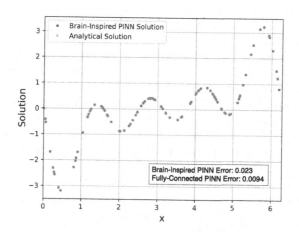

Fig. 3. Analytical and brain-inspired solutions of the differential equation: $d^2x(t)/dt^2 = \sin(t) + 4s \in (2t) + 9\sin(3t) + 16\sin(4t)$ with $x(0) = 0$ and $x(2\pi) = 0$.

The final PINN solution and its comparison with the analytical solution are presented in Fig. 3. The brain-inspired PINN can capture all the frequencies present in the solutions at a reasonable accuracy. To estimate the loss of performance due to large sparsity and modularity, we also run a fully connected neural network with two hidden layers of 21 neural units and calculate the error in the Euclidean norm compared to the analytical solution. The final error for the brain-inspired and fully-connected PINNs are 0.023 and 0.0094, respectively. As pointed out, architecture with high sparsity comes with a performance loss: in this example, the brain-inspired PINN has approximately twice the error of the fully connected MLP PINN.

4.1   Deriving Modular PINN Architectures

One of the advantages of brain-inspired neural networks is the possibility of deriving basic modules, the bare-minimum PINN architectures, that are small in scale and compact. These basic modules can be derived by solving simple problems, such as the logistic or Poisson equation in one dimension, with a simple archetypal source term, e.g., a sinusoidal source term with a single spectral component. Figure 4 shows different PINN bare-minimum architectures that can be derived by training brain-inspired PINN to solve simple archetypal problems with one and two hidden layers.

A few important points can be deduced by analyzing Fig. 4. For instance, it is striking that the solution of the logistic equation (top left panel of Fig. 4) requires only one neural unit in the hidden layer. Another important point comes up when analyzing the final brain-inspired PINN architecture obtained by training the neural network for source terms with higher frequency terms: to solve a low-frequency signal $\sin(t)$ requires only three neural units in the hidden layer. As we increase the frequency of the source term, we note that the number of neural units in the hidden layer increases. For instance, when solving the one-dimensional Poisson equation with $16\sin(4t)$, the brain-inspired converges to a fully connected MLP. In general, we observe that by increasing the frequency of the source term, more neural units are needed in the hidden layer to converge to a solution. This clearly manifests the spectral bias in the number of neural network connections needed to express higher-frequency components. Using a brain-inspired approach, we show that spectral bias is not only in the rate of convergence but also in the architecture of the PINN: higher frequency signal requires a larger number of neural units and layers to be accurately resolved.

As an additional note, we remark that bare-minimum architectures depend on the activation function that provides the basic basis function to express the solution. Using other activation functions leads to slightly different architecture modules for the problem presented here.

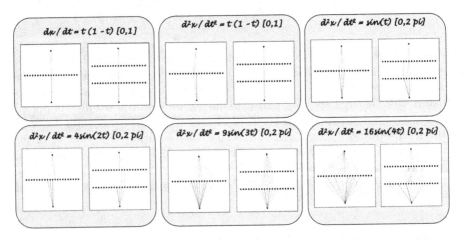

Fig. 4. Bare-minimum architectures derived running brain-inspired PINN for solving basic differential equations. Interestingly, some differential equations, such as the logistic equation, only require a neural unit in the hidden layer. Another important point is that source terms with higher frequencies result in denser PINN architectures as a manifestation of the spectral bias phenomenon.

An essential aspect of the experiments shown in Fig. 4 is that we can identify PINN bare-minimum architectures that we can use in larger architectures. These basic building blocks can be used similarly to convolutional [12] and attention [27] modules in established deep neural networks. Here, we can utilize a modular architecture based on the module derived from solving the Poisson equation in 1D for the $\sin(t)$ source term. For instance, we can use three modules as depicted on the right side of Fig. 1: the overall neural network consists of three building blocks with the same collocation point as input and whose output is combined by summing up the output of each building block. The modular PINN combines the modules into a larger NN using PyTorch superclasses and inheritance and is trained as traditional PINNs with an Adam optimizer. This is a simple example of modular PINN architecture, and larger modular architecture can be obtained by using a larger number of modules or using more advanced modules, e.g., modules obtained for two or more hidden layers or different source terms.

To understand the potential benefit of the modular compact PINN architecture, we solve the original problem expressed by Eq. 6 with a fully connected MLP PINN with one hidden layer and nine neural units; we compare the loss for the training data set and the error of the solution against the analytical solutions. We present the results in Fig. 5.

When analyzing Fig. 5, we note that the training loss values are similar, showing a comparable performance. However, the results of the two network architectures, regarding the test error and comparison of the analytical solution (bottom panel of Fig. 5), show a considerably higher performance of the modular PINN. We can see that the modular PINN has a test error (calculated as MSE) that is approximately two orders lower than the fully-connected MLP PINN. Compared to the analytical solution, the test dataset's final Euclidean error is 0.083 for the modular PINN vs. 0.57 for the fully-connected MLP PINN. We also note that a simple modular neural network performs better than the PINN networks (including the brain-inspired NN), whose results are previously presented in Fig. 3. The modular PINN, built on the top of a module obtained with BIMT, exhibits improved generalization properties in this use case and compact formulation.

5   Related Work

This work is at the intersection of PINN, brain-inspired neural networks, and modular structures for neural networks. PINNS have been extensively studied since their introduction by Raissi et al. in Ref. [25]. Subsequent research has focused on aspects such as convergence, stability, and numerical properties. PINNs have demonstrated versatility across various applications, including computational fluid dynamics [7], solid mechanics [3], molecular dynamics [5], and battery life cycle modeling [5]. However, previous PINN studies have primarily focused on fully connected architectures without incorporating sparsity or modularity.

Brain-inspired machine learning is gaining prominence as researchers seek inspiration from neural computations. Modular architectures have been a key focus in this domain. Convolutional Neural Networks (CNNs) [12] exemplify modular architecture, preserving symmetries and invariances under various

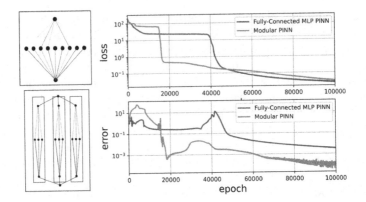

Fig. 5. Performance comparison between a fully connected MLP network with a hidden layer with nine neural units and a modular PINN using a network primitive obtained by solving $d^2x(t)/dt^2 = \sin(t)$. The modular PINN has three basic modules with three neural units. While the fully connected and modular PINNs have similar test loss values, the MSE against the analytical solution is two orders lower for the modular PINN.

transformations. Similarly, transformer neural networks [27] utilize attention mechanisms as fundamental building blocks. Additionally, graph neural networks [1] demonstrate structured architectures for processing graph data. These modular approaches enhance interpretability and scalability in neural network design.

Unlike previous PINN studies, this work introduces a novel approach where the PINN architecture evolves into a sparse and modular structure. By integrating brain-inspired techniques and modular design principles, this study explores new avenues for enhancing PINN architectures, potentially improving efficiency and interpretability in solving partial differential equations.

6   Discussion and Conclusion

This study investigated a fundamental aspect of PINN architectures, widely employed for solving differential equations. Conventionally, PINN methodologies have predominantly relied on fully connected MLP architectures. However, traditional numerical solvers for differential equations exhibit sparsity and a modular structure, wherein computations at a given point depend only on neighboring points, termed the stencil. This work merges established PINN techniques with a brain-inspired neural network approach to address these architectural limitations and enhance the solution of differential equations. Specifically, we leverage brain-inspired neural networks to achieve two primary objectives: first, to derive basic building blocks for constructing larger neural networks, and second, to obtain solutions for differential equations.

From a computational perspective, brain-inspired neural architectures offer significant advantages in terms of sparsity, leading to reduced computation and memory requirements. The examples presented in this work demonstrate remarkable sparsity, with simple equations requiring minimal neural units to achieve satisfactory accuracy. However, the degree of sparsity is influenced by various factors, including the characteristics of the differential equations, particularly those involving high-frequency components. While the achieved sparsity impacts memory storage needs, current support for sparse computations in mainstream deep-learning frameworks, such as PyTorch and TensorFlow, remains limited, necessitating advancements in this area to exploit computational benefits fully. Usage of frameworks and libraries for NN computation [6,20], directly in sparse formats, such as CSR/CSC/COO [11], is key to achieving both computational advantages, lowering the complexity costs of matrix multiply and memory requirements.

We observed that by increasing the frequency of the source term, more neural units are needed in the hidden layer to converge to a solution. This clearly manifests the spectral bias in the number of neural network connections needed to express higher-frequency components. Using a brain-inspired approach, we show that spectral bias is visible not only in the rate of convergence but also in the architecture of the PINN: higher frequency signal requires a larger number of neural units and layers to be accurately resolved. Understanding this phenomenon clarifies the challenges inherent in training PINNs for problems with high-frequency components and indicates the importance of developing architectures capable of accommodating such spectral biases.

This study proposed a novel contribution to PINNs architectures by introducing an approach to constructing modular architectures. By leveraging brain-inspired neural network techniques, we derived PINN bare-minimum architectures through BIMT on basic archetype problems. These modular building blocks (the bare-minimum architectures) exhibit potential for application in larger architectures reminiscent of the modular structures found in convolutional and attention-based modules within established deep NN. Modular PINNs offer promising possibilities for increasing the accuracy and efficiency of solving PDE problems, bridging the gap between traditional numerical methods and machine learning approaches [19]. Moreover, adopting modular architectures facilitates a transition from complex, fully connected Multilayer Perceptron (MLP) designs to simpler, more compact PINN architectures, thus reducing computational overhead and memory requirements. This study clarifies the architectural evolution of PINNs. It prepares for future research into developing and optimizing modular neural network frameworks for a wide range of scientific and engineering applications.

However, it is important to acknowledge that this study represents only an initial exploration of modular PINN architectures. Our investigation focused primarily on simple one-dimensional problems and utilized only a single building block derived from solving the Poisson equation with a sinusoidal source for the modular PINN. While our findings are promising, further research is needed to refine the composition of these building blocks and optimize the accuracy of compact and modular PINN architectures. Future studies could explore the incorporation of additional architectural primitives derived from a broader range of differential equations and problem domains. Additionally, research on enhancing the modularity and versatility of these architectures by incorporating multi-layer modules and exploring alternative activation functions could further improve their performance.

References

1. Bronstein, M.M., Bruna, J., LeCun, Y., Szlam, A., Vandergheynst, P.: Geometric Deep Learning: going beyond Euclidean data. IEEE Signal Process. Mag. **34**(4), 18–42 (2017)
2. Goodfellow, I., Bengio, Y., Courville, A.: Deep Learning. MIT Press (2016)
3. Haghighat, E., Raissi, M., Moure, A., Gomez, H., Juanes, R.: A physics-informed deep learning framework for inversion and surrogate modeling in solid mechanics. Comput. Methods Appl. Mech. Eng. **379**, 113741 (2021)
4. Hassabis, D., Kumaran, D., Summerfield, C., Botvinick, M.: Neuroscience-inspired artificial intelligence. Neuron **95**(2), 245–258 (2017)
5. Hassanaly, M., et al.: PINN surrogate of Li-ion battery models for parameter inference. Part I: implementation and multi-fidelity hierarchies for the single-particle model. arXiv preprint arXiv:2312.17329 (2023)
6. Ivanov, A., Dryden, N., Ben-Nun, T., Ashkboos, S., Hoefler, T.: Sten: Productive and efficient sparsity in PyTorch (2023)
7. Jin, X., Cai, S., Li, H., Karniadakis, G.E.: NSFnets (Navier-Stokes flow nets): physics-informed neural networks for the incompressible Navier-Stokes equations. J. Comput. Phys. **426**, 109951 (2021)
8. Karniadakis, G.E., Kevrekidis, I.G., Lu, L., Perdikaris, P., Wang, S., Yang, L.: Physics-informed machine learning. Nat. Rev. Phys. **3**(6), 422–440 (2021)
9. Kingma, D.P., Ba, J.: Adam: a method for stochastic optimization. arXiv preprint arXiv:1412.6980 (2014)
10. Kumar, S.K.: On weight initialization in deep neural networks. arXiv preprint arXiv:1704.08863 (2017)
11. Langr, D., Tvrdik, P.: Evaluation criteria for sparse matrix storage formats. IEEE Trans. Parallel Distrib. Syst. **27**(2), 428–440 (2015)
12. LeCun, Y., Bengio, Y., et al.: Convolutional networks for images, speech, and time series. Handb. Brain Theory Neural Netw. **3361**(10), 1995 (1995)
13. Lillicrap, T.P., Santoro, A., Marris, L., Akerman, C.J., Hinton, G.: Backpropagation and the brain. Nat. Rev. Neurosci. **21**(6), 335–346 (2020)
14. Liu, D.C., Nocedal, J.: On the limited memory BFGS method for large scale optimization. Math. Program. **45**(1–3), 503–528 (1989)
15. Liu, Z., Gan, E., Tegmark, M.: Seeing is believing: brain-inspired modular training for mechanistic interpretability. Entropy **26**(1), 41 (2024)
16. Liu, Z., Khona, M., Fiete, I.R., Tegmark, M.: Growing brains: co-emergence of anatomical and functional modularity in recurrent neural networks. arXiv preprint arXiv:2310.07711 (2023)
17. Loshchilov, I., Hutter, F.: Decoupled weight decay regularization. arXiv preprint arXiv:1711.05101 (2017)
18. Lu, L., Jin, P., Pang, G., Zhang, Z., Karniadakis, G.E.: Learning nonlinear operators via DeepOnet based on the universal approximation theorem of operators. Nat. Mach. Intell. **3**(3), 218–229 (2021)

19. Markidis, S.: The old and the new: can physics-informed deep-learning replace traditional linear solvers? Front. Big Data **4**, 669097 (2021)
20. Mishra, A., et al.: Accelerating sparse deep neural networks. arXiv preprint arXiv:2104.08378 (2021)
21. Mishra, S., Molinaro, R.: Estimates on the generalization error of physics-informed neural networks for approximating PDEs. IMA J. Numer. Anal. **43**(1), 1–43 (2023)
22. Paszke, A., et al.: Automatic differentiation in PyTorch (2017)
23. Paszke, A., et al.: Pytorch: an imperative style, high-performance deep learning library. Adv. Neural Inf. Process. Syst. **32** (2019)
24. Paul, A., Bandyopadhyay, R., Yoon, J.H., Geem, Z.W., Sarkar, R.: Sinlu: Sinu-sigmoidal linear unit. Mathematics **10**(3), 337 (2022)
25. Raissi, M., Perdikaris, P., Karniadakis, G.E.: Physics-informed neural networks: a deep learning framework for solving forward and inverse problems involving non-linear partial differential equations. J. Comput. Phys. **378**, 686–707 (2019)
26. Sirignano, J., Spiliopoulos, K.: DGM: a deep learning algorithm for solving partial differential equations. J. Comput. Phys. **375**, 1339–1364 (2018)
27. Vaswani, A., et al.: Attention is all you need. Adv. Neural Inf. Process. Syst. **30** (2017)
28. Wang, S., Yu, X., Perdikaris, P.: When and why PINNs fail to train: a neural tangent kernel perspective. J. Comput. Phys. **449**, 110768 (2022)
29. Xu, Z.Q.J., Zhang, Y., Luo, T., Xiao, Y., Ma, Z.: Frequency principle: fourier analysis sheds light on deep neural networks. arXiv preprint arXiv:1901.06523 (2019)
30. Yu, B., et al.: The deep Ritz method: a deep learning-based numerical algorithm for solving variational problems. Commun. Math. Statist. **6**(1), 1–12 (2018)
31. Zienkiewicz, O.C., Taylor, R.L., Zhu, J.Z.: The Finite Element Method: Its Basis and Fundamentals. Elsevier (2005)

NeRFlame: Flame-Based Conditioning of NeRF for 3D Face Rendering

Wojciech Zając[1] , Joanna Waczyńska[1] , Piotr Borycki[1] , Jacek Tabor[1] ,
Maciej Zieba[2] , and Przemysław Spurek[1,3(✉)]

[1] Faculty of Mathematics and Computer Science, Jagiellonian University, Krakow,
Poland
przemyslaw.spurek@uj.edu.pl
[2] Wroclaw University of Science and Technology, Wroclaw, Poland
[3] IDEAS NCBR, Warsaw, Poland

Abstract. Traditional 3D face models are based on mesh representations with texture. One of the most important models is Flame (Faces Learned with an Articulated Model and Expressions), which produces meshes of human faces that are fully controllable. Unfortunately, such models have problems with capturing geometric and appearance details. In contrast to mesh representation, the neural radiance field (NeRF) produces extremely sharp renders. However, implicit methods are hard to animate and do not generalize well to unseen expressions. It is not trivial to effectively control NeRF models to obtain face manipulation.

The present paper proposes a novel approach, named NeRFlame, which combines the strengths of both NeRF and Flame methods. Our method enables NeRF to have high-quality rendering capabilities while offering complete control over the visual appearance, similar to Flame. In contrast to traditional NeRF-based structures that use neural networks for RGB color and volume density modeling, our approach utilizes the Flame mesh as a distinct density volume. Consequently, color values exist only in the vicinity of the Flame mesh. Our model's core concept involves adjusting the volume density based on its proximity to the mesh. This Flame framework is seamlessly incorporated into the NeRF architecture for predicting RGB colors, enabling our model to represent volume density explicitly and implicitly capture RGB colors.

Keywords: NeRF · Flame · Avatar 3D

1 Introduction

Methods to automatically create fully controllable human face avatars have many applications in VR/AR and games [18]. Traditional 3D face models are based on fully controllable mesh representations. Flame [25] is a method used for mesh-based avatars [23,42]. Flame integrates a linear shape space trained using 3800

W. Zając and J. Waczyńska—Equal contribution.

L. Franco et al. (Eds.): ICCS 2024, LNCS 14832, pp. 346–361, 2024.
https://doi.org/10.1007/978-3-031-63749-0_24

human head scans with articulated jaw, neck, eyeballs, pose-dependent corrective blend shapes, and extra global expression blend shapes. In practice, we can easily train Flame on the 3D scan (or 2D image) of human faces and then manipulate basic behaviors like jaw, neck, and eyeballs. We can also produce colors for mesh by using textures. Unfortunately, such models have problems with capturing geometric and appearance details.

In contrast to the classical approaches, we can use implicit methods that represent avatars using neural networks. NeRFs [29] represent a scene using a fully connected architecture. As input, they take a 5D coordinate (spatial location $\mathbf{x} = (x, y, z)$ and viewing direction $\mathbf{d} = (\theta, \Psi)$) and return an emitted color $\mathbf{c} = (r, g, b)$ and volume density σ. NeRF extracts information from unlabelled 2D views to obtain 3D shapes. NeRF allows for the synthesizing of novel views of complex 3D scenes from a small subset of 2D images. Based on the relations between those base images and computer graphics principles, such as ray tracing, this neural network model can render high-quality images of 3D objects from previously unseen viewpoints. In contrast to the mesh representation, NeRF captures geometric and appearance details. However, it is not trivial to effectively control NeRF to obtain face manipulation.

There are many approaches to controlling NeRF, including generative models [21,38], dynamic scene encoding [22], or conditioning mechanisms [4]. However, our ability to manipulate NeRF falls short compared to our proficiency in controlling mesh representations.

This paper proposes NeRFlame, a hybrid approach for 3D face rendering that uses implicit and explicit representations; see Fig. 1. Our model is based on two components: NeRF and mesh, with only points in the mesh surroundings treated as NeRF inputs. Our method inherits the best features from the above approaches by modeling the quality of NeRF rendering and controlling the appearance as in Flame. We combine those two techniques by showing how to condition the NeRF model by mesh effectively. Our model's fundamental idea is conditioning volume density by distance to mesh. The volume density is non-zero only in the ε neighborhood of Flame

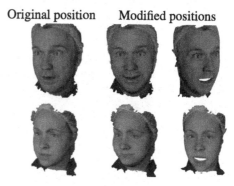

Original position   Modified positions

Fig. 1. Our model facilitates the manipulation of facial attributes in the context of human visage. NeRFlame use Flame as a conditioning factor in the NeRF-based model. We can produce novel views in training positions as well as in modified facial expressions.

mesh. Therefore, we use the NeRF-based architecture to model volume density and RGB colors only in the ε neighborhood of the mesh. Such a solution allows one to obtain renders of similar quality to NeRF and a level of control mesh similar to Flame (Table 1). In contrast to Dynamic Neural Radiance Fields, NeR-

Flame undergoes training using a single position of the human face rather than relying on sequences from various positions in movies. Despite this distinction, our model exhibits comparable functionality. We have the capability to generate a range of facial expressions and novel perspectives for newly encountered face positions facilitated by the Flame backbone. This enables us to model previously unseen facial expressions. Consequently, we conduct a comparison between our model and the traditional static Neural Radiance Fields.

The contributions of this paper are significant and are outlined as follows:

- We introduce NeRFlame – an innovative NeRF model conditioning be Flame that combines the best features of both methods, namely the exceptional rendering quality of NeRF and the precise control over appearance as in Flame.
- We demonstrate the ability to condition model volume density in NeRF by employing mesh representation, which represents a significant advancement over traditional NeRF-based approaches that rely on neural networks.
- We train our model on a single position of the human face rather than using entire movies, thereby highlighting the versatility and practicality of our approach.

Overall, our contributions offer a substantial advancement in the realms of 3D facial modeling and rendering, providing a foundation for future exploration and research in this domain.

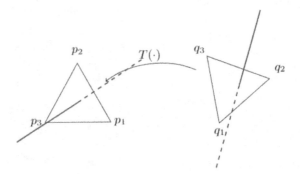

Fig. 2. Visualization of transformation in NeRFlame. We aim to aggregate colors along the ray during rendering in the new position (see the red line in the right image). NeRFlame uses Flame mesh, therefore we can localize the face's vertex, which is crossed with the ray p_1, p_2, p_3 and the corresponding triangle in the initial position mesh q_1, q_2, q_3. Thanks to such pairs of points, we estimate affine transformation T, which is used to find the ray in the initial position (see the red line in the left image).

2   Related Works

NeRFlame is a model of controllable human face avatars trained on a single 3D face represented by a few 2D images. Given our model's training on 2D facial

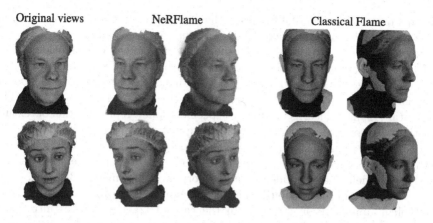

Fig. 3. Competition between NeRFlame and classical Flame fitting. NeRF-based model better fits human expression of the face.

images, we naturally refer to Static Neural Radiance Fields. However, our ability to adapt NeRF is noteworthy due to the utilization of Flame as a backbone. Consequently, in our exploration of related works, we incorporate considerations for Dynamic Neural Radiance Fields.

Static Neural Radiance Fields. 3D objects can be represented by using many different approaches, including voxel grids [10], octrees [19], multi-view images [3,27], point clouds [1,34,41], geometry images [35], deformable meshes [17,25], and part-based structural graphs [24].

The above representations are discreet, which causes some problems in real-life applications. In contrast to such apprehension, NeRF [29] represents a scene using a fully-connected architecture. NeRF and many generalizations [7,8,26,31, 32,36,39] synthesize novel views of a static scene using differentiable volumetric rendering.

One of the largest limitations is training time. To solve such problems in [14], authors propose Plenoxels, a method that uses a sparse voxel grid storing density and spherical harmonics coefficients at each node. The final color is the composition of tri-linearly interpolated values of each voxel. In [30], authors use a similar approach, but the space is divided into an independent multilevel grid. In [9], authors represent a 3D object as an orthogonal tensor component. A small MLP network, which uses orthogonal projection on tensors, obtains the final color and density. Some methods use additional information to NeRF, like depth maps or point clouds [6,12,32,40].

In our paper, we produce a new NeRF-based representation of 3D objects. As input, we use classical 2D images. However, RGB colors and volume density are conditioned by distance to Flame mesh.

Dynamic Neural Radiance Fields. Current solutions for implicit reprehension of human face avatars are trained on image sequences. We assume that we have external tools for segmenting frames in the movie. We often use additional information like each frame's camera angle or Flame representation.

In [15], authors implicitly model the facial expressions by conditioning the NeRF with the global expression code obtained from 3DMM tracking [37]. In [44], authors leverage the idea of dynamic neural radiance fields to improve the mouth region's rendering, which is not represented by the face model motion prior. The IMAvatar [42] model learns the subject-specific implicit representation of texture together with expression. In [16] authors use neural graphics primitives, where for each of the blend shapes, a multi-resolution grid is trained. In RigNeRF [5] authors propose a model that changes head pose and facial expressions using a deformation field that is guided by a 3D morphable face model (3DMM).

Diverging from the previously mentioned approach, certain researchers adopt an explicit apprehension strategy. In the case of [2], the authors introduced ClipFace, a method facilitating text-guided editing of textured 3D face models. In [23], a one-shot mesh-based model reconstruction is presented, while in [11,45], a model is proposed that draws upon a blend of 2D and 3D datasets.

Our method is situated between the above approaches. Similarly to the explicit approach, we use a single 3D object instead of movies to train. We also use Flame mesh to edit the avatar's shape end expressions. On the other hand, we use an implicit representation of the colors of objects.

3   NeRFlame: Flame-Based Conditioning of NeRF for 3D Face Rendering

In this subsection, we introduce NeRFlame - the novel 3D face representation that combines the benefits of Flame and NeRF models. We first provide the details about the Flame and NeRF approaches and further describe the concept of NeRFlame and how it can be used to control face mesh.

3.1   Flame

Flame (Faces Learned with an Articulated Model and Expressions) [25] is a 3D facial model trained from thousands of accurately aligned 3D scans. The model is factored in that it separates the representation of identity, pose, and facial expression, similar to the human body approach. It is represented by low polygon count, articulation, and blend skinning that is computationally efficient, compatible with existing game and rendering engines, and simple in order to maintain its practicality. The parameters of the model are trained by optimizing the reconstruction loss, assuming a detailed temporal registration of our template mesh with three unconnected components, including the base face and two eyeballs.

Formally, the Flame is a function from human face parametrization $\mathcal{F}_{Flame}(\beta, \psi, \phi)$ where β, ψ and ϕ describe shape, expression, and pose parameters to a mesh with n vertices:

$$\mathcal{F}_{Flame}(\beta, \psi, \phi) : \mathbb{R}^{k_\beta \times k_\psi \times k_\phi} \to \mathbb{R}^{n \times 3},$$

Original viwes   New renders obtain by NeRFlame

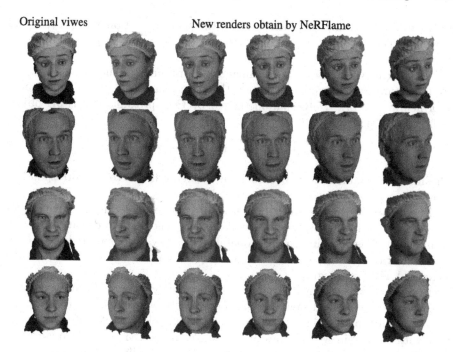

Fig. 4. Reconstruction of 3D object obtained by NeRFlame. As we can see, NeRFlame model the detailed appearance of the 3D face.

where k_β, k_ψ, and k_ϕ are the numbers of shape, expression, and pose parameters. In the classical version, we can fit our model to 3D scans or 2D images by using facial landmarks. Many strategies exist to choose landmarks and parameters for its training [25]. However, in the high level of generalization for input I – image 3D scan (or 2D image) and arbitrarily chosen method for estimation facial landmark points $\mathcal{LP}$ we minimize L2 distance

$$\min_{(\beta,\psi,\phi)} \|\mathcal{LP}(\mathcal{F}_{Flame}(\beta,\psi,\phi)) - \mathcal{LP}(I)\|_2$$

Such an approach is effective, but there are a few limitations. Localizing landmarks and choosing which parameters to optimize first is not trivial. On the other hand, for 2D images, the results are not well-qualified. We can use a pre-trained auto-encoder-based model DECA [13] for face reconstruction from 2D images to solve such problems.

In this paper, we train the Flame-based model in a NeRF-based scenario. As input, we take a few 2D images. As an effect, we obtain a correctly fitted Flame model and NeRF rendering model for new views.

3.2   NeRF

NeRF Representation of 3D Objects. NeRFs [29] are the model for representing complex 3D scenes using neural architectures. In order to do that, NeRFs take a 5D coordinate as input, which includes the spatial location $\mathbf{x} = (x, y, z)$ and viewing direction $\mathbf{d} = (\theta, \Psi)$ and returns emitted color $\mathbf{c} = (r, g, b)$ and volume density σ.

A classical NeRF uses a set of images for training. In such a scenario, we produce many rays traversing through the image and a 3D object represented by a neural network. NeRF parameterized by Θ approximates this 3D object with an MLP network:

$$\mathcal{F}_{NeRF}(\mathbf{x}, \mathbf{d}; \Theta) = (\mathbf{c}, \sigma).$$

The model is trained to map each input 5D coordinate to its corresponding volume density and directional emitted color.

The loss of NeRF is inspired by classical volume rendering [20]. We render the color of all rays passing through the scene. The volume density $\sigma(\mathbf{x})$ can be interpreted as the differential probability of a ray. The expected color $C(\mathbf{r})$ of camera ray $\mathbf{r}(t) = \mathbf{o} + td$ (where $\mathbf{o}$ is ray origin and d is direction) can be computed with an integral.

In practice, this continuous integral is numerically estimated using a quadrature. We use a stratified sampling approach where we partition our ray $[t_n, t_f]$ into N evenly-spaced bins and then draw one sample t_i uniformly at random from within each bin. We use these samples to estimate $C(\mathbf{r})$ with the quadrature rule [28], where $\delta_i = t_{i+1} - t_i$ is the distance between adjacent samples:

$$\hat{C}(\mathbf{r}) = \sum_{i=1}^{N} T_i(1 - \exp(-\sigma_i \delta_i))\mathbf{c}_i, \text{ where } T_i = \exp\left(-\sum_{j=1}^{i-1} \sigma_j \delta_j\right),$$

This function for calculating $\hat{C}(\mathbf{r})$ from the set of $(\mathbf{c_i}, \sigma_i)$ values is trivially differentiable.

We then use the volume rendering procedure to render the color of each ray from both sets of samples. Contrary to the baseline NeRF [29], where two "coarse" and "fine" models were simultaneously trained, we use only the "coarse" architecture. Our loss is simply the total squared error between the rendered and true pixel colors

$$\mathcal{L} = \sum_{r \in R} \|\hat{C}(\mathbf{r}) - C(\mathbf{r})\|_2^2 \qquad (1)$$

where R is the set of rays in each batch, and $C(\mathbf{r})$, $\hat{C}(\mathbf{r})$ are the ground truth and predicted RGB colors for ray $\mathbf{r}$ respectively.

3.3   NeRFlame

We introduce NeRFlame the 3D face model that combines the benefits of mesh representations from Flame and NeRF implicit representation of 3D objects. Thanks to the application of NeRFs, we can estimate the parameters of Flame directly from 2D images without using landmarks points. On the other hand, we obtain NeRF model, which can be editable similarly to Flame. In order to achieve that, we introduce the function that approximates the volume density using the Flame model.

Consider the distance function $d(\mathbf{x}, \mathcal{M})$ between point $\mathbf{x} = (x, y, z) \in \mathbb{R}^3$ and the mesh $\mathcal{M} := \mathcal{M}_{\beta,\psi,\phi}$ created by $F_{Flame}(\beta, \psi, \phi)$, with parameters β, ψ, ϕ. Note, that edges between vertices in Flame model can be directly taken from the template mesh (see [4] for details). We define the volume density function as:

$$\sigma(\mathbf{x}, \mathcal{M}) = \begin{cases} 0, & \text{if } d(\mathbf{x}, \mathcal{M}) > \varepsilon \\ (1 - \frac{1}{\varepsilon} d(\mathbf{x}, \mathcal{M})), & \text{otherwise,} \end{cases} \tag{2}$$

where ε is a hyperparameter that defines the neighborhood of the mesh surface. In practice, the values of the density volume function are non-zero only in the close neighborhood of the mesh.

The NeRFlame can be represented by the function:

$$\mathcal{F}_{NeRFlame}(\mathbf{x}; \beta, \psi, \phi, \Theta) = (\mathcal{F}_{\Theta}^{c}(\mathbf{x}), \sigma(\mathbf{x}, \mathcal{M})), \tag{3}$$

where $\mathcal{F}_{\Theta}^{c}$ is the MLP that predicts the color, similar to the NeRF model.

The model is trained in an end-to-end manner directly, optimizing the criterion (1) with respect to the MLP parameters Θ, and Flame parameters β, ψ, ϕ, which describe shape, expression, and pose. In NeRFlame, we utilize the original loss function used to train NeRF models. Therefore, the structure of colors on rays must be consistent. During training, the model modified the mesh structure to be consistent with the 3D structure of the target face. The simultaneous neural network $\mathcal{F}$ produces colors for the rendering procedure.

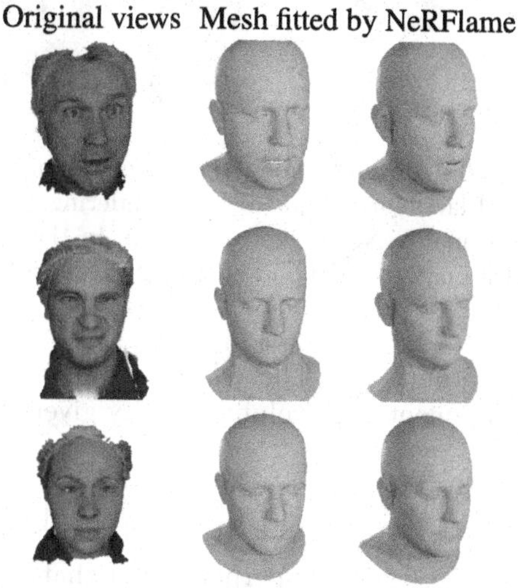

During training, we can see the trade-off between the level of mesh fitting and the quality of renders. The main reason is that we must train our model with small ε, and color must be encoded in a small neighborhood of the mesh. There-

Original views Mesh fitted by NeRFlame

Fig. 5. During the training of NeRFlame, we simultaneously model Flame mesh and NeRf dedicated to colors. In the above figure, we present the meshes fitted by NeRFlame.

fore, the quality of the model is lower than classical NeRF. With larger ε we obtain mesh which is not correctly fitted.

Therefore, we use the MLP $\mathcal{F}_{\Theta}^{\sigma}$ that predicts the volume density analogical to NeRF. In the first 10000 epoch we train our model with volume density given

Original position   Yawning

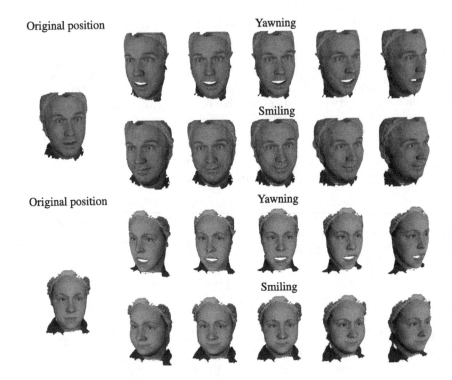

                                                    Smiling

Original position   Yawning

                                                    Smiling

Fig. 6. Our model allows producing manipulation of the human face. In NeRFlame, we use Flame for volume density rendering. Therefore, we can manipulate Flame features and modify NeRF representation. In the figure, we show two faces and their versions with open mouths and changing expressions from different views.

by formula (2). Then, Flame parameters are frozen, and we train only NeRF component with volume density given by

$$\sigma(\mathbf{x}, \mathcal{M}, \mathcal{F}_\Theta^\sigma) = \begin{cases} 0, & \text{if } d(\mathbf{x}, \mathcal{M}) > \varepsilon \\ \mathcal{F}_\Theta^\sigma(\mathbf{x}), & \text{otherwise,} \end{cases}$$

where ε is a hyperparameter that defines the neighborhood of the mesh surface and F_Θ^σ is MLP that predicts volume density. The value of ε increases gradually over time during training after 10 000 epochs. The value of ε increases up to $\varepsilon = 0.1$ at the end of the training procedure.

3.4   Controlling NeRF Models to Obtain Face Manipulation

The classical NeRF method is known to generate highly detailed and realistic images. However, it can be challenging to manipulate NeRF models to achieve precise facial modifications. Several techniques, such as generative models, dynamic scene encoding, and conditioning mechanisms, have been proposed to address this challenge. Nonetheless, controlling NeRF models to the same extent as mesh representations remains elusive. In contrast, the Flame is a straightforward model with three parameters, namely β, ψ, and ϕ, representing shape, expression, and pose, respectively. By performing simple linear operations on these parameters, it is possible to rotate the avatar, change facial expressions, and adjust facial features to a certain degree.

Our NeRFlame is built on the Flame model so that we can manipulate our Flame model to control density prediction σ. However, accurately predicting the RGB colors of the modified object is challenging. To solve this problem, we use a simple technique. To predict color after transformation, we return to the initial position where the color is known, see Fig. 2.

Let us consider NeRFlame model, which is already trained. We have parameters $\beta_1, \psi_1, \phi_1, \Theta, \varepsilon$, function

$$\mathcal{F}_{NeRFlame}(\mathbf{x}; \beta_1, \psi_1, \phi_1, \Theta)$$

and fitted mesh $\mathcal{M}_1$, created by the Flame from parameters β_1, ψ_1, ϕ_1. Let us assume that we apply some modification of Flame parameters, which means that we obtain new β_2, ψ_2, ϕ_2. Using these parameters, we can create the modified mesh $\mathcal{M}_2$, simply using the Flame model. Instead of retraining the NeRF model for new parameters requiring the 2D images for a new pose, we propose applying a simple transformation between the modified and original space.

Let's take a mesh $\mathcal{M}_2$ representing the new pose. We postulate using the affine transformation $T(\cdot)$, which transforms the point $\mathbf{x}_2$ on the mesh $\mathcal{M}_2$ to the original pose: $T(\mathbf{x}_2) = \mathbf{x}_1$, where $\mathbf{x}_1 \in \mathcal{M}_1$ is the corresponding point on original pose. After the transformation, we can identify an element in the original pose $\mathcal{M}_1$ for each element on the mesh $\mathcal{M}_2$. In practice, finding the transformation T is not a trivial task since it depends on the local transformation of the mesh.

However, for a given point $\mathbf{x}_1 \in \mathcal{M}_1$, we can find a face triangle described by vertices $(\mathbf{q}_1, \mathbf{q}_2, \mathbf{q}_3) \in V_1$ that contains $\mathbf{x}_1$, where V_1 is the set of vertices of the mesh $\mathcal{M}_1$. Because Flame model is shifting the vertices of the model keeping the connections unchanged, we can locate the corresponding triangle $(\mathbf{p}_1, \mathbf{p}_2, \mathbf{p}_3) \in V_2$ of the mesh $\mathcal{M}_2$. For each of the triangles that create the mesh, we define affine transformation:

$$T(\mathbf{p}_i) = \mathbf{q}_i, \text{ for } i = 1, 2, 3.$$

In such a situation, we assume that such transformations $T(\cdot)$ are affine, and we can use Least-Squares Conformal Multilinear Regression [33] to estimate the parameters. Practically, finding the transformation for each of the triangles is extremely fast, requires inverting a fourth-dimensional matrix, and can be parallelized. Having the parameters of transformations estimated, we can apply them directly to the $\mathbf{x}$ in NeRFlame model given by formula (3) and calculate the colors as in an unshifted pose.

Our approach is sensitive to some particular facial manipulations. When we open the mouth of our avatar, we obtain artifacts. Three main reasons cause such problems. First, it is difficult to fit the mouth around the mesh to images since it is very sensitive to perturbations. Additionally, the mesh lacks internal content, and it cannot represent the inside of the mouth and tongue. The third problem is that rays go through the open mouth, cut the mesh back of the head, and render some artifacts.

To solve such a problem, we remove rays through an open-mouth region. Such a solution is simple to implement since we can easily filter rays that do not cross mesh. On the other hand, allows for reducing most of the artifacts.

4   Experiments

In this section, we describe the experimental results of the proposed model. To our knowledge, it is the first model that obtains editable NeRF trains on a single object in one position. Most of the methods use movies to encode many different positions of the face. We can produce novel views in training positions and in modified facial expressions using knowledge only from one fixed position. Therefore, it is hard to compare our results to other algorithms. In the first subsection, we show that our model produces high-quality NeRF representations of the objects by comparing our model with our baseline classical NeRF and classical textured Flame. In the second subsection, we present meshes obtained by our model. Finally, we show that our model allows facing manipulations.

Table 1. Comparison of PSNR, SSIM, and LPIPS matrices between our model and NeRF and Flame baselines. While NeRF achieves better results, it lacks manipulation capabilities. In contrast, the Flame model produces inferior outcomes as it is trained solely on landmark points for mesh and texture.

	PSNR ↑			SSIM ↑			LPIPS ↓		
	NeRF	NeRFlame	Flame	NeRF	NeRFlame	Flame	NeRF	NeRFlame	Flame
Face 1	33.37	27.89	9.67	0.96	0.95	0.76	0.05	0.09	0.26
Face 2	33.39	29.79	12.44	0.96	0.96	0.80	0.05	0.06	0.24
Face 3	33.08	29.70	12.97	0.97	0.95	0.82	0.04	0.08	0.20
Face 4	31.96	25.78	12.51	0.96	0.92	0.79	0.04	0.10	0.23
Face 5	33.15	32.59	11.30	0.96	0.96	0.77	0.05	0.05	0.26
Face 6	32.42	29.18	11.45	0.96	0.95	0.76	0.06	0.07	0.26

Since the current literature does not provide suitable data sets for evaluating the NeRF-based model for modeling 3D face avatars, we created a data set using 3D scenes. We create a classical NeRF training data set. We construct 200 × 200 transparent background images from random positions.

4.1   Reconstruction Quality

In this subsection, we show that NeRFlame can reconstruct a 3D human face with similar quality as classical NeRF. Since we train our model on a single position, it is difficult to compare our model to dynamic neural radiance fields. Therefore we show that our model has a slightly lower quality than classical NeRF but allows dynamic modification. On the other hand, we show that we obtain better results than textured Flame, which cannot capture the geometry and appearance details of the human face.

In Fig. 3, we compare NeRFlame and textured Flame. As we can see, NeRFlame can reproduce facial features and geometry. On the other hand, Flame produces well-suited textures, but the mesh is not well-suited. In Table 1, we present a numerical comparison. We compare the metric reported by NeRF called PSNR (*peak signal-to-noise ratio*), SSIM (*structural similarity index measure*), LPIPS (*learned perceptual image patch similarity*) used to measure image reconstruction effectiveness. As we can see NeRF gives essentially better since do not allow manipulation. On the other hand, Flame model gives inferior results since we train mesh and texture only on landmark points. In Fig. 4, we present new renders of the model obtained by NeRFlame. As we can see, NeRFlame model the detailed appearance of the 3D face.

4.2   Mesh Fitting

The RGB colors generated by NeRF are present only in the ε vicinity of the mesh. This approach allows for a precise fitting of the mesh to the human face, which is critical for generating animated models. In Fig. 5, we present the rendered faces and corresponding meshes produced by our NeRFlame approach. The results demonstrate that our method can accurately capture the underlying mesh structure.

4.3   Face Manipulation

Our approach enables the manipulation of human facial features. Leveraging Flame as a backbone, NeRFlame offers the ability to manipulate Flame features and modify NeRF representations. In Fig. 6, we showcase three faces and their modification including open mouths and changing expressions, that can be manipulated using our model in a manner similar to the classical Flame model.

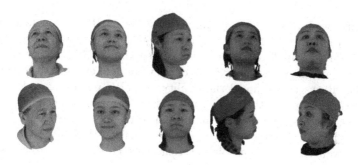

Fig. 7. The renders obtained by NeRFlame on MoFaNeRF dataset.

NeRFlame simultaneously train mesh and NeRF components for color. After training, we can exchange the produced mesh in our model to obtain a modification of the final look of the avatar, see Fig. 1.

4.4   Comparison with MoFaNeRF [43]

Comparison between MoFaNeRF [43] and NeRFlame is not possible directly. MoFaNeRF is trained on a large dataset. MoFaNeRF is a generative model witch ability to control face position and expression. NeRFlame is a classical NeRF-based model trained individually on each element of the dataset separately.

This experiment shows that our NeRFlame reconstruct objects with a PSNR value similar to that of MoFaNeRF. In Fig. 7, we present faces trained on elements from the MoFaNeRF dataset. In Table 2, we present numerical experiments, but it should be noted that the models were trained entirely differently on different parts of the data set. It only states that we have similar render quality.

Table 2. Quantitative evaluation of representation ability. Results are from MoFaNeRF experiments [43]. * Comparison between MoFaNeRF [43] and NeRFlame is not possible directly. It should be highlighted that models were trained entirely differently on different dataset parts. It only states that we have similar-quality of renders.

Model	PSNR(dB)	SSIM	LPIPS
FaceScape	27.96 ± 1.34	0.932 ± 0.012	0.069 ± 0.009
i3DMM	24.45 ± 1.58	0.904 ± 0.014	0.112 ± 0.015
MoFaNeRF	31.49 ± 1.75	0.951 ± 0.010	0.061 ± 0.011
MoFaNeRF-fine	30.17 ± 1.71	0.935 ± 0.013	0.034 ± 0.007
NeRFlame*	29.57	0.955	0.060

5   Conclusions

In this work, we introduce a novel approach called NeRFlame, which combines NeRF and Flame to achieve high-quality rendering and precise pose control. While NeRF-based models use neural networks to model RGB colors and volume density, our method utilizes an explicit density volume represented by the Flame mesh. This allows us to model the quality of NeRF rendering and accurately control the appearance of the final output. As a result of offering complete control over the model, the quantitative performance of our approach is marginally inferior to that of a static NeRF model. We believe that future work should prioritize advancements in mesh fitting techniques. By doing so, we can maximize the potential for extensive modifications.

Acknowledgments. The work of P. Spurek was supported by the National Centre of Science (Poland) Grant No. 2021/43/B/ST6/01456. The work of M. Zieba was supported by National Science Centre of Poland, grant no 2020/37/B/ST6/03463. We gratefully acknowledge Polish high-performance computing infrastructure PLGrid (HPC Centers: ACK Cyfronet AGH) for providing computer facilities and support within computational grant no. PLG/2023/016357. The research of P. Borycki was funded by the program Excellence Initiative-Research University at the Jagiellonian University in Kraków.

References

1. Achlioptas, P., Diamanti, O., Mitliagkas, I., Guibas, L.: Learning representations and generative models for 3D point clouds. In: International Conference on Machine Learning, pp. 40–49. PMLR (2018)
2. Aneja, S., Thies, J., Dai, A., Nießner, M.: ClipFace: text-guided editing of textured 3D morphable models. arXiv preprint arXiv:2212.01406 (2022)
3. Arsalan Soltani, A., Huang, H., Wu, J., Kulkarni, T.D., Tenenbaum, J.B.: Synthesizing 3D shapes via modeling multi-view depth maps and silhouettes with deep generative networks. In: Proceedings of the IEEE Conference on Computer Vision and Pattern Recognition, pp. 1511–1519 (2017)
4. Athar, S., Shu, Z., Samaras, D.: FLAME-in-NeRF: neural control of radiance fields for free view face animation. arXiv preprint arXiv:2108.04913 (2021)
5. Athar, S., Xu, Z., Sunkavalli, K., Shechtman, E., Shu, Z.: RigNeRF: fully controllable neural 3D portraits. In: Proceedings of the IEEE/CVF Conference on Computer Vision and Pattern Recognition, pp. 20364–20373 (2022)
6. Azinović, D., Martin-Brualla, R., Goldman, D.B., Nießner, M., Thies, J.: Neural RGB-D surface reconstruction. In: Proceedings of the IEEE/CVF Conference on Computer Vision and Pattern Recognition, pp. 6290–6301 (2022)
7. Barron, J.T., Mildenhall, B., Tancik, M., Hedman, P., Martin-Brualla, R., Srinivasan, P.P.: Mip-NeRF: a multiscale representation for anti-aliasing neural radiance fields. In: Proceedings of the IEEE/CVF International Conference on Computer Vision, pp. 5855–5864 (2021)
8. Barron, J.T., Mildenhall, B., Verbin, D., Srinivasan, P.P., Hedman, P.: Mip-NeRF 360: unbounded anti-aliased neural radiance fields. In: Proceedings of the IEEE/CVF Conference on Computer Vision and Pattern Recognition, pp. 5470–5479 (2022)
9. Chen, A., Xu, Z., Geiger, A., Yu, J., Su, H.: TensoRF: tensorial radiance fields. In: Avidan, S., Brostow, G., Cissé, M., Farinella, G.M., Hassner, T. (eds.) ECCV 2022. LNCS, vol. 13692, pp. 333–350. Springer, Cham (2022). https://doi.org/10.1007/978-3-031-19824-3_20
10. Choy, C.B., Xu, D., Gwak, J.Y., Chen, K., Savarese, S.: 3D-R2N2: a unified approach for single and multi-view 3D object reconstruction. In: Leibe, B., Matas, J., Sebe, N., Welling, M. (eds.) ECCV 2016. LNCS, vol. 9912, pp. 628–644. Springer, Cham (2016). https://doi.org/10.1007/978-3-319-46484-8_38
11. Daněček, R., Black, M., Bolkart, T.: EMOCA: emotion driven monocular face capture and animation. In: 2022 IEEE/CVF Conference on Computer Vision and Pattern Recognition (CVPR), pp. 20279–20290. IEEE (2022)
12. Deng, K., Liu, A., Zhu, J.Y., Ramanan, D.: Depth-supervised NeRF: fewer views and faster training for free. In: Proceedings of the IEEE/CVF Conference on Computer Vision and Pattern Recognition, pp. 12882–12891 (2022)

13. Feng, Y., Feng, H., Black, M.J., Bolkart, T.: Learning an animatable detailed 3D face model from in-the-wild images. ACM Trans. Graph. (ToG) **40**(4), 1–13 (2021)
14. Fridovich-Keil, S., Yu, A., Tancik, M., Chen, Q., Recht, B., Kanazawa, A.: Plenoxels: radiance fields without neural networks. In: Proceedings of the IEEE/CVF Conference on Computer Vision and Pattern Recognition, pp. 5501–5510 (2022)
15. Gafni, G., Thies, J., Zollhofer, M., Nießner, M.: Dynamic neural radiance fields for monocular 4D facial avatar reconstruction. In: Proceedings of the IEEE/CVF Conference on Computer Vision and Pattern Recognition, pp. 8649–8658 (2021)
16. Gao, X., Zhong, C., Xiang, J., Hong, Y., Guo, Y., Zhang, J.: Reconstructing personalized semantic facial nerf models from monocular video. ACM Trans. Graph. (TOG) **41**(6), 1–12 (2022)
17. Girdhar, R., Fouhey, D.F., Rodriguez, M., Gupta, A.: Learning a predictable and generative vector representation for objects. In: Leibe, B., Matas, J., Sebe, N., Welling, M. (eds.) ECCV 2016. LNCS, vol. 9910, pp. 484–499. Springer, Cham (2016). https://doi.org/10.1007/978-3-319-46466-4_29
18. Grassal, P.W., Prinzler, M., Leistner, T., Rother, C., Nießner, M., Thies, J.: Neural head avatars from monocular RGB videos. In: Proceedings of the IEEE/CVF Conference on Computer Vision and Pattern Recognition, pp. 18653–18664 (2022)
19. Häne, C., Tulsiani, S., Malik, J.: Hierarchical surface prediction for 3D object reconstruction. In: 2017 International Conference on 3D Vision (3DV), pp. 412–420. IEEE (2017)
20. Kajiya, J.T., Von Herzen, B.P.: Ray tracing volume densities. ACM SIGGRAPH Comput. Graph. **18**(3), 165–174 (1984)
21. Kania, A., Kasymov, A., Ziba, M., Spurek, P.: HyperNeRFGAN: hypernetwork approach to 3D NeRF GAN. arXiv preprint arXiv:2301.11631 (2023)
22. Kania, K., Yi, K.M., Kowalski, M., Trzciński, T., Tagliasacchi, A.: CoNeRF: controllable neural radiance fields. In: 2022 IEEE/CVF Conference on Computer Vision and Pattern Recognition (CVPR), pp. 18602–18611. IEEE (2022)
23. Khakhulin, T., Sklyarova, V., Lempitsky, V., Zakharov, E.: Realistic one-shot mesh-based head avatars. In: Avidan, S., Brostow, G., Cissé, M., Farinella, G.M., Hassner, T. (eds.) ECCV 2022, Part II. LNCS, vol. 13662, pp. 345–362. Springer, Cham (2022). https://doi.org/10.1007/978-3-031-20086-1_20
24. Li, J., Xu, K., Chaudhuri, S., Yumer, E., Zhang, H., Guibas, L.: GRASS: generative recursive autoencoders for shape structures. ACM Trans. Graph. (TOG) **36**(4), 1–14 (2017)
25. Li, T., Bolkart, T., Black, M.J., Li, H., Romero, J.: Learning a model of facial shape and expression from 4D scans. ACM Trans. Graph. **36**(6), 1–17, 194 (2017)
26. Liu, L., Gu, J., Zaw Lin, K., Chua, T.S., Theobalt, C.: Neural sparse voxel fields. In: Advances in Neural Information Processing Systems, vol. 33, pp. 15651–15663 (2020)
27. Liu, Z., Zhang, Y., Gao, J., Wang, S.: VFMVAC: view-filtering-based multi-view aggregating convolution for 3D shape recognition and retrieval. Pattern Recogn. **129**, 108774 (2022)
28. Max, N.: Optical models for direct volume rendering. IEEE Trans. Visual Comput. Graphics **1**(2), 99–108 (1995)
29. Mildenhall, B., Srinivasan, P.P., Tancik, M., Barron, J.T., Ramamoorthi, R., Ng, R.: NeRF: representing scenes as neural radiance fields for view synthesis. In: Vedaldi, A., Bischof, H., Brox, T., Frahm, J.-M. (eds.) ECCV 2020. LNCS, vol. 12346, pp. 405–421. Springer, Cham (2020). https://doi.org/10.1007/978-3-030-58452-8_24

30. Müller, T., Evans, A., Schied, C., Keller, A.: Instant neural graphics primitives with a multiresolution hash encoding. ACM Trans. Graph. (ToG) **41**(4), 1–15 (2022)
31. Niemeyer, M., Barron, J.T., Mildenhall, B., Sajjadi, M.S., Geiger, A., Radwan, N.: RegNeRF: regularizing neural radiance fields for view synthesis from sparse inputs. In: Proceedings of the IEEE/CVF Conference on Computer Vision and Pattern Recognition, pp. 5480–5490 (2022)
32. Roessle, B., Barron, J.T., Mildenhall, B., Srinivasan, P.P., Nießner, M.: Dense depth priors for neural radiance fields from sparse input views. In: Proceedings of the IEEE/CVF Conference on Computer Vision and Pattern Recognition, pp. 12892–12901 (2022)
33. Schmid, K.K., Marx, D.B., Samal, A.: Tridimensional regression for comparing and mapping 3D anatomical structures. Anat. Res. Int. **2012** (2012)
34. Shu, D.W., Park, S.W., Kwon, J.: Wasserstein distributional harvesting for highly dense 3D point clouds. Pattern Recogn. **132**, 108978 (2022)
35. Sinha, A., Bai, J., Ramani, K.: Deep learning 3D shape surfaces using geometry images. In: Leibe, B., Matas, J., Sebe, N., Welling, M. (eds.) ECCV 2016. LNCS, vol. 9910, pp. 223–240. Springer, Cham (2016). https://doi.org/10.1007/978-3-319-46466-4_14
36. Tancik, M., et al.: Block-NeRF: scalable large scene neural view synthesis. In: Proceedings of the IEEE/CVF Conference on Computer Vision and Pattern Recognition, pp. 8248–8258 (2022)
37. Thies, J., Zollhofer, M., Stamminger, M., Theobalt, C., Niessner, M.: Face2Face: real-time face capture and reenactment of RGB videos. In: 2016 IEEE Conference on Computer Vision and Pattern Recognition (CVPR), pp. 2387–2395. IEEE Computer Society (2016)
38. Trzciński, T.: Points2NeRF: generating neural radiance fields from 3D point cloud. arXiv preprint arXiv:2206.01290 (2022)
39. Verbin, D., Hedman, P., Mildenhall, B., Zickler, T., Barron, J.T., Srinivasan, P.P.: Ref-NeRF: structured view-dependent appearance for neural radiance fields. In: 2022 IEEE/CVF Conference on Computer Vision and Pattern Recognition (CVPR), pp. 5481–5490. IEEE (2022)
40. Wei, Y., Liu, S., Rao, Y., Zhao, W., Lu, J., Zhou, J.: NerfingMVS: guided optimization of neural radiance fields for indoor multi-view stereo. In: Proceedings of the IEEE/CVF International Conference on Computer Vision, pp. 5610–5619 (2021)
41. Yang, F., Davoine, F., Wang, H., Jin, Z.: Continuous conditional random field convolution for point cloud segmentation. Pattern Recogn. **122**, 108357 (2022)
42. Zheng, Y., Fernández Abrevaya, V., Bühler, M., Chen, X., Black, M.J., Hilliges, O.: IM avatar: implicit morphable head avatars from videos. In: 2022 IEEE/CVF Conference on Computer Vision and Pattern Recognition (CVPR), pp. 13535–13545. IEEE (2022)
43. Zhuang, Y., Zhu, H., Sun, X., Cao, X.: MoFaNeRF: morphable facial neural radiance field. In: Avidan, S., Brostow, G., Cissé, M., Farinella, G.M., Hassner, T. (eds.) ECCV 2022, vol. 13663, pp. 268–285. Springer, Cham (2022). https://doi.org/10.1007/978-3-031-20062-5_16
44. Zielonka, W., Bolkart, T., Thies, J.: Instant volumetric head avatars. arXiv preprint arXiv:2211.12499 (2022)
45. Zielonka, W., Bolkart, T., Thies, J.: Towards metrical reconstruction of human faces. In: Avidan, S., Brostow, G., Cissé, M., Farinella, G.M., Hassner, T. (eds.) ECCV 2022, Part XIII. LNCS, vol. 13673, pp. 250–269. Springer, Cham (2022). https://doi.org/10.1007/978-3-031-19778-9_15

Dynamic Growing and Shrinking of Neural Networks with Monte Carlo Tree Search

Szymon Świderski and Agnieszka Jastrzębska[✉]

Warsaw University of Technology, Warsaw, Poland
A.Jastrzebska@mini.pw.edu.pl

Abstract. The issue of data-driven neural network model construction is one of the core problems in the domain of Artificial Intelligence. A standard approach assumes a fixed architecture with trainable weights. A conceptually more advanced assumption is that we not only train the weights, but also find out the optimal model architecture. In this paper, we present a new method that realizes just that. We show how to create a neural network with a procedure that allows dynamic shrinking and growing of the model while it is being trained. The decision-making mechanism for the architectural design is governed by a Monte Carlo tree search procedure which simulates network behavior and allows to compare several candidate architecture changes to choose the best one. The solution utilizes a Stochastic Gradient Descent-based optimizer developed from scratch to realize the task of network architecture modification. The paper is accompanied with a Python source code of the prepared method. The proposed approach was tested in visual pattern classification problems and yielded highly satisfying results.

Keywords: neural network · changing architecture · training · Monte Carlo tree search · shrinking · growing · Stochastic Gradient Descent

1 Introduction

Neural network training algorithms development is an essential theoretical and practical problem of Artificial Intelligence. The underlying task is network architecture design. Typical methods assume that a programmer specifies subsequent components of the network and uses an optimization algorithm of choice to find out weight values. At the same time, there is a pressing need to deliver effective methods that relieve programmers from the task of neural network architecture specification. One trend is to use predefined designs, tested by others on some benchmark datasets. A more conceptually advanced scenario is to offer training algorithms that optimize the network architecture during the training procedure. In this scenario, the training algorithm is responsible not only for setting the weights but also for modifying the model design.

This research was supported by the Warsaw University of Technology within the Excellence Initiative: Research University (IDUB) programme.

L. Franco et al. (Eds.): ICCS 2024, LNCS 14832, pp. 362–377, 2024.
https://doi.org/10.1007/978-3-031-63749-0_25

The idea of delegating neural model design to an optimization algorithm has been present in the literature domain for some time now. Unfortunately, up to this day, the practical use of this approach is quite limited. This is first and foremost due to the modest effectiveness of the available approaches. Most of the existing studies, such as the ones of Zhang et al. [16], were delivered for plain feed-forward neural networks. The demands of contemporary data analysis, especially in the field of image classification, are not matched when such models are used. There are some studies, such as the very recent paper by Evci et al. [5], that offer more insights. The aforementioned paper shows an approach when a network is grown/shrunk neuron-by-neuron. Comprehensive theoretical and empirical studies are overall rare.

In light of the facts outlined in the previous paragraph, in this paper, we contribute a novel method for neural network training. We are publishing it as an open-source package. We chose the name *growingnn*. It was uploaded to PyPi. Its description is under https://pypi.org/project/growingnn/. The delivered method uses error backpropagation as a base for weight update. The action of architecture design is carried out by enabling a scheduler that after each K epochs allows the neural architecture to change. The change can be realized by adding or removing a layer of a predefined type from the network. The current implementation covers three types of neural layers: (i) a plain, feed-forward layer, (ii) a convolutional layer, and (iii) a layer with residual connections. These three layer types are typical in contemporary advanced models for visual pattern recognition.

The key novelty of the presented work is the use of Monte Carlo tree search to simulate network performance. We use it to determine the optimal decision with regard to the design change. To achieve that, the neural model evolution strategy is represented as a tree. The network is redesigned in such a way that a change in the structure has a minimal impact on the already-learned weight values. In this manner, instead of offering a trivial wrapper solution that trains from scratch a new model using a library algorithm after an architecture update, we develop a novel optimizer that performs continual (progressive) training and reuse of the already-learned neural connections. Furthermore, the changes to the network architecture concern entire layers, not single neurons.

The proposed approach was empirically evaluated in applied visual pattern recognition problems involving standard benchmark datasets: MNIST [3] and FMNIST [14]. The outcomes of the proposed model were compared with baseline strategies for neural architecture change orchestration: random and greedy approaches. In both cases, the new method involving the Monte Carlo tree search yielded much better results.

The remainder of this paper is structured as follows. Section 2 addresses relevant literature positions in the domain of architecture-changing neural network training methods. Section 3 outlines the theoretical background of our approach. Section 4 shows the results of empirical tests of the new method. Section 5 concludes the paper.

2 Literature Survey

In recent years, the topic of dynamic neural network architecture change has attracted noticeable attention. We shall start this discussion by mentioning the method known as GradMax [5]. It is a method capable of growing a neural architecture during the training procedure without costly retraining. The idea is very similar to the one discussed in this paper, but the methods that handle each change are very different from our methods. GradMax operates on the level of a single neuron. In our algorithm, there is a very wide spectrum of possible changes for a network that allows architecture to grow and shrink. GradMax maximizes gradients for new weights and efficiently initializes them using singular value decomposition (SVD). This approach makes new neurons not impact existing knowledge which is contradictory to our method for which neural network has a short period of instability. The idea of changing the structure using gradient information is relatively common in the literature. One of the first of this kind was a model called resource-allocating network [11]. In this method, when a given pattern was unrecognizable, new neurons were added. An analogous idea was published by Fahlman and Lebiere under the name Cascade-Correlation Architecture [6]. Miconi [10] proposed an approach that uses information about gradient value to adjust the number of layers and layers' size. However, his approach works only for residual levels. Neural networks can grow not only by adding new neurons but also by splitting existing neurons. The article by Kilcher et al. [9] about escaping flat areas via structure modification introduces a new strategy of this kind. When the change of loss is slowing down and the network encourages a flat error surface, the proposed method adds new neurons by splitting the existing ones. This work has two important elements in common with our method. The changes to the structure are made when the network's ability to learn decreases and we also believe that their method of splitting the neuron is in a few ways similar to our method that uses quasi-identity matrices. A very important question in the field of neural networks that can change their structure is how much the neural network can adapt to the problem. A study about Convex Neural Networks [1] shows that these networks can adapt to diverse linear structures by adding neurons in a single hidden layer in each step of training, which also forces the network to grow.

3 The Method

The proposed algorithm consists of two components. The first component performs weight adjustment. The second component is the orchestrator, which launches a procedure to change the network architecture. The change takes place each K epoch and it works in a guided manner. Its decisions are made based on the outcome of the Monte Carlo tree search. A rough outline of the new routine is given in the Algorithm 1.

The algorithm is fundamentally built upon Stochastic Gradient Descent (SGD) [4]. It relies on low-level computations, avoiding elaborate tools used in some contemporary training algorithms. This simplicity makes it ideal for research that focuses on fundamental machine-learning principles. For us, a model is a main structure that stores layers as nodes in a directed graph structure, it operates on the identifiers of layers. The layer is assumed to be an independent structure that contains information about incoming and outgoing connections. The default starting structure is a graph that has an input and output layer and one connection between those. In each generation, a new layer may be added or an existing layer may be removed. As the structure grows, each layer gains more incoming and outgoing connections. In the propagation phase, the layer waits until it receives signals from all input layers. Once received, the signals are averaged, processed, and propagated through all outgoing connections.

Algorithm 1. Training Algorithm

1: **Input:** Dataset
2: **Output:** Model
3: **Initialization:**
4: SimSet ← Create simulation set
5: Model ← Create a model with basic structure
6: **for** each generation **do**
7:     GradientDescent(Model, Dataset, epochs)
8:     **if** canSimulate() **then**
9:         Action ← MCTS.simulation(Model)
10:        Model ← Action.execute(Model)
11:    **end if**
12: **end for**

The function *canSimulate()* called in Line 8 in Algorithm 1 represents a module that is later referred to as *simulation orchestrator*. The orchestrator determines the point in the training procedure when a simulation is executed during the learning process to change a current neural architecture. At the end of each generation, the simulation orchestrator checks if a simulation is needed. The moment at which the simulation is executed is very important because it helps maintain a balance between the exploration and exploitation of potential structures. A model may retain a particular architecture for several epochs, or it may require frequent changes. Too frequent changes may prevent a specific architecture from being fully trained, while infrequent changes may lead to constantly running into local minima and significantly increase learning time, rendering the method inefficient. In the section dedicated to parameter exploration, we examined various approaches, but ultimately, we decided to use a method known as *progress check*. In this method, after each generation, we check whether there has been an improvement in the model's learning. If there is no improvement, then the simulation is run.

In our algorithm, the learning rate plays a crucial role. We implemented a custom modification of the progressive learning rate in line with the work of

Schaul et al. [12]. In this approach, the progressive learning rate ensures that the learning rate is very close to zero in the first epoch after the structure change. Thereafter, the learning rate grows to a constant value through the training process. When the maximum constant value of the learning rate is reached, the learning rate slowly decreases before the next action. This minimizes the negative impact of introduced changes on the already learned information in the network.

3.1 Neural Architecture Design Changes

The algorithm draws inspiration from the achievements of ResNet-50 [8] and ResNeXt [15], particularly from the success of residual connections. In the presented algorithm, the model's structure is treated as a graph, in which layers are nodes and connections between layers are directed edges. The structure built from combining residual and sequential connections shapes a directed acyclic graph, which also functions as a flow network [7]. The algorithm allows adding and removing layers from the model without losing the residual structure of layers. All layers operate asynchronously, and signals move through the network in a manner akin to recursion. Such a structure has key properties, as it allows for unlimited network size, ensures that data always flows through the network without supervision, prevents deadlocks, and prohibits cyclic or unnecessary layers.

In general, a network has a tendency to add new layers, which allows the network to grow and learn new features. Changes to the network structure are added in the form of actions. An action concerns essentially either an addition or a removal of a layer. The algorithm generates all of the possible candidate connections for a given type of layer. Each possible connection for a given layer type in structure defines one single action that can be run on the current model.

The current implementation supports four different types of layers.

1. sequential dense layer;
2. residual dense layer;
3. sequential convolution layer;
4. residual convolution layer.

Dense layers can be connected to and from any layer kind as long as a residual structure is preserved. Convolution layers have a specific rule they can only be added after another convolution layer as input. In our experiments, the initial layer is always convolutional as the model is primarily developed to deal with computer vision tasks.

3.2 Monte Carlo Tree Search

Monte Carlo Tree Search (MCTS) [13] is a search algorithm used in decision processes, particularly in games and simulations. It builds a tree structure by simulating different possible moves, evaluating their outcomes, and expanding the tree. Each iteration in this simulation is divided into four parts.

1. **Selection.** Starting from the root of the tree, one child is selected. The main difficulty is to maintain a balance between exploration and exploitation. This balance is controlled by upper confidence bound applied to trees [2]:

$$a^* = \arg \max_{a \in A(M_s)} \left(Q(M_s, a) + C_s \cdot \sqrt{\frac{\ln N(s)}{N(M_s, a)}} \right). \tag{1}$$

For a given set of actions $A(s)$ generated for a current model structure M_s, the formula selects the action chosen in the child node during the selection. $Q(M_s, a)$ denotes the average result of scores from the rollout phase. $N(s)$ is a number denoting how many times model structure M_s has been analyzed. $N(M_s, a)$ denotes the number how many times action a has been processed for model structure M_s. Initially, the root node consists of a model structure for the current generation.

2. **Expansion.** The algorithm adds new children of a node. It executes all possible actions for a model structure, creating a set of children nodes, each having a different model structure.

3. **Rollout** or playoff. For a given node which is a leaf, in the simulation tree, the algorithm is trying to play a random game. In our adaptation, the rollout executes n random actions on a given model, to simulate future possible changes after a given action. After the rollout, the resulting structure is passed to the score function. The score function trains the resulting structure on the simulation dataset and the resulting accuracy represents the score from the rollout.

4. **Backpropagation.** All nodes are updated according to the score function after the rollout.

MCTS in this implementation is time-limited. After a user-specified time, the simulation returns a single action. The assumption is that the action performed on the current structure should set the stage for subsequent actions to converge toward an optimal structure. In each generation, MCTS identifies changes in the structure toward an optimal configuration. This behavior is analogous to the gradient descent algorithm, which in each epoch determines the change in weight for improvement.

For each structure, the algorithm has seven possible action types to generate

1. **Action: Add a dense sequential layer.** Executing this action adds a sequential layer between two other layers. Unlike residual layers, the sequential layer does not have the ability to determine the initial state of weights in the layer.

2. **Action: Add a dense residual random layer.** A residual layer between two layers does not need to be added between layers that are directly connected. A residual layer can be added between any two layers between which there is a path in the same direction. The initial state of the weights is very important because it determines what will happen to the network's knowledge after adding a new layer.

3. **Action: Add a dense residual "zero" layer.** This action involves adding a residual layer, for which the initial value of all weights is zero. Since the weights are zero, the residual layer should have almost no effect on the network output in the first forward propagation execution after adding this layer. The assumption is that the network may be at a point in space where it cannot move towards the global minimum, as it lacks a dimension in which it could move.

4. **Action: Add a dense residual identity layer.** The idea in adding this layer is that the value of the input layer to this layer is enhanced. The weight matrix in this layer is an identity matrix and the bias values are zero, which means that the output from this layer is the same as the output from the input layer to the newly added layer because it is a residual layer.

5. **Action: Add a sequential convolution layer.** Adding a sequential convolution layer works the same as it was in a dense layer, but convolution layers can be only connected to another convolution layer as input, as output it can be a convolution or dense layer. In the convolution type of actions, there are no predefined weights initial state.

6. **Action: Add a residual convolution layer.** Adding a residual convolution layer works the same as it was in a dense residual layer, with the same constraints as it was with convolution sequential action.

7. **Action: Remove a layer.** Removing a layer cannot change the main principles in the structure, the graph that the layers create must be directed and acyclic. The algorithm allows to removal of any layer except the initial input and output layer of the model. The algorithm may create additional connections to maintain the established structure.

Before executing the method, a default number of neurons in layers, denoted as def_{neu}, can be set. This default value does not force all layers to have the same number of neurons, but rather most of them will align with it. Initially, the output layer will have the number of incoming connections set to def_{neu}, which subsequently influences most layers to adopt this neuron count. This alignment occurs because each added layer adapts to the layers it connects to.

3.3   Training Scheme – The Complete Algorithm

In the presented algorithm, iterations for training the model are divided into generations and then into epochs. In every generation, the structure of the model can change. In every epoch, the model changes its weights. For each generation, the algorithm runs a gradient descent algorithm for a specified number of epochs. While training learning rate changes progressively. For the first and last epoch in a single generation, the learning rate is close to zero. In the middle of training, the learning rate gets to some maximal value, this approach makes it easier for the structure to adapt to new changes in the network. Although the key properties of gradient descent are preserved, there is a big change in data flow in forward and backward propagation. If a layer has more than one input signal, it waits until all information is gathered, after that all input signals are averaged and then processed.

The biggest advantage of the residual structure is that there is no need to supervise the data flow. When a signal is sent to the input layer of the model, it is guaranteed that the signal will travel through the network up to the final output layer and then return to the caller.

The model starts forward propagation by sending a signal to the input layer as a forward signal. Similarly, after calculating the loss, backward propagation starts by sending a signal to the output layer as a backward signal. Forward propagation for a given layer waits until the layer receives inputs from all incoming links. The layer assumes that the received input is in the correct form. When one layer sends the input to another layer, the latter converts it to the desired size. These steps are summarized in the Algorithm 2.

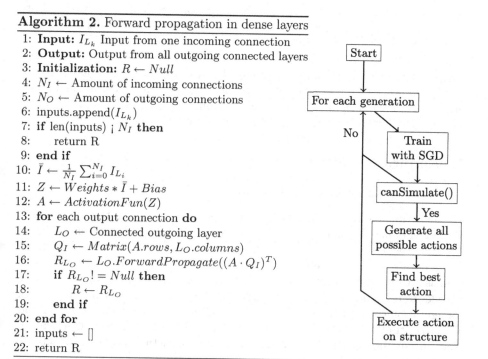

Algorithm 2. Forward propagation in dense layers

1: **Input:** I_{L_k} Input from one incoming connection
2: **Output:** Output from all outgoing connected layers
3: **Initialization:** $R \leftarrow Null$
4: $N_I \leftarrow$ Amount of incoming connections
5: $N_O \leftarrow$ Amount of outgoing connections
6: inputs.append(I_{L_k})
7: **if** len(inputs) ¡ N_I **then**
8:     return R
9: **end if**
10: $\bar{I} \leftarrow \frac{1}{N_I} \sum_{i=0}^{N_I} I_{L_i}$
11: $Z \leftarrow Weights * \bar{I} + Bias$
12: $A \leftarrow ActivationFun(Z)$
13: **for** each output connection **do**
14:     $L_O \leftarrow$ Connected outgoing layer
15:     $Q_I \leftarrow Matrix(A.rows, L_O.columns)$
16:     $R_{L_O} \leftarrow L_O.ForwardPropagate((A \cdot Q_I)^T)$
17:     **if** $R_{L_O}! = Null$ **then**
18:         $R \leftarrow R_{L_O}$
19:     **end if**
20: **end for**
21: inputs $\leftarrow$ []
22: return R

After all input signals are received, the layer calculates output using learned weights and activation function. This output is adjusted to each layer on the outgoing connection by Quasi-Identity matrices (Q_I). A quasi-identity matrix is created to mimic identity matrices as closely as possible, by resizing the identity matrix to fit a specific size, we can quickly adjust a vector to a different size while maintaining its essential characteristics. It is an efficient way to adjust vector size without losing their fundamental features.

At this point in the algorithm, a layer can send input signals to all connected outgoing layers. This happens by running forward propagation recursively on each of the connected layers. Because the structure of the graph of connections between layers is a directional non-cyclic graph built on residual or sequential connections, we know that the last outgoing layer will return the output from the output layer of the whole model.

3.4   Complexity Analysis

Monte Carlo three search is time limited but it must analyze all actions at a depth of one in the searched tree. For a structure with k hidden layers, it is possible to generate $k * (k - 1)/2$ actions to add sequential layers, as this is the maximum number of edges that can be added in a directed graph with k vertices without introducing cycles. The maximum number of actions for adding residual layers is $(k - 1)!$, as the highest number of possible residual connections occurs in a graph resembling a path, where each vertex can connect to all subsequent vertices except itself. This means that time complexity for one generation is $O(((k - 1)! + k * (k - 1)/2) * n * e)) + C_{SGD} = O(k! * n * e) + C_{SGD}$, where C_{SGD} denotes the complexity of training the model using SGD, n is the size of simulation set, e is the number of epochs for training the model in simulation. In our experiments simulation set had 10 examples per class which means that $n = 100$ and the training time in the simulation was 10 epochs.

4   Empirical Analysis

4.1   Datasets and Empirical Setup

The experiments reported in this paper were conducted on the widely used MNIST [14] and Fashion MNIST [14] datasets. MNIST is a dataset of handwritten digits, while Fashion MNIST contains images of various items of clothing. These datasets are suitable for testing both convolutional and non-convolutional models. We have deliberately reduced the size of the initial neural architecture in order to efficiently evaluate the validity of the algorithms. The initial network configuration consists of a single convolutional input layer with a 3×3 filter and a dense output layer with 10 hidden neurons. In addition, we conducted experiments using different random seeds to ensure the robustness and generalizability of our results across different training scenarios.

In what follows, we address the overall quality of the designed algorithm. Furthermore, we inspect the most critical parameters present in the method. In the conducted experiments, the training set was divided into training and testing subsets, with the testing set comprising 20% of the training data. The distribution of images per class was even. More specifically, we used stratified sampling implemented in the *train_test_split* function in the sklearn library. As a result, the testing set for MNIST comprised 8,400 images, with the training set consisting of 33,600 images. For FMNIST, the testing set included 12,000 images, and the training set comprised 48,000 images.

4.2   Classification Quality of the New Approach

The conducted experiment aimed to empirically validate two hypotheses:

RH1. The first hypothesis states that the algorithm induces neural network growth that leads to an optimal structure.

RH2. The second hypothesis states that the Monte Carlo simulation can identify optimal changes in the network structure.

To verify the first hypothesis RH1, the experiments were conducted on the discussed MNIST and FMNIST datasets, with the initial network structure deliberately reduced to stimulate network growth.

The second hypothesis RH2 was verified by comparing the Monte Carlo algorithm with greedy and random approaches. The random approach randomly selects a change to execute on the network. The greedy approach evaluates each potential change in a single step. The learning process using the Monte Carlo simulation is characterized by a stable and persistent drive to improve quality, each subsequent network change was selected to optimize the model's overall performance. After introducing a change to the network, there are a few epochs during which the network is unstable, but it quickly returns to a stable state and achieves a higher score than before the change. Because the presented data analysis problem is relatively simple and the network has a very small number of neurons, it quickly falls into a state of procrastination. In this state, the network has learned everything it could within its current structure.

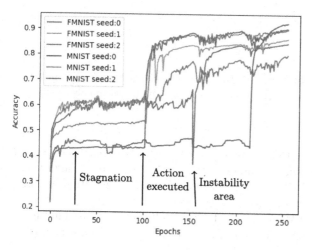
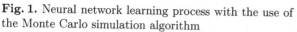

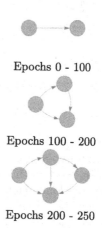

Fig. 1. Neural network learning process with the use of the Monte Carlo simulation algorithm

Fig. 2. Graphs representing history of structure for FMNIST seed 2

In Fig. 1, we observe the model learning process using our algorithm that employs Monte Carlo simulation. The graphs illustrate classification accuracy, with architectural changes in the neural network introduced every 50 epochs. Epoch number is indicated on the OX axis.

Visible "jumps" (see Fig. 1) in the learning process occur shortly after the introduction of a modification to the network. The perturbations and instabilities during learning represent a transition phase in which the learning rate gradually changes in the first epochs after the modification. Each generation lasts for 50 epochs, so the observed phases of stability and procrastination are prominent in full cycles. The structure in each generation learns all possible features it can acquire with the structure available in the given generation. A modification was needed to extend the structure so that it could learn a new feature in the existing data set. Figure 1 shows that when the model was unable to learn more, it fell into the stability region. After the Monte Carlo simulation selects the best action, it allows the network to be better fitted. The results obtained confirm the hypothesis that the algorithm induces a growth of the neural network that leads to an optimal structure (Fig. 3).

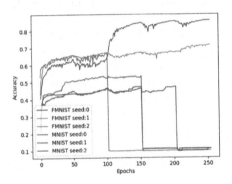 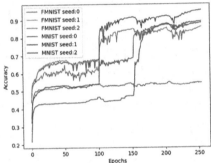

Fig. 3. Ablation study: learning process with a random simulation algorithm

Fig. 4. Ablation study: learning process with a greedy simulation algorithm

Subsequently, we address the experiments conducted to verify the second research hypothesis concerning the justifiability of using the MCTS algorithm. We test two what-if scenarios. In the first scenario, we replace the MCTS method with a random architecture modification. In the second scenario, we test a greedy search method in place of the MCTS method. The results are illustrated in Figs. 3 and 4, depicting the learning progress in terms of accuracy across epochs during training.

In the plot concerning a randomized scenario (Fig. 3), we observe that the learning process is highly unstable. It is possible that the algorithm may eventually reach a well-working structure. However, it is much more likely that the chosen changes will cause a total collapse of the predictive power. Observable sudden drops in the learning process indicate that the selected change prevented further learning.

Table 1. Classification accuracy reported in a comparative study assessing the impact of a simulation algorithm. The procedure was repeated nine times with different seeds

algorithm	FMNIST			MNIST			mean
	seed 0	seed 1	seed 2	seed 0	seed 1	seed 2	
greedy	83%	67%	82%	86%	55%	91%	77.67%
Monte Carlo	82%	83%	79%	90%	87%	87%	88.35%
random	83%	70%	9%	11%	11%	11%	11.43%

In Fig. 4, we illustrate the learning process for the greedy approach. It shows significantly better results than the random simulation.

Subsequently, let us examine the numerical quality scores of different processing pipelines. These are summarized in Table 1. It becomes apparent that the overall score of the greedy simulation is worse than the score achieved by the Monte Carlo simulation. Changes introduced by the greedy algorithm mostly had a positive impact on the algorithm's performance but did not lead it to achieve structures as good as those obtained through the Monte Carlo simulation.

Since the Monte Carlo simulation was able to look further into the future compared to the greedy simulation, chosen actions had a better long-term impact on the network structure. The Monte Carlo simulation selects the best change in the current generation but also considers subsequent ones, thereby determining the direction of changes in each generation leading to an optimal structure. In contrast, the greedy simulation only analyzes all possible steps in a single generation and does not consider potential future changes.

The random simulation effectively illustrates that not all changes are good for the structure, choosing the best action is crucial for the learning process, emphasizing the necessity of simulation. It is evident that random changes to the network result in the development of a structure that is unfortunately not favorable for the presented problem. This intuitively indicates that there exists an optimal structure for the given problem, and the Monte Carlo simulation is the best for discovering this structure. The obtained results confirm the hypothesis that Monte Carlo simulation can identify optimal changes in the network structure.

4.3 Parameters of the Method and Their Impact on the Procedure

The development of this algorithm required some design decisions that were made based on theoretical analyses and empirical results. In this section, we discuss the most important parameters that were fixed in the discussed experiments. Below, we present the results from three experiments aimed at determining key parameters for this algorithm. The first parameter determines the mode of evaluating simulation results, the second parameter dictates the initiation moment of the simulation, and the third parameter governs how the learning rate parameter should work. All experiments in this section were run on the MNIST dataset with three different seeds.

The first parameter to be tested was the mode of operation of the *score function* during the algorithm simulation. The algorithm can use either the *loss* or *accuracy* parameter to score a model after a particular action. The choice of this parameter is particularly important when using Monte Carlo Tree Search. On the one hand, accuracy values are immediately normalized and fit well with the established and effective patterns in Monte Carlo Tree Search. On the other hand, using the loss value as an evaluation criterion has the potential to convey more information about the quality of the model after a specific action has been performed.

In that regard, we inspect and compare the impact of using loss and accuracy as two distinct evaluation criteria. The experiments were conducted on both datasets. They were repeated three times with different seeds.

Table 2. Classification accuracy depending on various score function modes

score function	seed 1	seed 2	seed 3	mean
accuracy	89.5%	85.0%	53.7%	76.0%
loss	44.9%	89.5%	55.7%	63.3%

The experiments have shown that the mean score of the models that used accuracy in grading actions was bigger than that of those that used loss, which is shown in Table 2. We stipulate that it is because UCB1 (UCB stands for Upper Confidence Bounds) in Monte Carlo simulation works better with normalized values like accuracy.

Subsequently, we examined different operational modes for the simulation orchestrator. There are several strategies one can design for this task. We chose to investigate three specific modes: *constant*, *progress check*, and *overfit*.

Table 3. Classification accuracy and the number of changes of network architecture for various simulation orchestrator working modes

orchestrator	accuracy				number of simulations			
	seed 1	seed 2	seed 3	mean	seed 1	seed 2	seed 3	mean
overfit	89.3%	86.6%	84.6%	86.3%	18	18	18	18
progress check	83.6%	88.0%	90.6%	87.4%	10	7	9	8.6
constant	89.9%	88.2%	87.1%	88.3%	18	18	18	18

Table 3 showcases the average classification accuracy on test sets for various orchestrator operation modes. The "overfit" method displayed slightly worse performance compared to the other two. In this mode, the algorithm triggers simulations when there is a high probability of overfitting based on the model's learning history. The "constant" mode conducts a simulation in each generation, while the "progress check" mode verifies if the model has achieved improved accuracy compared to the previous generation; if not, a simulation is initiated. All these methods show promise, yielding similar results. Despite the "constant" method having the best mean score, the "progress check" mode attains the highest maximum score. Consequently, we analyzed the extent of changes induced by these orchestrators. The primary objective of this module is to find a balance between exploiting and exploring the model structure and adjusting the number of changes to the model's learning progress. Table 3 details the number of changes caused by the orchestrator. Both the "overfit" and "constant" orchestrators made changes in each generation. Perhaps the "overfit" mode should be better parameterized for a given problem to be less or more sensitive. The "progress check" made changes in approximately half of the possible generations. We reduced the number of neurons in the layers to close to 10, explaining why adding more layers consistently yielded high scores in all orchestrators. However, the "progress check" orchestrator, with fewer changes, emerged as the most stable and efficient method, making it the likely best choice for future experiments (Table 4).

Table 4. Classification accuracy depending on different learning rate scheduler modes

Learning rate scheduler	seed 1	seed 2	seed 3	mean
constant mode	54.3%	83.9%	81.9%	73.3%
progress mode	86.5%	87.9%	85.2%	86.5%

Subsequently, we compared the use of a progressive learning rate with a constant learning rate. Progressive means that the learning rate increases up to a certain point during one generation and then decreases almost to zero before starting the next generation. This happens so that the changes introduced by the algorithm to the model have the least impact on the features learned so far. With a constant learning rate, the same value applies throughout all generations. From the conducted experiments, we inferred that the progressive mode yields higher results.

5   Conclusion

The paper has brought forward a new approach to dynamic neural network training procedures. The proposed procedure relies on a simulation orchestrator that launches an MCTS procedure. The outcome of this procedure is a decision concerning a change in the neural architecture. The addressed solution works on the level of a layer: after each simulation, we may decide to add a layer, remove a layer, or keep the current architecture intact. The detailed formalism was presented for convolutional, plain sequential (dense) layers, and residual sequential layers. The new method, in contrast to the approaches existing in the literature, empowers more flexible model design through the use of a wide variety of neural layers.

The validity of the use of the MCTS algorithm for design was tested in an ablation and substitution study. We have replaced the Monte Carlo simulations with a random decision-making algorithm and with a greedy algorithm. The latter performed substantially worse. The random algorithm was unacceptable.

An indispensable component of the delivered study was the prepared source code. It is openly available as a Python package named *growingnn*. It was uploaded to PyPi: https://pypi.org/project/growingnn/. It contains the implementation of the training algorithm prepared from scratch in Python. We want to underline the scarcity of open-source codes in the domain of dynamic neural topology adjustment methods and we believe that our work would bring practical value to the researchers working in this area.

References

1. Bach, F.R.: Breaking the curse of dimensionality with convex neural networks. J. Mach. Learn. Res. **18**, 1–53 (2017). https://jmlr.org/papers/volume18/14-546/14-546.pdf
2. Bagaria, V., Baharav, T.Z., Kamath, G.M., Tse, D.N.: Bandit-based Monte Carlo optimization for nearest neighbors. IEEE J. Sel. Areas Inf. Theory **2**(2), 599–610 (2021). https://doi.org/10.1109/JSAIT.2021.3076447
3. Deng, L.: The MNIST database of handwritten digit images for machine learning research. IEEE Sig. Process. Mag. **29**(6), 141–142 (2012)
4. Dogo, E.M., Afolabi, O.J., Nwulu, N.I., Twala, B., Aigbavboa, C.O.: A comparative analysis of gradient descent-based optimization algorithms on convolutional neural networks. In: 2018 International Conference on Computational Techniques, Electronics and Mechanical Systems (CTEMS), pp. 92–99 (2018). https://doi.org/10.1109/CTEMS.2018.8769211
5. Evci, U., Vladymyrov, M., Unterthiner, T., van Merriënboer, B., Pedregosa, F.: GradMax: growing neural networks using gradient information. In: Proceedings of ICLR 2022 (2022). https://iclr.cc/virtual/2022/poster/7131
6. Fahlman, S.E., Lebiere, C.: The Cascade-Correlation Learning Architecture, pp. 524–532. Morgan Kaufmann Publishers Inc., San Francisco (1990)
7. Goldberg, A.V., Tardos, É., Tarjan, R.E.: Network Flow Algorithms. Princeton University, Princeton (1989). https://www.cs.princeton.edu/techreports/1989/216.pdf
8. He, K., Zhang, X., Ren, S., Sun, J.: Deep residual learning for image recognition. In: 2016 IEEE Conference on Computer Vision and Pattern Recognition (CVPR), pp. 770–778 (2016). https://doi.org/10.1109/CVPR.2016.90
9. Kilcher, Y., Bécigneul, G., Hofmann, T.: Escaping flat areas via function-preserving structural network modifications. In: Proceedings of ICLR 2019 (2019). https://openreview.net/forum?id=H1eadi0cFQ
10. Miconi, T.: Neural networks with differentiable structure. CoRR abs/1606.06216 (2016). http://arxiv.org/abs/1606.06216

11. Platt, J.: A resource-allocating network for function interpolation. Neural Comput. **3**, 213–225 (1991). https://doi.org/10.1162/neco.1991.3.2.213

12. Schaul, T., Zhang, S., LeCun, Y.: No more pesky learning rate guessing games. J. Mach. Learn. Res. **28**(3), 343–351 (2013). https://proceedings.mlr.press/v28/schaul13.html

13. Swiechowski, M., Godlewski, K., Sawicki, B., Mandziuk, J.: Monte Carlo tree search: a review of recent modifications and applications. Artif. Intell. Rev. **56**(3), 2497–2562 (2023). https://doi.org/10.1007/s10462-022-10228-y

14. Xiao, H., Rasul, K., Vollgraf, R.: Fashion-MNIST: a novel image dataset for benchmarking machine learning algorithms. CoRR abs/1708.07747 (2017). http://arxiv.org/abs/1708.07747

15. Xie, S., Girshick, R., Dollár, P., Tu, Z., He, K.: Aggregated residual transformations for deep neural networks. In: 2017 IEEE Conference on Computer Vision and Pattern Recognition (CVPR), pp. 5987–5995 (2017). https://doi.org/10.1109/CVPR.2017.634

16. Zhang, X., Yang, T., Wang, L., Liu, S., Yan, J., He, Z.: Architecture growth of dynamic feedforward neural network based on the growth rate function. In: 2022 IEEE 11th Data Driven Control and Learning Systems Conference (DDCLS), pp. 1190–1195 (2022). https://doi.org/10.1109/DDCLS55054.2022.9858492

Author Index

L. Franco et al. (Eds.): ICCS 2024, LNCS 14832, pp. 379–380, 2024.
https://doi.org/10.1007/978-3-031-63749-0

-

Printed in the United States
by Baker & Taylor Publisher Services